Microcomputer Interfacing and Applications

Microcomputer Interfacing and Applications

Second Edition

Mustafa A. Mustafa, BSc, PhD, AMIEE

Newnes
An imprint of Butterworth-Heinemman Ltd
Linacre House, Jordan Hill, Oxford OX2 8DP

A member of the Reed Elsevier plc group

OXFORD LONDON BOSTON MUNICH NEW DELHI
SINGAPORE SYDNEY TOKYO TORONTO WELLINGTON

First published as *Digital and Analogue Interfacing for Microcomputers*
by Blackwell Scientific Publications 1991.
Second edition 1994

British Library Cataloguing in Publication Data
Mustafa, Mustafa A.
Microcomputer Interfacing and
Applications. – 2Rev.ed
I. Title
004.616

ISBN 0 7506 1752 7

Library of Congress Cataloguing in Publication Data
Mustafa, M. A. (Mustafa Ahsen)
Microcomputer interfacing and applications/Mustafa A. Mustaf. –
2nd ed.
p. cm.
Rev.ed. of: Digital and analogue interfacing for microcomputers.
1990.
Includes index.
1. Computer interfaces. 2. Digital-to-analog converters.
3. Analog-to-digital converters. I. Mustafa, M. A. (Mustafa
Ahsen). Digital and analogue interfacing for microcomputers.
II. Title
TK7887.5.M87
004.6′16–dc20

94–25743
CIP

Typeset by Vision Typesetting, Manchester
Printed in Great Britain by Hartnolls, Bodmin

Contents

Contents

Preface

The continuous increase in the performance/price ratio of microcomputers makes them very attractive as intelligent controllers in domestic and industrial applications. They are available at relatively low prices and users can select the most suitable microcomputer for their needs. Such a microcomputer may not provide the users with all the required features. Any microcomputer-controlled system requires an interface between the microcomputer and the rest of the equipment. Users may be able to purchase additional hardware and possibly associated software to satisfy the interface requirements. They can select from a variety of add-on cards and devices available from various suppliers. Commercial add-ons, however, may not fully satisfy the requirements. Users may want to design and produce interfaces tailored to their needs and to be able to modify those interfaces should the need arise for interfacing additional equipment. The cost of an add-on may prove to be too high for a company requiring a large number of such add-ons for inclusion in its products, or for computer owners aiming to expand their microcomputers and use them as intelligent controllers. In most situations, communication links between the microcomputer and the outside world are easy to design and apply once the designer develops an understanding of how to approach the interfacing problem.

When people inexperienced in the subject of interfacing attempt to expand their microcomputers, they may, for example, aim to introduce analogue input/output lines. The circuits they use may be perfectly suitable for many applications and may have been presented in a book dealing with interfacing, but they may not fully satisfy the requirements. They could, for instance, prevent users from running a few tasks in parallel. They may end up redesigning the circuit or they may introduce additional unnecessary hardware. This book emphasizes the importance of defining the objectives of the interface, comparing solutions and producing the most suitable interface. Such an interface is based not only on selection of the optimum input/output devices, but also on its effect on the operation of the microcomputer. Add-on circuitry should not overload the microcomputer or degrade its performance. The book emphasizes the importance of achieving a balance between carrying out various calculations and maintaining a suitable sampling rate of the interface signals. This aids in achieving the most effective use of any microcomputer. A large number of examples are included, they present to the reader many of the problems that can arise when attempting to expand a microcomputer. They provide information on how these problems can be solved and, whenever possible, more than one solution is given to each problem.

The book describes how the requirement of a given application can be met by the use of any of several microcomputers. For instance, the timing requirements of a demanding application are not necessarily best satisfied by employing a fast and expensive microcomputer. Instead, a simple microcomputer can be utilized in association with additional hardware. The book is not confined to one programming language; software solutions are presented in the form of procedures. Each procedure consists of a group of sequential steps which are easy to understand and can be easily translated to any of the low-level or high-level programming languages used in microcomputers. This approach makes the book useful to a wide range of readers. It will help them to achieve the optimum interface for whatever microcomputer they decide to use and whatever programming language they decide to adopt.

The first chapter begins by considering the main parts of a microcomputer. It explains the principle of operation of microprocessors and illustrates the advantages of using microcontrollers and coprocessors. It describes how tasks are executed and how interrupts are handled. Chapter 2 explains how devices can be interfaced to processors. It gives practical examples illustrating the difference in requirements between different types of processors. Chapter 3 begins with an explanation of personal and industrial microcomputers. It then proceeds to consider a simple task and illustrates the existence of several solutions. It explains that each solution is best suited to a certain type of application. Chapter 4 presents examples of the use of input and output ports to exchange digital signals between a microcomputer and the outside world. It illustrates the difference between meeting interfacing requirements with software or hardware. Section 4.6 describes the behaviour of different types of loads and serves as an introduction to the solutions of some of the examples presented in later chapters.

Chapter 5 starts by explaining the principle of multiplexing and storing signals and gives examples of commercial ICs. It proceeds to describe the principles of popular methods of data conversion and the types of errors occurring in practical data converters. Chapters 6 and 7 consider interfacing single-channel and multichannel data converters with microcomputers through input and output ports. They present examples illustrating techniques of constructing and interfacing data converters. Chapter 8 deals with interfacing commercial single-channel and multichannel data conversion ICs with microcomputers. It presents examples demonstrating how such ICs can be operated under the control of a microcomputer. The examples also include interfacing conversion ICs directly with the data bus of a microcomputer.

Chapter 9 deals with the operation of counters and timers and their interfacing with microcomputers. Such devices play a very important role in interfacing applications, particularly those dealing with real-time events.

Chapters 10 to 16 present examples illustrating microcomputer-controlled applications. These examples emphasize the importance of obtaining a balance between carrying out various calculations and maintaining a high sampling rate in order to achieve the most effective use of a microcomputer. Consequently, some of the presented solutions are based on the inclusion of additional circuits to perform some of the tasks.

Chapter 10 explains the operation of a few types of switching device and gives examples of their interfaces with microcomputers in various applications. Chapter 11 considers a different field of applications. It describes a few optical sources and sensors and gives examples of their utilization in interfacing applications, e.g. displays, isolators and encoders. Chapter 12 considers the use of a microcomputer to produce different types of waveform. It presents a few methods which include providing manual control of the generated waveforms.

Chapter 13 describes the implementation of controlled operations by a microcomputer. It considers applying real-time control strategies by a microcomputer where a robotic mechanism is taken as an example. It describes how the path of motion can be specified and how the movements of single- and multi-degree of freedom robotic mechanisms can be commanded. The last section gives examples on the incorporation of microcomputers in sequential processing applications.

Chapter 14 starts with a description of the use of microcomputers in temperature-control applications. It then proceeds to introduce a few types of temperature sensors and their interfaces with microcomputers. It presents a few methods of controlling temperature, explaining the benefits and limitations of each method. Chapter 15 includes a brief description of d.c. and a.c. motors. It presents examples of controlling their operation from microcomputers, taking into account the effect of the solution on the operation of the microcomputer. Chapter 16 presents a few types of applications, starting with interfacing keyboards and then explaining data transfer applications and the use of microcomputers as programmable testers. It also describes how an existing interface may be modified to meet the requirements of a new application. Section 16.7 considers sources of noise glitches and provides hardware and software solutions to minimize their effects.

Chapter 17 begins with an explanation of the utilization of a microcomputer as a monitoring and a data-capturing device. It highlights the limitations imposed by the storage capability of the microcomputer. The chapter then proceeds to explain the operating systems of microcomputers, and then describes the capabilities of microcomputers and different ways of upgrading them. The last section deals with programming languages and tools employed during the development of microcomputer-based projects.

The book gives engineering and computer/science students information on how to approach practical design problems in order to produce the most suitable interface. It provides design and development engineers with alternative solutions to interfacing problems. The book will also be attractive to senior engineers and managers aiming to develop an understanding of microcomputers and their interfacing with the outside world.

Mustafa A. Mustafa

List of Trademarks

The following is a list of trademarks, probable trademarks and the probable owner of each trademark mentioned within this book. No responsibility can be taken by the author or the publisher for any omissions or errors.

Alpha is a trademark of Digital Equipment Corporation.
Analog Devices is a trademark of Analog Devices Incorporated.
GEC Plessey Semiconductors is a trademark of GEC Plessey Semiconductors Limited.
IBM, AT, PS/2 and OS/2 are registered trademarks of International Business Machines Corporation (IBM).
Intel, Intel 386, Intel 387, Intel486, Intel486 SX, Intel486 DX, Intel486 DX2 and Pentium are trademarks of Intel Corporation.
Microsoft, MS and MS-DOS are registered trademarks of Microsoft Corporation.
Motorola 68000 is a trademark of Motorola Incorporated.
National Semiconductor is a registered trademark of National Semiconductor Corporation.
PC/XT is a trademark of International Business Machines Corporation.
Texas Instruments is a trademark of Texas Instruments Incorporated.
UNIX is a registered trademark of Unix Systems Laboratories.
Windows is a trademark of Microsoft Corporation.
Zilog Z80 is a trademark of Zilog Corporation.

The operation and interfacing of many devices are explained within this book. Details of these devices have originated from the devices' data sheets (which are copyrighted to the manufacturer and where further information can be found). No manufacturer's name is mentioned for devices available from a few sources.

Chapter 1

Operation of a microcomputer

1.1 Introduction

Microcomputers are attractive for use in numerous applications, as they offer digital programmability, flexibility and processing power. Recent technological advances have improved the performance and storage capabilities of microcomputers and reduced their cost. This has led to the availability of powerful microcomputers at affordable prices.

The availability of a wide range of commercial software packages and a variety of hardware eases the use of microcomputers and increases their suitability for applications in offices, in laboratories, on shop floors, etc. Nowadays microcomputers are offered in a variety of sizes and shapes with prices proportional to their processing power and the features they offer. Each type is best suited for a certain range of applications. For example, a microcomputer incorporated as a part of an industrial controller may possess much processing power but be small in physical size. It may be designed to satisfy the requirements of a specific application without the need for means of displaying information. On the other hand, a microcomputer designed for use in data-monitoring or graphics-design applications is likely to have a higher storage capability and be provided with a monitor, an LCD display or a video interface. Both of these microcomputers are represented in block diagram form in Fig. 1.1.

The parts of a microcomputer can be categorized into three areas; these are: processing, storage and input/output.

1.2 Processor

This is the fundamental part of a microcomputer, and contains most of the control and decision-making logic. It can be either a microprocessor or a

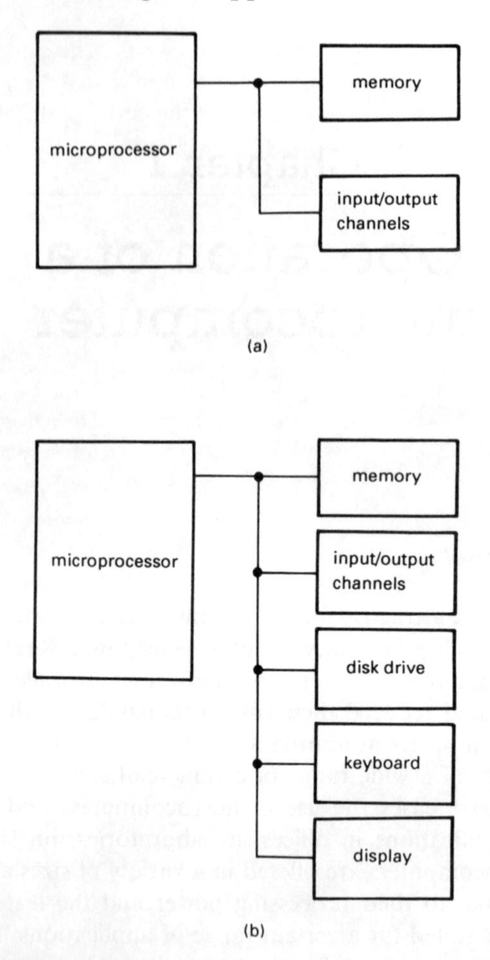

Fig. 1.1 Parts of a microcomputer: (a) with I/O channels; (b) with I/O channels, disk drive and user interface.

microcontroller. Arithmetic and logic operations are performed in a processor by the central processing unit (CPU), which also controls the transfer of data. Most processors contain a single CPU, but very high performance can be achieved by the inclusion of two CPUs in the same processor, where both CPUs operate in parallel, as in the Intel Pentium™ processor.

The CPU performs the desired system functions by executing a program stored in memory. The program consists of a group of instructions written in low-level or high-level language (Section 17.7.1). These instructions are translated by development tools (e.g. assemblers or compilers) into a code specific to the processor which will be stored in and executed from memory. The internal operation of the processor is based on the use of 'microcoded instructions', or 'hardwired instructions'. The performance of a processor can be improved by the inclusion of hardwired instructions, since they are faster to execute. This, however, increases the complexity of the processor.

The interface between the processor and the rest of the system is through three buses. These are:

- the address bus
- the data bus
- the control bus

Each of these buses consists of several lines, the number of which depends on the processor. Let us consider a few examples:

- The data bus in an 8-bit microprocessor, such as the Zilog Z80™, consists of 8 lines, whereas its address bus consists of 16 lines.
- In a 16-bit microprocessor, such as the Intel 8086, Intel 80186 and the Motorola 68000™, the data bus consists of 16 lines, whereas the number of address lines is not the same in all of these processors. The address bus in the 8086 and 80186 consists of 20 lines, while in the 68000 the address bus consists of 23 lines.
- A 32-bit microprocessor, such as the Intel386™ DX, Intel486™ SX and Intel486™ DX, possesses 32 data lines and 32 address lines.
- A slightly different type of processor is that possessing an internal structure of a different resolution to the external data bus. The Intel 8088 and Intel 80188 are examples of processors with a 16-bit internal structure and an 8-bit data bus. They provide a 20-line address bus and an 8-bit external data bus to allow the use of 8-bit hardware with a 16-bit processor (e.g. operation from a single 8-bit memory device). Similarly, the Intel386™ SX microprocessor consists of a 32-bit internal structure and a 16-bit data bus.
- The Intel Pentium processor has a 64-bit data bus. Internally, this bus is applied to two pipelines, allowing the execution of two 32-bit instructions simultaneously.
- The Digital Alpha™ is based on a 64-bit architecture capable of 64-bit addressing and can handle data types from bytes to quadwords (8 bytes).

The address bus is an output bus (unidirectional). Each combination of the logical states of the address bus lines refers to a specific address which can be a specific memory location, an enable signal to an interface device, or an unused address. On the other hand, the data bus is an input and output bus (bidirectional). It serves as an input when the processor reads information from memory or from an input device. It becomes an output bus when the processor writes information to memory or to an output device. Normally, the processor controls the data transfer through the bus by the use of its control lines (the control bus). In some systems, however, the processor is allowed to transfer the control of the bus to another device for a short period of time. This is done in certain situations to permit an external device to transfer information directly to memory (Section 16.3.3).

Many devices can be connected to the data bus, but only one device can place information on that bus at any instant of time. This is made possible by the use of '3-state' devices. As the name implies, each of these devices possesses three states. They are a logic high (logic 1), a logic low (logic 0) and a high-impedance state (off). Such a device can therefore be employed to permit the transfer of information to, or from, the processor through the data bus only when instructed by the processor to do so (Chapter 2).

Microcomputers with an 8-bit data bus transfer 1 byte of information per instruction. On the other hand, a microcomputer with a 16-bit data bus normally transfers 2 bytes of information with one instruction. Consequently, a microcomputer possessing a 16-bit data bus has the ability to communicate with suitable devices and memory elements faster than one possessing an 8-bit data bus. A designer can utilize such a feature to achieve high performance from a 16-bit microcomputer. Although 8-bit processors deal with bytes, they can also handle 16-bit operations by considering each 16-bit word to consist of an upper byte and a lower byte. Sixteen-bit mathematical and data-transfer operations will therefore take longer than when they are handled by a 16-bit processor. The transfer of 16-bit information through an 8-bit data bus is achieved in two steps, one byte at a time.

The need for more processing power has led to the development of high-performance 32-bit and 64-bit processors. The capability of such processors to exchange multibytes (4 or 8 respectively) of information in one step makes them very attractive for applications where high-resolution information needs to be processed and transferred at a high rate. One of the main areas using fast high-resolution processors is that of personal microcomputers where (in the majority of applications) a huge amount of data is required to be transferred to and from the processor at the highest possible rate.

High resolution is always preferred, but other factors (such as sampling rate and cost) are given a higher priority in applications such as real-time control. This is because high-resolution processors are more costly and high-resolution calculations can take longer than those of low resolution (this depends on the type and resolution of the processor).

1.2.1 Microcontrollers

The requirement for a higher performance/cost ratio is satisfied in a very large number of applications by incorporating a microcontroller rather than a microprocessor. The former is an IC containing a CPU plus a number of peripherals. Generally, microcontrollers are designed with specific application areas in mind. The availability of microcontrollers in various forms, with different processing capabilities and with different peripherals, has provided designers with a very wide choice. This enables them to optimize their design to achieve the best balance between performance, cost and board size.

An example of a microcontroller is the Intel 80C196KB. It contains a 16-bit CPU, RAM, EPROM (or ROM) and can accept interrupts from 28 sources. It contains many peripherals, such as an 8-channel 10-bit analogue/digital (A/D) converter, a pulsewidth modulation (PWM) channel (which can be utilized as a digital/analogue (D/A) converter), a serial port, high-speed input and output channels, two 16-bit timers/counters, and input/output lines. The number of usable input and output lines depends on which of the peripherals are used. This is because the flexibility of the Intel 80C196KB (as well as that of many other microcontrollers) has been increased by making many of the pins multifunctional. They can be defined by software to be input, output, or connected to a specific internal unit. The inclusion of peripherals within the same IC simplifies the circuit, reduces cost and can improve performance. This advance in technology made it possible to increase the capabilities and execution speed of microcontrollers, making them very attractive for use in a vast number of applications.

It is often desirable to design the software so that it will fit within the internal memory of a microcontroller. This allows one to configure the microcontroller for operation in a single-chip mode. This approach eliminates the need for additional memory devices. It also removes the need for address and data lines (unless needed by external peripherals). This gives the designer the ability to configure these lines as digital-input or digital-output lines. The availability of a higher number of these lines simplifies interfacing tasks considerably and can reduce the number of components and board size. In some applications, however, the size of the internal memory is not large enough to contain the software. A microcontroller can therefore be configured to use external memory, or a combination of both internal and external memory. Memory size is not the only limitation, since many interfacing applications impose the requirement of using a large number of lines from the microcontroller. If this is not possible, then one may be forced to introduce additional ports, or special techniques to meet the requirements with the available number of lines, as shown throughout this book.

1.2.2 Coprocessors

The demand for high processing and sampling rates in many applications may exceed the ability of a single processor. Some of the tasks can be transferred to a second processor which operates in parallel with the main processor. Depending on the type of system and the requirements, the operation of the second processor is normally controlled by demand signals supplied from the main processor. In this way, the operation of both processors is synchronized. The main processor is referred to as the 'master', while the second processor is called the 'slave'. An example of a dual-processor system is an industrial controller using two processors (e.g. microcontrollers). The slave processor is employed to send and receive signals through multichannel input and output lines. After reading input signals, it performs calculations and then conditions and scales signals into a form acceptable to the master processor. It also scales signals received from the master processor before supplying them to the outside world through an interface circuit. The main processor carries out the rest of the tasks and uses the slave processor as a source and destination of communications signals.

Processors are widely incorporated to manipulate data represented by integer numbers. Additional software can be written or purchased to enable the processor to manipulate floating-point numbers. The execution of such software requires more processing time than when dealing with integer values. In situations where fast results are required, or when processor time becomes insufficient to handle various tasks and carry out mathematical operations, a special type of coprocessor is incorporated to assist the main processor. It is referred to as a 'math-coprocessor'. The high price of such a coprocessor is justified in a large number of applications, since the master processor offloads the time-consuming mathematical functions to the coprocessor and receives the results much faster than if the calculations were handled by software. Clearly, the improvement in performance depends on the type of application. Such a coprocessor can, for example, be extremely beneficial in a graphics-based application, but not in a data-transfer application. Many microcomputers (e.g. personal computers) are designed with a socket for a math-coprocessor, either occupied or left empty to give the user the option to include it if and when the need arises. A coprocessor has to be of the same type and of the same frequency as the main processor. For instance, an Intel387™ DX-33 Math-coprocessor can be added to a microcomputer incorporating an Intel386 DX-33MHz as the main microprocessor. The increase in the use of math-based applications, particularly in graphics, has led to the design of processors with an on-chip math-coprocessor, e.g. the Intel486 DX microprocessor.

In applications where the cost of a math-coprocessor is considered to be too high, the calculation requirements may be satisfied by the use of a second, lower-cost processor. For example, a fast microcontroller can be a very attractive solution, since it can be employed to handle the time-consuming mathematical operations using software, and, at the same time, introduce additional input/output ports, timers, etc. The choice of processor will depend on the application. The designer should keep in mind, however, that the time taken to execute various mathematical functions varies from one processor to another. Performing the mathematical operations using software will obviously need longer execution time than if a math-coprocessor is incorporated. If, however, the software is correctly arranged, then the effect of delays introduced by the additional calculations can be reduced considerably. This is because, while the second processor is busy calculating the required values, the main processor can carry on executing other tasks. For a vast number of applications, the inclusion of a second processor (rather than a math-coprocessor) is the optimum solution. That is because it satisfies the requirements more economically even though software needs to be written to perform these tasks, thereby increasing the development time. When comparing both approaches, the difference in performance depends on the capability of both processors as well as the type of calculations, the rate of their repeat and the way they communicate with each other.

1.2.3 Demultiplexing

In some processors, such as the 68000 and the Z80, the address and data lines appear on separate pins. On the other hand, the 8086 and 80186 use a multiplexed bus structure where both the address and the data lines appear on the same set of pins at different times. The latter approach reduces the pin count of the IC, but introduces the requirement for demultiplexing the address and the data buses outside the processor. The demultiplexing process separates the address bus from the data bus to give the processor the ability to read from, or write to, memory and other peripherals (there are some devices capable of direct connection to a multiplexed data bus).

A way of demultiplexing an 8-bit bus is shown in Fig. 1.2(a), where a single 74LS373 IC (octal transparent latch) is incorporated to latch address lines A0 to A7. The circuit of Fig. 1.2(b) employs two 74LS373 ICs to latch address lines A0 to A15 of a microprocessor (or a microcontroller) possessing a 16-bit data bus. The IC has two control lines: OE (output enable) and LE (latch enable); both are active-low. The former is shown connected to the 0-V line, thus permanently enabling the output of the IC. The separation of the address and data signals is accomplished by using the address latch enable (ALE) signal supplied by the processor. It is connected to the LE line of the latch. When the LE line is at a logic-high state, the output of the IC is in the same state as its input. Changing the state of that line to a logic low instructs the IC to latch the information. Consequently, demultiplexing is performed when the processor takes three consecutive actions. First, it sets the ALE to a logic-high state and supplies the address signals through the multiplexed bus lines. Second, it changes the state of the ALE line, thus commanding the latching of the address. Third, it supplies the data signals through the bus lines. During such operations, the processor activates the appropriate control lines, identifying the type of operation and controlling the transfer of information.

In Chapter 2, a description is given of how the data, address and control lines can be used to interface devices to processors. It shows how devices

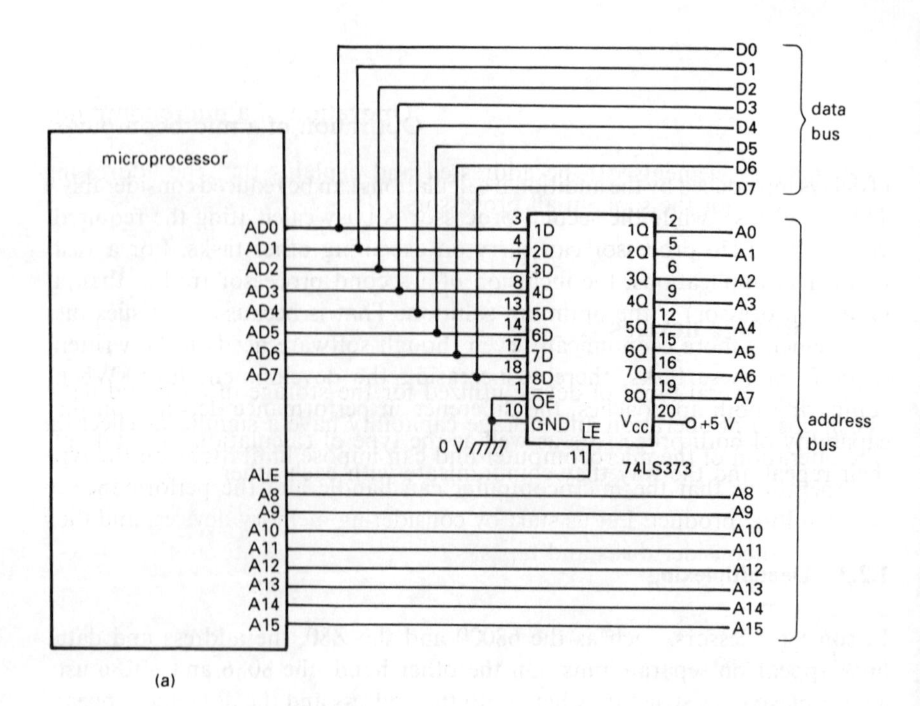

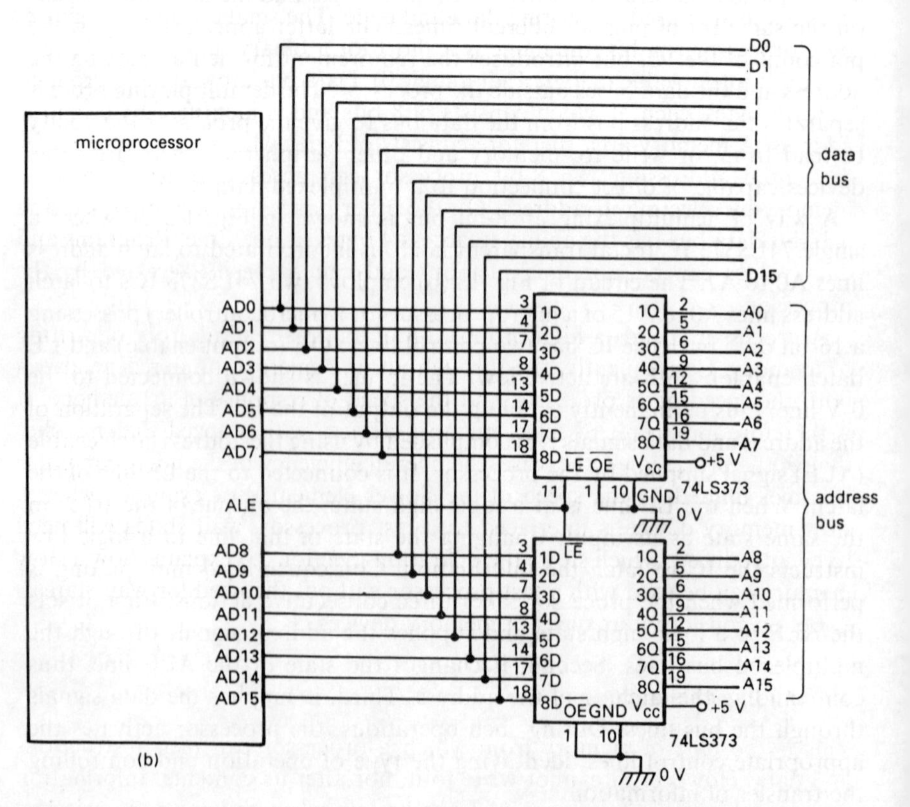

Fig. 1.2 Demultiplexing data and address buses: (a) 8-bit data bus; (b) 16-bit data bus.

and memory elements can be addressed and explains that the addressing capability is not the same in all processors.

1.3 Storage devices

There are several types of device utilized for the storage of code and data. Their speed of operation and storage capability have a significant effect on the operation of the microcomputer and can impose limitations on the type of operations that the microcomputer can handle and the performance of the resulting product. Let us start by considering memory devices, and then proceed to consider disks and tapes.

1.3.1 Memory

This is the area used for storing data and code. The smallest unit recognized by a processor is a bit (the name is a short for a binary digit). Each bit can store a Boolean value (a logic 1, or a logic 0). Data bits are normally grouped to form bytes, each of which consists of 8 bits and is identified by a unique address. A memory area therefore consists of bytes which are numbered by sequential addressing. An 8-bit processor can access any single byte in memory by selecting its address. On the other hand, a processor possessing a 32-bit data bus has the ability to access 4 bytes at one time. Programs and data values can be stored in memory and can be easily accessed by the processor.

A processor spends a high proportion of its time reading from or writing to memory. Consequently, the throughput of a microcomputer depends on both the speed of its processor and the speed of response of its memory. It is therefore important to select memory devices with a speed that matches that of the processor. Otherwise, the overall efficiency of the microcomputer becomes limited by the speed of the slower element. For example, when a slow-memory device is interfaced to a fast processor, wait states will need to be introduced, which slows the operation (Section 1.5 explains how a slow memory can be used with a fast processor without the need for wait states).

Let us consider four types of memory devices.

ROM (Read Only Memory)

This is the part of memory from which a processor can read instructions and data. However, it cannot write to it, nor alter its contents. Information saved in this type of memory will not be lost when the power is switched off. Consequently, instructions and fixed data are permanently stored in

ROM and remain intact even when not supplied with power. This type of memory can therefore be used to store the program which operates the microcomputer. There are three types of ROMs, as follows:

- *Masked ROM*. In this type of memory, information is stored in ROM ICs during manufacture. If the contents of the masked ROM are to be altered, then a new ROM will need to be manufactured, and may take months to obtain. Since ROMs are manufactured for a specific program, they need a large initial investment, but with the production of large quantities individual ROMs become inexpensive.
- *PROM*. These are Programmable ROMs. They can be programmed only once and are often referred to as 'OTP' (One Time Programmable) devices. They will retain the stored information, as they cannot be erased. On a per unit basis, PROMs are more expensive than masked ROMs. However, they can be programmed individually, thus eliminating the need for the large initial investment required for the masked ROM.
- *EPROM*. These are Erasable Programmable ROMs. They can be programmed, erased and reprogrammed thousands of times and will therefore not be scrapped when a program alteration is required. They were originally used only during program development, but as their price has dropped they have become the most popular program-storing devices. When the contents of an EPROM need to be altered, the IC needs to be removed from the equipment, erased by the use of ultraviolet light and then reprogrammed electronically.

RAM (Random Access Memory)

This type of memory is normally used for temporary storage during the execution of the program. The RAM retains its contents as long as it is supplied with power. Once the power is removed, the RAM loses its contents. Some applications have the requirement to maintain the contents of the RAM even when the power to the microcomputer is switched off. They use a battery (normally rechargeable) to provide the RAM with power. A large number of microcomputers do not use a battery back-up and the RAM is reloaded with information every time the power is applied.

The RAM in a microcomputer can be either of the following:

- *SRAM*. This type of memory is called the Static RAM. Each bit of information is stored in a flip-flop. Once written, the bits will be latched until the power is turned off, or other information is written to it.
- *DRAM*. This is the Dynamic RAM; it stores each bit of information as a charge on a capacitor. When compared to static RAM, this type of memory is slower, cheaper and uses less power and space. It is therefore

used to provide large areas of memory in computers. However, it needs refreshing regularly to prevent discharging the capacitors.

EEPROM (Electrically Erasable and Programmable ROMs)

This type of memory can be programmed and reprogrammed electronically without the need for its removal from the equipment. EEPROMs can therefore be soldered to the board. Rather than erasing all the information stored in an EEPROM, an alteration to its contents can be achieved by erasing and programming selected bytes individually. This makes EEPROMs attractive for a large number of applications. They are, however, slower and more expensive than EPROMs and there is a limitation to the number of times an EEPROM device can be reprogrammed, e.g. 10 000 times.

Flash EPROM

Similar to EEPROMs, programming and reprogramming this type of memory can be achieved electronically while the device is in circuit. It is not, however, possible to reprogram individual bytes of this type of memory. The availability of high-density flash EPROM devices makes them preferable to EPROMs in applications where the programs are likely to change.

Some applications impose the requirement of frequent on-line collection and storage of information which should not be lost if the microcomputer is turned off or forced to reset due to a fault. For example, if the microcomputer is employed to control the movement of a robot arm, then the microcomputer needs to know the exact position of each degree of freedom and the end-point of that arm. Position transducers are connected to each joint in order to provide the microcomputer with feedback signals. Assume that each transducer provides a signal corresponding to the angle moved, but not the actual position (Section 11.8.1). This implies that the microcomputer needs to calculate the actual position from the feedback signals and stores it in memory. A decision is needed on what type of memory to use for the storage of such continuously updated information. The obvious choice is RAM, but if the power to the microcomputer is lost, or a reset is activated, then the contents of the RAM will be erased, leaving the microcomputer without a record of the latest position of the robot arm. Consequently, the feedback information needs to be stored in a memory device which is protected against power failure. One may consider using an EEPROM for such storage, since it permits the reprogramming of individual bytes electronically and the stored information will not be lost after a reset or a power failure. But EEPROMs are slower than RAM and allow a limited number of reprogramming operations. One can then resort to the use of a battery-backed RAM. The increase in the use of battery-backed

RAM has led to the availability of a wide choice of batteries, and ICs containing both the RAM and the battery in a DIL (dual-in-line) package.

Memory devices are included in microcomputers and allowance has been made (in many microcomputers) for upgrading by the inclusion of additional devices. The following are examples of popular forms:

- DIL (dual-in-line) ICs are the most popular memory devices that can be added to a microcomputer. Factory-fitted devices may be of the surface-mounted type.
- SIMMs (single-in-line memory modules) are small PCBs (printed circuit boards) containing a few surface-mounted RAM ICs. They can be inserted in special connectors on the microprocessor board (normally referred to as the 'motherboard' in a multiboard microcomputer). The increase in the use of high-resolution microcomputers has led to a sharp increase in the utilization of such modules. For example, in IBM and compatible personal computers (PCs) each SIMM uses nine single-bit memory devices. This provides storage of a single byte plus a parity-check bit. The latter is included to detect corrupted data. PCs using 32-bit processors require four such SIMMs to store the 32 bits of information. Upgrading the memory of such computers means adding or replacing SIMMs in groups of four.
- Memory is also available in the form of a 'memory cartridge' or a 'memory card'. These are used by some microcomputers to provide an easily interchangeable solution. They include a means of data storage such as ROM, EPROM, EEPROM or battery backed-up RAM. Each of these storage devices can include useful data, or a program to operate the microcomputer in a special manner. When the need to run a different program arises, the operator can replace the storage device with another containing different data or a new program. This is much easier than spending time reloading the information. Some microcomputers are provided with the capability of writing to such a device, thus using it as a means of extending the data-storage area of the microcomputer, or as a means of data exchange with a similar microcomputer.

1.3.2 Disks

A disk is a device dedicated to the storage of information. Such information can be programs or data which, in many cases, are more valuable than the microcomputer! Many microcomputers incorporate one or more disk drives within the microcomputer casing. Use can also be made of external disk drives. One cannot define the optimum storage device without knowing the capability of the microcomputer and the type of application. For example, fast storage may be given the highest priority in data-monitoring and data-collecting applications. On the other hand, the increase in the use of colour graphics and the integration of sound and video in multimedia applications has led to the need for massive storage capacities. There are many other applications where security and portability of data are given the ultimate priority.

Floppy disks

This type of disk is a reliable low-cost storage medium which usually comes in either of two physical sizes, $5\frac{1}{4}$ inches and $3\frac{1}{2}$ inches. Such a disk consists of tracks, each track being divided into several sectors in which digital information is stored. When a new disk is to be used, it needs to be formatted by a microcomputer, which writes information into the disk in order to identify its sectors. This allows quick storing and accessing of data. A floppy disk formatted by one type of microcomputer will not work in an incompatible microcomputer. If the need arises, one can reformat the disk and use it as a new disk, thereby losing the information which was originally on the disk.

Floppy disks can be damaged as a result of incorrect storage or use. They are available in low-density and high-density forms. The continuous increase in the size of programs has meant that there is more demand for the higher-density type. Floppy disks do not have a capacity large enough to store very large programs or data. However, the combination of their limited storage capacity, very low cost and portability makes them useful for data exchange between microcomputers as well as for the back-up of useful data and programs. Large programs or data can be stored on a few disks and/or use is made of 'compression software' which arranges the information so that it occupies a much smaller area. The information will then need to be decompressed before use. The use of multidisk software means that one has to insert the appropriate disk in the disk drive every time such information is required. The software will inform the user to insert the required disk when the information is needed. This can be considered tedious in applications where frequently used information is stored on different disks. On the other hand, software packages are supplied on floppy disk. The software is likely to be compressed to fit within the minimum number of disks. All the user needs to do is type 'install' and the stored information will be decompressed automatically and installed on a suitable area on the hard disk without intervention from the user.

Hard disks

These are fixed disks possessing a much larger storage capacity than floppy disks. Many microcomputers, e.g. IBM-compatible PCs include or accept the inclusion of one or more such disks within the microcomputer casing.

The microcomputer can access information within a hard disk in a relatively short time. The access time is shorter than that of a floppy disk, but longer than that of memory, leading to a preference to store frequently needed temporary information in a memory area rather than on a disk. If the memory area is not large enough, some programs use a hard disk as an additional storage medium for such information.

The 'average access time' is normally used as a measure of the performance of a hard disk. For instance, hard disks are advertised in computer magazines indicating their storage capacity (in megabytes) and access time (in milliseconds). The latter represents the time needed for the disk to retrieve stored information. As programs get bigger, with frequent access needed to stored information, the speed of a hard disk as well as the rate of transferring data become important factors in selecting a hard disk. They can affect the overall performance of the computer. Consequently, it is not advantageous to spend a large sum of money on a very fast machine and include a slow hard disk.

In addition to the performance of a hard disk, its price is mainly dependent on its storage capacity. When buying or upgrading a microcomputer, it is not easy to estimate the optimum capacity of a hard disk. The fast and continuous development of more sophisticated and easier-to-use software means that there is an ever increasing need for more storage capacity. These days, only a limited amount of serious work can be done efficiently in PCs without using a hard disk. Some programs will not run without a hard disk, while others will need to go through a tedious procedure. Hard disks in old microcomputers were slower and generally of lower capacity. In time, software applications became more sophisticated. This increased the size of the programs and increased the demand for hard disks with larger and larger capacities. Nowadays, it is not unusual for a single software application program to occupy more than 10 megabytes on a hard disk. A fast hard disk with a high storage capacity connected through a fast bus will be extremely useful in real-time data acquisition applications, especially if high resolution is used.

The type of hard disk and its performance are closely related to the type and characteristics of the disk controller used. Two of the main types of controllers employed in personal computers are as follows:

- IDE (Integrated Drive Electronics) controllers provide a high rate of data transfer and low access time. If upgrading from an older type of disk to an IDE type, one needs to replace the controller as well as the disk. Such a controller is simple, because the control circuit is located inside the disk drive. This makes the controller cards simple and of extremely low cost, but they are limited to controlling two hard-disk drives (which is perfectly adequate for the majority of applications).

- SCSI (Small Computer Systems Interface) is another type of disk controller. It gives a very high rate of transfer and a low access time (a high rate of throughput). This feature makes such a controller very attractive. They are normally incorporated when several drives are employed (including CD-ROM, tape drives, etc.). Up to seven drives can be daisy-chained together and controlled by a single SCSI controller (they obviously need to be SCSI-compatible devices). When compared to IDE drives and controllers, SCSI drives and controllers are more expensive. But when a high rate of data transfer is required and/or a large number of drives are to be employed, then the extra price is justified.

Depending on the type of microcomputer and what it includes, it may not be possible to add disk drives within its casing. Although more costly than internal drives, external drives offer an expandable and transportable solution. Nowadays external drives have the additional advantage of being small, and are light enough to be carried in a briefcase, or stored in a safe place to secure important data. If space is available within the computer casing and security of data is required, then use can be made of removable hard disks. These use removable compact hard-disk cartridges to provide a good method of secure data storage.

Information stored in disks needs to be backed-up in case the disk breaks down or is destroyed in a fire, or the information is overwritten due to operator error. The back-up can be accomplished by the use of floppy disks or tapes.

Optical disks

Different types of optical disks are available from several manufacturers. Their operation is based on using a laser to store information on, and retrieve information from, a removable disk. Let us consider three types of removable high-storage optical disks.

- *WORM* (*Write Once Read Many times*). This type of disk allows a single write and has an average access time of tens of milliseconds. It offers an extremely high storage capacity, allowing the storage of massive amounts of data. Prevention of overwriting makes it an excellent storage medium for valuable data which need to be stored permanently.
- *CD-ROM*. This is a read-only type of optical disk. It possesses a massive storage capacity and has an average access time of hundreds of milliseconds. It is mainly employed in the distribution of massive amounts of data (hundreds of megabytes). It may include mathematical conversion tables, various wave shapes, graphical images, or various information for multimedia applications.

- *Rewriteable optical disks.* This type of optical disk offers a high storage capacity as well as the ability to overwrite the stored information. It is thus attractive for use in applications where massive amounts of data are to be stored and updated from time to time.

  Nowadays optical drives have some limitations. Their access time and prices are higher than those of floppy-disk and hard-disk drives. However, they possess massive storage capability. The continuous increase in the demand for large storage devices makes optical disks attractive for use in many applications.

RAM disk

This is not a physical disk, but it is part of the RAM configured as a disk. The configuration is performed by the operating system of the microcomputer in response to instructions within the program. Once configured, it can be used as a disk to store and retrieve information. The utilization of such a disk can eliminate the need for continuously accessing the hard disk. Since the access time of RAM is much lower than that of a hard disk, storing and retrieving information is much faster than when using a hard disk or a floppy disk. One should, however, remember that the information is not fully secured while in RAM. Turning the power off or resetting the microcomputer will cause the loss of this information. Consequently, one should periodically store any useful information on the hard disk.

One of the useful applications for RAM disks is to load them with repeatedly accessed information, e.g. program-development tools. Such tools are normally stored in a hard disk and can be easily transferred to the RAM disk by software. Since only a copy of the tools is stored in RAM, there is no risk of losing information if the power is turned off. The tools (which are used for running and debugging programs) will run much faster and the hard disk can then be accessed only to store the developed or modified work.

1.3.3 Tapes

Tapes form an attractive solution to some data storage problems, since they provide a large medium of storage (hundreds of megabytes) at a relatively low cost. Reel-to-reel magnetic tapes are used in large mainframe computers, while cartridge tapes are employed in many microcomputers. However, access time is measured in seconds, which limits the use of tapes to such applications as transferring massive amounts of data, or backing-up information in various types of computers. Obviously, in order to be able to use the tapes, hardware (a tape drive) needs to be purchased. One has the choice between an external unit or one which can be fitted within the casing of a computer. The cost of such a drive can be considered high for many microcomputer-based applications. But the high-storage capability, reusability and low cost of tapes make them attractive to use, particularly to back up information stored on hard disks in preference to the use of floppy disks (a single tape can include the same amount of information as hundreds of floppy disks).

1.4 Input and output devices

The input and output devices of a microcomputer can be divided into two categories. The first provides the microcomputer with the ability to interface to a user, and the second allows the microcomputer to communicate with other devices and equipment.

1.4.1 User interface

This is accomplished by the use of input and output devices. An example of a user-interface input device is the keyboard of a microcomputer. It allows a user to supply commands and information to enable the microcomputer to fulfil a certain requirement. Clearly, not all microcomputers possess the same type of user-interface devices. Some, such as those used as controllers in industry, possess a few switches to permit an operator to supply commands such as start, stop, move left, move right, etc. Other microcomputers, such as PCs, include a keyboard with all the alphanumeric characters and several control and function keys.

An example of a user-interface output device is the display. It enables a microcomputer to supply the user with various information. Means of displaying information vary from one microcomputer to another. An industrial microcomputer may, for instance, incorporate a few LEDs (light-emmitting diodes) to indicate the state of operation and to inform the operator of a fault when and if it occurs. Many other microcomputers supply information through the use of 7-segment or alphanumeric displays. Others have a built-in monitor, or a connector to a suitable display.

Other types of user-interface device are joysticks, mice, tracking balls and light pens.

1.4.2 Device interface

This book is concerned with interfacing microcomputers to the outside world. It is therefore mainly concerned with using devices such as input and output ports to receive and transmit signals into and out of microcomputers. For example, through an input port a microcomputer can detect whether switches are closed or opened, and through an output port it can supply information and issue commands to close relays or to start industrial controllers. Each port is assigned a unique identifying address. The designer of an interface needs to decide on the number of required input and output ports, and also the number of lines per port. Suitable ICs can then be selected accordingly.

If the microcomputer is based on a microcontroller rather than a microprocessor, then all the interfacing requirements may be accomplished by the use of the input and output ports within the microcontroller itself without the need for additional devices. This is not always possible and external devices will need to be provided even with microcontrollers (many of the examples given in this book provide means of optimizing the number of input and output lines of a microcomputer in a given application).

Chapter 2 explains the introduction of input and output ports to a microcomputer and the transfer of information through them. Not every interface of a microcomputer to an external device needs the introduction of a port, as some devices include one or more 3-state buffers and can be directly connected to the data bus. Examples of the implementation of both approaches are given throughout the book.

1.5 Cache memory

This is a special part of RAM dedicated to use as a temporary storage area between one device (e.g. a fast processor) and a slower device (e.g. a hard disk, a monitor, or a slower area of RAM). It is designed into some microcomputers in order to speed up their operation. Such microcomputers incorporate a fast and powerful processor, the computational speed of which is not the main source of delays in many applications. Such a processor is capable of executing instructions at a high rate, but a delay can occur if the processor communicates with a slower device. Ideally, a processor should be allowed to transfer data at the highest possible speed through a fast bus to and from peripherals, but in practice there are limitations. Let us consider three areas where a cache memory can help.

1.5.1 Memory cache

The memory of a PC is an example of an area where cache memory can improve performance economically. Such computers include a large capacity of RAM employed for temporary storage. DRAM is used because it is cheaper than SRAM, but it is slower. Wait states can be introduced within the processor in order to extend its cycle, thereby allowing the slower memory to respond. Since wait states slow the operation, higher performance can be achieved by incorporating a faster memory (SRAM) as a buffer between the fast processor and the slow-memory devices. The processor deals with the fast memory without the need for wait states. Information is then transferred between the fast and the slow memory without delaying the processor. The percentage improvement in performance depends on the capacity of the memory cache and the rate at which information is stored in and retrieved from the RAM.

1.5.2 Disk cache

Let us now examine another use of cache memory. Consider a situation where a fast processor uses a hard disk to store and retrieve information. Before the write or read operation can take place, the hard disk moves its internal mechanical parts to the area where information is to be stored in, or retrieved from. The time it takes to perform these actions depends on the type of the hard disk and it is much longer than the storage or retrieval time of RAM. The time needed to read from and write to RAM is the same for different bytes. This is not the case for a disk, where the time depends on the position of the head relative to the storage area of interest. Furthermore, the processor can access a single byte of RAM if it needs to, but disk information is accessed as sectors. For these reasons, faster rate of transfer of information between a processor and a hard disk can be realized if an area of RAM is utilized as a buffer for temporary storage between the processor and the disk drive. The processor can dump information and carry on with its other activities while the information is transferred to the disk drive. If a controller keeps a record of the contents of the cache RAM, then if the information is needed it can be retrieved without accessing the disk. An important thing to remember, however, is that cache is employed for temporary storage. Consequently, if information is written to the cache and the power is turned off before it had the chance to write it to the disk, then that information will be lost.

1.5.3 Video cache

Sending information to a display is another example where cache memory can help to improve the performance of microcomputers. For instance, the operation of the processor of a PC is much faster than the rate at which information can be transferred to a monitor. The processor can either introduce wait states, or deliver the information to a memory buffer (cache memory), and then continues with its other activities while the information is being transferred. The size of the buffer, however, should be chosen relative to the information to be transferred. This depends on the type of monitor and the mode of operation. The increase in the use of graphics, particularly in high-resolution colour, has led to the need to transfer more data, thereby increasing the need for a buffer of a higher capacity. Further information is given in Section 17.5.4.

1.6 Storage limitations

A microcomputer possessing a large area of RAM allows large programs to be loaded and run. It also permits storage and retrieval of large amounts of data. This is very attractive, since accessing information stored in RAM is much faster than when it is stored on a disk or a tape. When purchasing a microcomputer, there is usually a choice of capacity of RAM and disk drive. The user needs to decide on what size is required and what can be afforded. Normally these can be upgraded at a later stage, but there are limitations as described in Chapter 17. With the increase in the use of larger and better programs, the RAM area within a microcomputer is not always large enough for the storage of calculated or collected data. Microcomputers employed for such operations can be programmed to store some of the data into a disk as fast as the system allows. However, in many situations the access time of the disk is much greater than the rate at which data is collected. This can impose a restriction on the amount and/or the rate of collecting sequential data, even when a large area of cache memory is employed. This is further discussed in Chapter 17.

1.7 Operation of a microcomputer

Under normal operation, a processor works through its instructions sequentially except when responding to an interrupt, calling a subroutine, or branching in response to a jump instruction. Each instruction requires a known duration of time for its execution which is a multiple of the clock period of the processor. Some instructions require longer execution times than others; this depends on the complexity of the instruction.

1.7.1 Program counter

The processor uses a register called the 'program counter' to keep track of the next instruction location in memory. After fetching an instruction, the contents of the program counter are incremented. If the flow of the program is to be altered (such as due to a jump instruction), then the contents of the program counter are altered accordingly.

1.7.2 Temporary branching

During a subroutine call or while responding to an interrupt, the execution sequence of the program is diverted temporarily and the contents of the program counter are stored in a register called the 'stack' before activation of the temporary branch. This is followed by execution of the subroutine, or the interrupt-servicing routine. When the temporary branch has been completed, the processor retrieves the contents of the program counter from the stack and resumes the execution of instructions from where it branched.

1.7.3 Stack

The stack is a group of sequential memory locations whose operation is based on the principle 'first-in last-out', i.e. it can be accessed from one end. The allowable depth of the stack depends on the capacity of the RAM area. It will therefore grow within the RAM as more bytes are stored in it. If care is not taken, then the stack can grow and overwrite information stored in the RAM. The designer should therefore consider the worst possible case and ensure that the area reserved for the stack is sufficient. This does not represent the only use of the stack, as it can also be employed to store information such as the contents of general-purpose registers used by a temporary branch.

1.7.4 Execution speed

The speed at which a processor steps through its program depends on the frequency of its clock. High performance can therefore be achieved when the processor is driven from a clock possessing the highest acceptable frequency.

Manufacturers have produced different versions of the same processor to satisfy various requirements. For instance, the Intel 8086 processor is offered in 5-, 8- and 10-MHz clock rate versions, whereas the Intel 80186 processor is available in 8-, 10- and 12.5-MHz versions (the 8-MHz version of the 80186 is designed to give twice the performance of the 5-MHz 8086). One can, for example, increase the performance of a PC incorporating a 33-MHz Intel486 DX microprocessor by using the 50-MHz version of the same microprocessor (clearly, this is only possible if the memory and interface devices are capable of operating at the higher clock frequency). Many computer boards include links to allow the user to change the frequency.

In some processors the clock is generated by an external device, while many other processors contain an on-chip oscillator whose frequency is determined by an externally connected crystal. The designer can therefore define the frequency of operation of the processor by selecting the crystal (the choice is obviously limited by the capability of the processor, which is stated in its data sheet). For example, the Intel 80186 microprocessor contains an on-chip clock generator which requires a crystal frequency twice that of the CPU clock. Therefore, operation at 10 MHz is achieved by using a 20-MHz crystal. The 80186 also allows the use of an external oscillator instead of an internal one.

Designers of microcomputers normally operate the processor at the highest specified speed. This gives the highest performance, and if this is not required then one can use a lower-frequency version of the processor which is normally of a lower cost (if available). An attractive way of increasing performance without altering the hardware is to use a processor with a frequency-doubling capability, such as those within the Intel486™ family of microprocessors (referred to as the Intel486™ DX2 microprocessors). For example, operation at 66 MHz can be achieved by replacing a 33-MHz Intel486 DX microprocessor with the Intel486 DX2-66 microprocessor. One should keep in mind, however, that the improvement is only within the processor, not in the external circuit.

1.7.5 Memory read/write

To read from a memory device, the processor provides an address on the address bus and pulses the read line of that device. This commands the device to transfer data to the data bus. When writing to a memory device, the processor provides an address and pulses the write line of that device after placing data on the data bus. This is further described in Chapter 2.

1.7.6 Reset

The operation of a processor can be reset by activating the reset line for a specified number of machine cycles. A simple circuit can be included external to the processor to activate an automatic reset after the power is switched on. The active state is not the same for all processors. For instance, the reset line of the 8086 is active-high, whereas in the Z80 the reset line is active-low. After a reset, the processor aborts the execution path and starts executing instructions beginning at a specific location, e.g. location 0 in the Z80 and FFFF0H in the 8086. The program can therefore be arranged so that after a reset, the processor executes an initialization procedure. It defines the address of the stack pointer, enables the acceptance of certain interrupts, specifies lengths of working buffers, sets locations corresponding to various variables to their default values, etc.

1.8 Execution of tasks

A microcomputer can be programmed to execute a variety of tasks. During its operation, the processor steps through the program, executing one instruction at a time. This introduces a limitation that the processor can handle only one operation at any instant of time (this is not the case for all processors, as the Intel Pentium processor can execute two instructions simultaneously). Clearly, this limitation does not mean that a microcomputer cannot handle the simultaneous execution of several tasks over a period of time. Multitask operation is possible because the microcomputer can be programmed to adopt the time-sharing approach where a repetitive loop can be generated to execute the tasks in sequence. An alternative way of implementing tasks is to use interrupts (Section 1.9).

For example, a microcomputer may need to read values from two input devices, add them and supply the sum to an output device. Let us assume that the microcomputer performs these actions in 10 µs and that the minimum acceptable rate of repeating this operation is 10 kHz. If the microcomputer is dedicated to the execution of this task, then it can repeat its execution once every 10 µs, i.e. an update rate of 100 kHz. This is higher than the minimum acceptable rate and will therefore satisfy the requirement. However, executing the task at the highest possible rate implies that the microcomputer does not have time for other activities. This clearly is an inefficient way of using the microcomputer, which can be provided with the ability to implement other tasks. The first step is to reduce the computation time spent on the above-described operation. This can be achieved by performing these operations at a lower rate. The above specifies a minimum acceptable updating rate of 10 kHz, i.e. the output can be updated once every 100 µs. Since the task under consideration can be processed in 10 µs, the requirements can be satisfied by using only 10% of the computation time. The rest can be utilized for implementing other tasks as long as their execution does not take more than 90 µs. Let us assume that the microcomputer will be used to execute two additional tasks, each of which is executed in 30 µs. In this case, the execution of the three tasks is arranged in a loop whose duration is 70 µs (which is within the design specification). This represents an example of a 'software loop' which can be processed within a fixed time interval. This is not always the case in software loops. The operation of numerous microcomputer-based systems is arranged such that the microcomputer interrogates several external flags and performs the tasks whose corresponding flag is active. In such systems, the duration of the software loop depends on the number of active flags. Microcomputers, therefore, deal with samples of input and output signals and the maximum sampling rate depends on the time required to execute various tasks within a software loop. The execution time depends on the type of processor used, the way it is programmed and the frequency of its clock. The last of these indicates that if a program containing a software loop is transferred from one microcomputer to another possessing a faster clock, then that loop will be executed faster. In applications having no critical timing requirement, the increase in the execution speed can contribute to speeding the response of the system. The increase in speed is not always favourable. If, for example, it is required that the repetition rate be the same in both microcomputers, then software delays will need to be introduced in the faster microcomputer.

Rather than executing all the tasks at the same rate, the requirements of many applications can be met successfully by arranging different rates of execution, as shown in the following example.

Example 1.1: **Write a procedure for executing two tasks. The microcomputer takes 10 μs to execute the first task, which needs to be repeated at a rate of 10 kHz. The second task takes 20 μs and is to be run at half the rate of the first task.**
A repetition rate of 10 kHz means that the task must be executed once every 100 μs. One way of satisfying the requirement is to arrange the execution of the tasks in a software loop represented by:

1 EXECUTE TASK 1
2 EXECUTE TASK 2
3 WAIT FOR 70 MICROSECONDS
4 EXECUTE TASK 1
5 WAIT FOR 88 MICROSECONDS
6 GO TO STEP 1

The procedure is in the form of a continuous loop. Once the microcomputer finishes executing one task, it begins executing the other task or waits for a specified interval. The execution time of steps 1 and 2 is 30 μs. Therefore, a 70-μs delay is introduced in step 3 to obtain the required 100-μs duration between the beginning of step 1 and the end of step 3. Similarly, the time from the beginning of step 4 until step 1 is also 100 μs (where the execution time of step 6 is assumed to be 2 μs). Consequently, the procedure executes task 1 once every 100 μs and task 2 once every 200 μs.

An alternative way of executing tasks at different rates is to use 'software counters', one per task. Such a counter is a memory location (e.g. a byte) which contains a number inversely proportional to the required rate of execution. Before executing a task, a software loop decrements the content of the counter. If the result is 0, then the counter is reloaded with the full count and the task executed. Otherwise, the task is bypassed. This approach provides flexibility and makes it easy to include many tasks within the same software loop, but execute them at different rates.

In applications having the requirement of passing information from one task to another, use can be made of predefined memory locations. These can either be individual locations corresponding to single variables, or a block of memory which can be reserved for data storage and retrieval.

A microcomputer can be programmed to use its decision-making capability to determine the execution sequence of tasks or of steps within a task. The microcomputer can, for instance, take a decision after examining the state of switches or sensors as shown in the following example.

Example 1.2: **Consider an application where a microcomputer-controlled robot arm is employed to pick objects from either of two locations (x and y) and place them at a defined destination (z). The robot controller receives three digital commands. The first two (X and Y) provide information on whether to pick an object from location x or location y respectively. If both commands are received simultaneously, then location x is given the priority. The operation should continue as long as the third command line (W) is at a high logic state.** Let us assume that, at start, the end-point of the robot is at location z. The application requirements can therefore be satisfied when the microcomputer executes the following procedure, where a byte 's' is employed to store a value identifying the destination to which the end-point of the robot needs to move:

```
 1  READ INPUTS X AND Y
 2  IF INPUT X IS ACTIVE THEN LET s = x AND GO TO STEP 5
 3  IF INPUT Y IS ACTIVE THEN LET s = y AND GO TO STEP 5
 4  GO TO STEP 9
 5  MOVE END-POINT AS DEFINED IN s
 6  PICK OBJECT
 7  MOVE END-POINT TO z
 8  PLACE OBJECT
 9  IF INPUT W IS ACTIVE THEN GO TO STEP 1
10  END PROCEDURE
```

In this procedure, the program consists of a sequential loop in which the program-flow path is determined by the state of the digital-input signals. If neither input X nor input Y is active, then the microcomputer executes steps 1, 2, 3, 4 and 9. If input W is not active, then step 10 is executed to terminate the procedure. If, on the other hand, input W is active, then the execution of steps 1, 2, 3, 4 and 9 continues until input X, or input Y, becomes active. This informs the microcomputer of the location from which the robot arm should pick the next object. Note that in order to give input X a higher priority than Y, both inputs are only read in step 1 and input X is tested before Y.

An operator or a controlling device can terminate the procedure by altering the state of input line W. The detection of a change of state by the microcomputer stops the pick-and-place operation only when the end-point is located at point z. This means that when the robot starts next time, it will be ready to move to point x or y depending on the command. This implies that only two paths of motion (Section 13.6) need to be stored in the memory of the microcomputer, (x to z) and (y to z), i.e. there is no need to move the end-point from point x to point y.

The method of execution described so far is known as 'polling'. It is simple to apply and is best suited for situations where the microcomputer is dedicated to the implementation of a single task or a sequence of operations. The execution process may include intervals where the microcomputer waits for a device to process data or to take an action. This can take a fair amount of the processing time, during which the processor is unavailable for other, possibly more important, tasks. This may considerably reduce the throughput and efficiency of the microcomputer. Examples are given within the book illustrating that, for such a situation, the burden can be minimized by correctly arranging the program or by assigning dedicated hardware to perform the time-consuming tasks under the control of the processor.

1.9 Interrupts

As explained earlier, a processor executes tasks by stepping through its instructions sequentially. In some situations, the processor may be executing a task and information becomes available from an external device. It may be available for a very short period and has to be read by the processor before it is lost. In such a situation, an interrupt can be generated by the external device to ask the processor to read the information. In a similar manner, interrupts can be utilized to provide a processor with the ability to supply information to an external device at regular intervals. This can be achieved by permitting a timing device to interrupt the processor periodically. The use of interrupts eliminates the need for executing unnecessary software delays. If correctly organized, the use of interrupts can greatly contribute to the design of high-performance systems. However, the use of interrupts is not always needed, as the requirements of many applications can be best satisfied by the polling approach. Many of the examples given in this book compare both methods in order to select the optimum solution for an interfacing problem.

1.9.1 Processor handling of interrupts

When an interrupt occurs, the processor finishes executing its current instruction and then the following procedure is adopted:

1 The contents of the program counter are stored in the stack.
2 The program execution vectors to a specific memory location corresponding to that particular interrupt.
3 The interrupt-service routine is executed.
4 At the end of the interrupt, a return instruction is executed. This commands the processor to reload the program counter with the information it has stored in the stack.
5 The execution of the program resumes from the point it left when the interrupt appeared.

The interrupt routine may use some registers, RAM locations, or flags used by the main program. These need to be restored to their original contents after the completion of the interrupt routine. This is achieved by including instructions in the interrupt routine to store the valuable contents in the stack at the beginning of the interrupt routine (e.g. the PUSH instruction) and to restore them at the end of the interrupt and before returning to the main program (e.g. the POP instruction). The programmer should keep in mind that the operation of the stack is based on the principle first-in last-out. Therefore, taking information out of the stack should be performed in an order opposite to that of their storage.

Operations involving the use of interrupts (such as diverting the execution path, storing and retrieving information, and returning to the main program) take time and impose a restriction on the use of interrupts. This becomes significant and needs to be taken into serious consideration when the interrupt is repeated at a high rate. Attention should also be paid to the time needed to process the interrupt-service routine. If the processing of this routine introduces unacceptable delays in to the main program, then the designer can, for example, write the interrupt-service routine in such a way that it consumes as short a time as possible. It handles the time-critical parts and leaves the more time-consuming functions to be performed by the main program. A short interrupt-service routine may start by reading from, or writing to, one or more memory locations; it then returns to the main program after setting a flag to indicate that an action has been taken.

The number of interrupts available in a microcomputer for use by an interfacing circuit, their type and their priorities depend on many factors. The main factors are the processor upon which the microcomputer is based, whether the microcomputer includes an interrupt controller or not, and the number of interrupts used for the operation of the microcomputer itself. Some microcomputers provide the user with access to some, or all, of its processor interrupt lines. Others provide a few interrupt lines and specify conditions for their use. Information on such interrupts is provided in the technical manual of the microcomputer. Examples 1.3 and 4.8 consider a situation where an interface circuit generates interruption signals which need more interrupt lines than those available in a microcomputer.

When using interrupts, one should keep in mind that:

- Noise on interrupt lines can cause random interruptions of operation which could lead to undesirable actions (the subject of noise and its reduction is dealt with in Section 16.7).
- Care should be taken when connecting interrupt signals to microcomputers, since connecting the wrong signal, or one with the wrong voltage level, could damage the microcomputer (especially if the interrupt line is connected directly to the processor without buffers or protection devices).

1.9.2 Types of interrupts

One way of categorizing interrupts is according to their source:

- 'Internal interrupts' are those generated from within the microcomputer, e.g. as a result of detecting an error within the microcomputer, or when an internal counter reaches a specified count.
- 'External interrupts' are produced by devices external to the microcomputer, e.g. when an interface circuit signals the microcomputer to synchronize the execution of tasks with the occurrence of an external event.

An alternative way of categorizing interrupts is according to the degree of control they offer to the programmer:

- A 'maskable interrupt' is one which provides the programmer with the ability of enabling or disabling it (e.g. line INTR in the Intel 8086 microprocessor).
- A 'non-maskable interrupt' (e.g. line NMI in the 8086 microprocessor) cannot be disabled. It is therefore used to perform time-critical tasks of the highest priority, such as indicating serious faults, or resetting the microcomputer as a result of detecting a catastrophic error.

The ability to mask and unmask interrupts is extremely useful, especially in programs containing time-critical parts which cannot tolerate delays resulting from processing the interrupt-service routine. Rather than ignoring the interrupt-based approach and adopting the polling approach, a programmer may choose to disable the interrupts during the execution of a time-critical part of the program. This approach leads to the formation of a non-interruptable section of the program. The interrupts can then be re-enabled after completing the execution of that part of the program. The programmer should, however, ensure that the interrupting device will not lose the information before re-enabling the interrupt (Example 4.8 illustrates a method of using external devices to store the source of interrupt until recognized by the microcomputer).

The method of enabling and disabling interrupts depends on the type of processor used. In a large number of processors, this task is performed by setting, or clearing, one or more bits (flags) in a specified register. For example, a processor containing N maskable interrupts may provide a user with N flags to permit the user to enable or disable each of the interrupts individually. For instance, the maskable interrupt-request line (INTR) in the Intel 8086 microprocessor can be masked by using software to reset the interrupt status flag bit. Instead of using individual flags to enable/disable interrupts, the Motorola 68000 microprocessor possesses seven interrupts (numbered 1 to 7). The enabling of interrupts is performed according to the contents of 3 bits in the status register. The processor only responds to a request from an interrupt if its number is higher than the binary combination of the 3 bits (this is further described in the next section).

Interrupts can also be categorized according to the type of signal required for their activation to:

- level-triggered or
- edge-triggered

For example, both the maskable and the non-maskable interrupts of the Zilog Z80 microprocessor are active-low. The former is level-triggered, whereas the latter is edge-triggered. On the other hand, the maskable interrupt-request line in the Intel 8086 is activated by a high-level signal, whereas the non-maskable interrupt is activated by a rising edge.

1.9.3 Interrupt priorities

When using a few interrupts, it is possible that two or more interrupt sources might be activated simultaneously. The processor, however, can only execute one interrupt-service routine at a time. This implies that a decision is needed on which of the interrupts the processor should respond to, and whether to allow one interrupt to interrupt another. Such decisions can be taken when interrupts are given 'priorities', where a high-priority interrupt can interrupt another of a lower priority. Attention should be paid to the effect of one interrupt on another. When two interrupts occur simultaneously, the servicing of the interrupt of the highest priority takes precedence over responding to the other interrupt or attending to other background tasks.

Both the Z80 and the 8086 microprocessors have two hardware interrupt lines, one maskable and the other non-maskable. The latter possesses a higher priority than the former (the 8086 also possesses internal interrupts). The 68000 microprocessor utilizes three of its pins (named IPL2, IPL1 and IPL0) to accept a 3-bit digital number representing one of seven interrupt levels, or a single non-interrupt level. The latter is represented by the digital number 000, whereas numbers 001 to 111 represent the seven interrupts. Each interrupt is represented by a number equal to its priority, for example, level 7 possesses the highest priority and level 1 possesses the lowest. As explained in the previous section, an interrupt can be accepted only if its number is higher than the contents of 3 bits in the status register. Once an interrupt is accepted, the status register is saved and the number of the accepted interrupt is written into the three status bits. This only allows a high-priority interrupt to interrupt another of a lower priority. Interrupt level 7 forms a special case; it is non-maskable.

Rather than having to encode the hardware interrupt requests individually, use can be made of an 8-to-3-line encoder such as the 74LS148 IC. Figure 1.3 shows the use of this IC to produce seven prioritized interrupt-request lines. Note that the EI (enable input) line of the encoder is connected to the 0-V line, thus permanently enabling the IC to accept interrupt requests.

Interrupt priorities can be made programmable by the use of a 'priority interrupt controller'. This can be an individual IC, such as the Intel 8529 employed in the IBM and compatible PCs. Interrupt controllers are also designed within microcontroller ICs. Not all such controllers are the same, but, in general, one can say that a controller of this type can be programmed through writing to three registers. The state of each bit within the first register defines whether each of the interrupts is level-triggered or edge-triggered. Consequently, a register of a single byte is reserved to give a programmer the ability to define the state for eight individual interrupts. In a similar way, each bit of the second register permits the definition of whether each interrupt is to be masked or unmasked. The third register allows setting the priority of each interrupt. This register provides the designer with a powerful tool for setting and altering priorities by software. Setting the priorities of eight individual levels requires 3 bits per interrupt. If there are eight individual interrupts, then 4 bytes are reserved for this purpose, where each nibble of a byte defines a priority level of a specific interrupt. Another register is also included to identify the active interrupt level during any instant of time (where the background is given the lowest priority). This gives the controller the ability to compare the level of a newly requested interrupt with the active interrupt.

Interrupt priorities are particularly useful when using the microcomputer in a real-time multitask system. The microcomputer can, for instance, be programmed to receive commands from operators, communicate with external devices and at the same time maintain the real-time nature of the system. For example, a microcomputer can be programmed to respond to any of two interrupts generated from a system. The first interrupt is employed to command a fast termination of a process, whereas the second interrupt commands the microcomputer to display a message indicating the state of specified input lines. In such a case, the first interrupt should be given a higher priority than the second. It may also be possible to use the latter as a background task rather than as an interrupt (within a software loop).

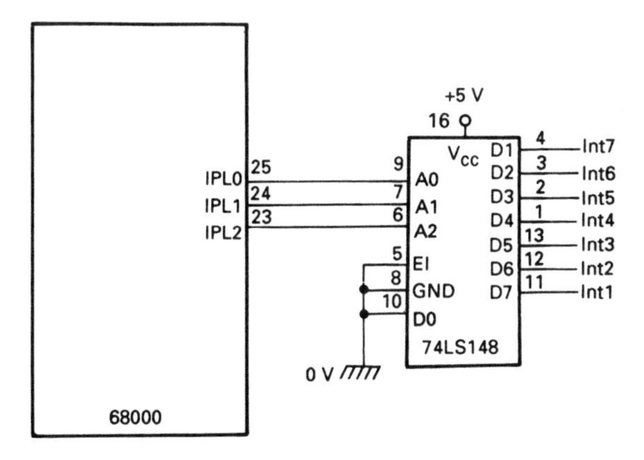

Fig. 1.3 Interrupt-encoding circuit.

1.9.4 Operation of a multi-interrupt system

Let us consider a situation where two interrupts are utilized. One is given a higher priority than the other. The program is therefore split into three levels, as follows.

Level 1

This is the high-priority interrupt. It includes the parts of the program which require immediate response, such as responding to a request from an input device when an event occurs. The interrupt-service routine may take the form of reading a value or executing a task. When the microprocessor handles the high-priority interrupt, the execution of other parts of the program stops until completion of the execution of the interrupt-service routine. In many applications, the high-priority interrupt can occur at random and will delay the execution of a lower-priority interrupt as well as background tasks. This should be taken into consideration by the designer.

Level 2

This is the low-priority interrupt. It includes the part of the program which needs to be serviced when an event occurs, but can tolerate execution delays occurring as a result of servicing a higher-priority interrupt.

Level 3

This is the background program. It contains all parts of the program to which no immediate response is essential. The designer should ensure that there will be sufficient time for the execution of the background tasks efficiently, even if all the interrupts are requested at the same time.

1.9.5 Applications considerations

Many computerized systems are designed to be interrupt-driven, i.e. the microcomputer waits for an interrupt to occur before taking a corresponding action. Between interrupts, the microcomputer can either be kept in a standby state or, preferably, employed to provide additional functionality, which makes the system more attractive and easier to use. An example of an interrupt-driven system is an industrial controller designed to wait for two types of interrupts. The first is an indication from a sensor, while the second is a command from an operator or an external device. Each interrupt is an independent task and there is little or nothing to do in the way of background. Such a controller can be designed to spend most of its time servicing the interrupts efficiently, thereby leaving a very short idling interval. This is not the case for a controller whose interrupts are used to update a few memory locations, leaving most of the processing to be performed in the background. In this type of controller, the percentage time spent on servicing the interrupts is limited by the required rate of servicing the background tasks. This is because there should be sufficient time to use the values delivered by an interrupt before they are overwritten by the occurrence of the next interrupt.

An interrupt-driven system makes it easy for the microcomputer to execute many independent tasks. In this way, adding new tasks or deleting others will neither create the need to alter the way of attending to the rest of the tasks nor introduce delays. Such alterations or delays may be needed in a sequential time-sharing system. For instance, if task 2 is to be removed from the procedure of Example 1.1, then the time delay following the execution of task 1 must be increased by 20 μs, or the procedure rewritten. One should keep in mind, however, that the occurrence of an interrupt diverts the execution of the program. Consequently, one should examine the critical paths within that program and estimate the effect of the diversions and the resulting delays in the execution of sequential steps. One should also keep in mind that in the case of using several interrupts, a low-priority interrupt can be interrupted by another of a higher priority. Operation should therefore be designed so that any part of a background task, or a low-priority interrupt, should be capable of tolerating the resulting delays. This may require more frequent servicing of certain background tasks, particularly those dealing with external devices. In some situations, the designer may need to introduce additional hardware, or to write the interrupt service routine in such a way that it consumes minimum execution time.

When using interrupts, one of the main causes of faults is the overflow of the stack. This occurs when the information saved on the stack occupies more memory than its allocated length. Its overflow causes writing over RAM locations which may be used by the program. This leads to unpredictable results. The designer should therefore allocate a large enough area for the stack. This is not difficult in systems with a large capacity of RAM, but in many systems (particularly those whose total memory area is the internal memory within a low-cost microcontroller) the stack area needs to be optimized. This can be achieved by examining the program and estimating the maximum length of information that will be stored in the stack at any instant of time. The maximum length is likely to be reached when a high-priority interrupt appears during servicing another of a lower priority, where each of these interrupts stores information in the stack. A development tool, such as an in-circuit emulator (Section 17.7), allows a designer to terminate the execution of the program at a selectable location or instant and examine the contents of the memory area. This helps the designer to observe the information placed in the stack and determine the optimum depth of the stack.

Example 1.3: **A microcomputer-controlled robot is employed to perform any of three preprogrammed operations. The selection of an operation is performed through one of three active-low lines. The microcomputer is so involved with controlling the robot that it cannot spend time monitoring these lines. How can the information be supplied to the microcomputer, assuming that only one interrupt-request line is available?**

The application imposes the requirement that the microcomputer should spend as little time as possible on the execution of the user-interface task. The designer is faced with a problem that only one interrupt signal should be generated from the three interruption sources. One solution to the problem is presented in Fig. 1.4, where a 3-input NAND gate is employed to multiplex the signals (the 74LS10 IC contains three such gates; see Appendix). The output line of the NAND gate can be connected to the interrupt line of the microcomputer.

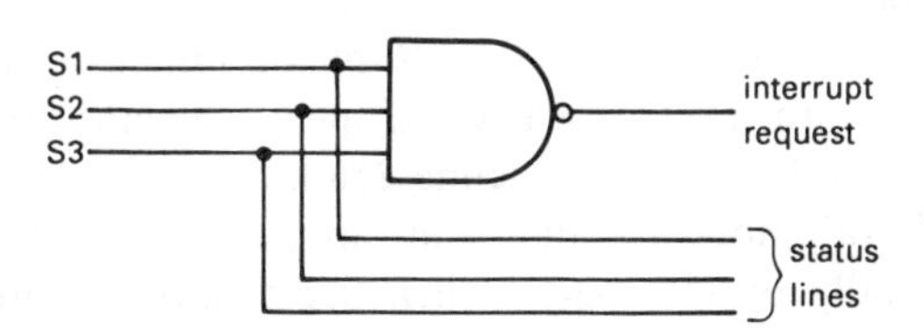

Fig. 1.4 Multiplexing an interrupt-request line.

The NAND delivers a logic-low signal when all of its input lines are at a logic-high state (not active). When any of the three sources (S1, S2 or S3) is activated, it supplies the NAND gate with a logic-low signal. The state of the output line will change to high, thereby interrupting the microcomputer. The interrupt signal, however, only tells the microcomputer that one (or more) of the three lines is activated. It provides no information on which line it is. Consequently, the three interruption signals are also connected to three digital-input lines of the microcomputer. The interrupt-service routine starts by testing the state of the lines to determine the source of the interrupt. The successful detection of an active request line results in the setting of a software flag corresponding to that particular request. The change in the state of the flag will be recognized by the background program, which will branch to execute the appropriate routine at a later instant.

If the interruption signals carry time-dependent signals of different importance, then software can be used to prioritize the requests. This can be accomplished by testing the request lines in sequence. In this way, the first line tested possesses the highest priority.

The solution assumes that the source of the interrupt will be there long enough for the microcomputer to examine and recognize it. In some situations, however, interrupts are requested by very narrow pulses. Such pulses may disappear before the microcomputer has had the chance of examining which of the sources caused the interrupt. If it was possible for this to happen, then the above solution could be replaced by that of Example 4.8, where the source of the interrupt is stored until read and cleared by the microcomputer.

1.10 Quantization and resolution

A digital word consists of several bits, each of which is given a weight which is different from the weight of the rest of the bits. The MSB (Most Significant Bit) has a weight of half of the full scale, whereas the second MSB has a weight of one-quarter of the full scale. In other words, the *n*th bit has a weight of $\frac{1}{2}$ to the power n. Consequently, in a 16-bit word, the LSB (Least Significant Bit) has a weight of 1/65 536 of the full scale. Representing an analogue value by an *n*-bit digital word means placing it within 2 to the power n distinct levels, e.g. 8 bits define 256 levels and 16 bits define 65 536 levels. Figure 1.5(a) shows an analogue ramp signal. If this signal is to be represented digitally by four levels, then the resulting representation is as shown in Fig. 1.5(b), where the steps are large. On the other hand, Fig. 1.5(c) gives an 8-level representation of the same signal, where the steps are smaller than those of Fig. 1.5(b). Consequently, the difference between the actual signal and the digital representation decreases with increasing a number of levels. In order to get a digital representation which exactly equals

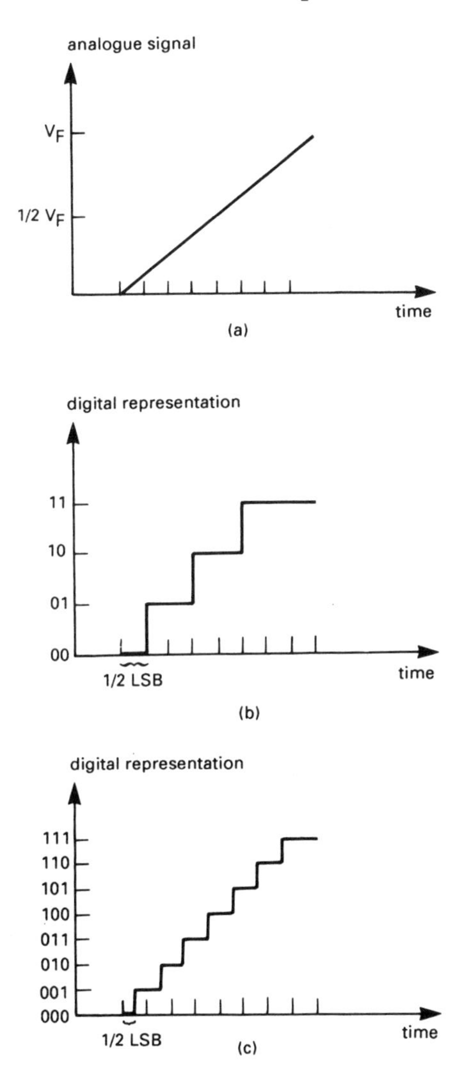

Fig. 1.5 A ramp signal: (a) analogue representation; (b) 4-level digital representation; (c) 8-level digital representation.

the analogue signal, an infinite number of levels is required. Since this is not practically possible, there will always be an error which is inversely proportional to the number of bits representing the function. This error is known as the 'quantization error', and the digital representation of the analogue signal 'quantizes' the analogue value into the nearest digital level.

The size of each level is defined by the size of 1 LSB which, in turn, is defined by the resolution of the conversion device. When converting an analogue value into a digital form, the resolution represents the smallest change in the analogue input which will be recognized by the conversion device. Similarly, when converting a digital value into an analogue form, the resolution of the conversion device represents the smallest change in the analogue signal that the conversion device can produce. The definition of the resolution assumes the use of an ideal conversion device. One can therefore say that an *n*-bit converter provides a resolution of $\frac{1}{2}$ to the power n. Resolution can also be defined in terms of bits of the conversion device, e.g. 8-bit resolution.

One can also see from Fig. 1.5 that the maximum difference between the analogue value and the digital representation is $\pm\frac{1}{2}$ LSB, i.e. a digital system will not recognize a change in the analogue value when it is less than $\frac{1}{2}$ LSB. The relationship given in Fig. 1.5 assumes that V_F represents the full-scale value of the conversion device. The relationship is true for an ideal converter. More information is provided in Section 5.7, together with a description of errors which occur in a practical converter.

Chapter 2

Microcomputer interfacing techniques

2.1 Introduction

The main goal in the design of numerous microcomputer-based systems is to produce a reliable, high-performance, easy-to-use product at the lowest cost. An important part of the design is the interface between the microcomputer and the equipment it monitors or controls. It depends on whether or not the microcomputer possesses digital and/or analogue input and output lines for implementing the interface. If it does, then a decision is required on how to utilize these lines to satisfy the interface requirements efficiently. In many situations, existing input and output lines need to be saved for other uses, thereby creating the need for introducing additional hardware to implement the interface. The type and complexity of the hardware depend on both the microcomputer and the add-on device. They also depend on the type, direction and length of information to be transferred. The term 'add-on' is used in this book to refer to a device interfaced, or to be interfaced, to a microcomputer. It can be a port (input or output), an IC, a circuit, or an external system.

Some add-ons can be connected directly to the data bus; they possess one or more 3-state buffers through which signals are transferred to, or from, a microcomputer (see Section 1.2.1). Other add-ons do not include 3-state buffers, so their interface to a microcomputer introduces the need for using 3-state devices (ports). In either case, the interface circuit can be designed to allow the microcomputer to behave as a master which will produce signals to control the transfer of information through the data bus. A microcomputer can be programmed to communicate with an add-on regularly, or following the occurrence of an event. Alternatively, the communication can commence when the microcomputer receives a request from an add-on.

This chapter presents examples of interfacing add-ons to a few microprocessors. It concludes by designing two types of interface circuits for use in many of the examples presented within the book. Interfacing microcontrollers follows the same methods, although in most applications, microcontrollers use their own internal parts (memory and peripherals). This is because interfacing through data and address lines means introducing additional devices and losing a large proportion of the input/output ports of the microcontroller.

2.2 Device-selection signals

When interfacing add-ons to a microcomputer, each add-on needs to be given a unique address. A logic circuit is required to produce a pulse only when that address is supplied by the microcomputer. This pulse selects (enables) an add-on to communicate with the microcomputer. During communication, the selected add-on supplies information to the microcomputer, or reads information from it.

It is possible to interface many add-ons to the data bus (this may introduce the need for improving the drive capability of the bus, as described in Section 2.5). During its operation, a microprocessor only deals with one device at a time. It makes no difference if that device is an internal IC or an add-on. The microprocessor selects the device it wants to communicate with by supplying a suitable 'device-selection' pulse to the device's enable line. The interface of add-ons to a microprocessor creates the need for a flexible decoding arrangement to produce an efficient interface and to permit the use of additional add-ons in future. Suitable interfacing devices need to be chosen by the designer to prevent bus contention, which occurs when one device is selected while another is undergoing deselection (the bus contention results when fast decoders are used with devices possessing slow turn-off times).

There are normally many addresses available for the designer to use in deriving device-selection signals. One method of decoding is *memory mapping*, in which the microcomputer deals with a device as one or more memory locations. With this method, the communication is achieved by executing instructions such as 'move to' or 'load from' memory. Another method of decoding is *input/output mapping*, which is supported by some, but not all, microprocessors. With this method, the microprocessor executes IN and OUT instructions to produce an address of an input or an output device rather than an address of memory.

There are different ways of deriving selection signals. They depend on several factors, such as the type of microprocessor and the available (unused) addresses. A designer seeking the production of the optimum interface should compare solutions in order to arrive at the most suitable design for the application. The following sections illustrate ways of generating device-selection signals by combining two types of signals produced by a microprocessor. The first is the *device-control* signal, which allows the microprocessor to control the communication process (Section 2.2.1). The second signal is the *addressing* signal, which allows the microprocessor to identify a device (which could be an input/output or a memory device) with which it wishes to communicate (Section 2.2.3).

Remember that:

device-control signal + addressing signal = device-selection signal

2.2.1 Deriving device-control signals

This section considers the formation of device-control signals by combining microprocessor-generated control signals. To achieve this task, we consider some of the control lines available in microprocessors.

Read and write signals

We start by considering two signals which identify the direction of transfer of information, the read and write signals. They permit a microprocessor to indicate whether it wishes to read from an input source or to write to an output destination. During communications, the microprocessor generates either of these two signals together with an addressing signal (Section 2.2.3) to identify the device it wishes to communicate with.

In some microprocessors, such as the Intel 8086 and 80186 and the Zilog Z80, the read and write signals appear on two separate active-low lines. In other microprocessors, such as the Motorola 68000, both the read and write signals are issued from one line. A logic 1 on that line indicates a read command, and a logic 0 indicates a write command. If required, the two signals can be separated by the use of an inverting gate, as shown in Fig. 2.1.

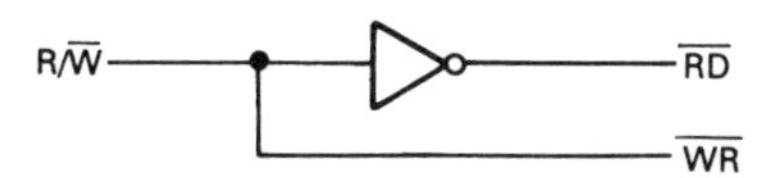

Fig. 2.1 Separating the read and write signals.

Using the read and write signals

Figure 2.2 shows a block diagram representing an example of interfacing input and output devices to a microcomputer. A decoding circuit is used to produce a device-selection signal which enables either of the two devices for communication with the microcomputer. The decoding circuit can be designed to consist of two parts. The first produces an active signal only if the address is correct (addressing signal). The second part produces an output whose state depends on the state of the control lines (device-control signals). The decoding circuit combines the output of both parts to form a unique device-selection signal per device.

The read and write signals can be included as part of the device-control circuit. This is not always the case in interfacing circuits, as certain devices accept the read and/or write signals as inputs. The interface of such devices can therefore be accomplished without supplying the read and write signals to the decoding circuit. Examples of devices accepting read and/or write signals are memory ICs. A ROM IC allows a microcomputer to read from it, but not write to it. It therefore contains an output-enable line (active-low) which can be driven from the read line of a microprocessor. Such a line is also available in a RAM IC to allow a microprocessor to read information from it. A RAM, however, is a read-write device. It therefore contains another active-low input line, called the write line. It can be connected to the write line of the microprocessor to be activated whenever the microprocessor wants to send information to the IC.

Let us assume that the read and write signals are supplied to the input of the decoding circuit of Fig. 2.2. The microprocessor reads information from device A and writes information to device B. Signal CS1 of Fig. 2.2 can therefore be activated by supplying a unique address and activating the read line. This signal enables device A to supply information to the data bus. Device A can be the source of the information, or it can act as carrier of externally generated information. The decoding circuit can activate signal CS2 in the same way except for activating the writing rather than the read

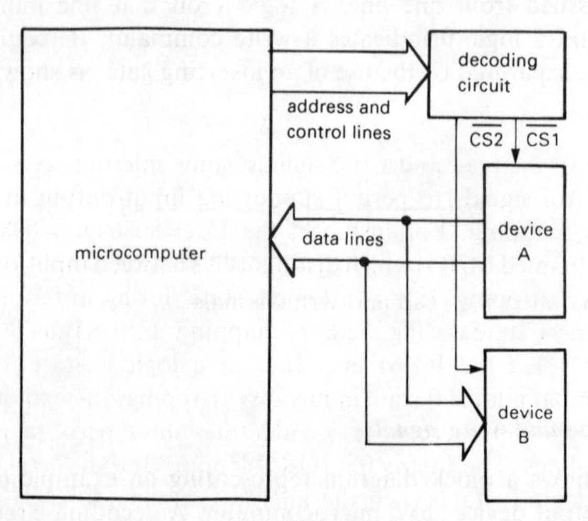

Fig. 2.2 Interfacing devices to a microcomputer.

line. This results in enabling device B to accept information from the data bus. This information can be used by device B, or passed on to an external circuit or system.

Note that it is possible to combine an addressing signal with the read line to form a device-selection signal for an input device, and to combine the same addressing signal with the write line to form a device-selection signal for an output device. In this case, both the input and the output devices share the same address, but only one of the devices can be enabled at any instant of time (depending on the state of the read and write signals). If the microcomputer aims to read from the input device, it executes an instruction which generates the required address and activates the read line. Similarly, if the microprocessor wants to transfer information to the output device, it issues the same address except, this time, it activates the write line. The simple circuit of Fig. 2.3 can be used to produce such signals. The circuit consists of two OR gates employed to produce two device-selection signals (CS1 and CS2) from a single addressing signal (CS). Line CS1 activates communications with an input device, while line CS2 activates communication with an output device.

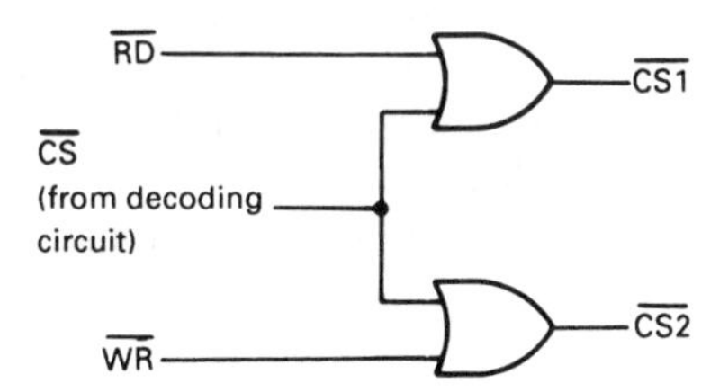

Fig. 2.3 Generating device-selection signals.

Other control signals

In addition to the read and write signals, some microprocessors include one or two control signals to permit supporting input/output mapping as well as memory mapping. For instance, the Intel 8086 and 80816 include an output line (named M/IO) which distinguishes between input/output mapping and memory mapping. It is at a logic 0 state during input/output mapping and at a logic 1 state during memory mapping. In the Intel 8088 and 80188, this line is called the IO/M line. It is at a logic 1 state in input/output mapping and at a logic 0 state in memory mapping. Instead of using a single line, the Zilog Z80 microprocessor identifies the type of mapping by using two output lines. The first is the IORQ (Input/Output ReQuest) line which is set to a logic 0 state when the microprocessor addresses a device using input/output mapping. The second line is the MEMRQ (MEMory ReQuest) line, which is set to a logic 0 when the microprocessor addresses a memory-mapped device. These signals can be included as inputs to a circuit generating device-selection signals.

The Motorola 68000 microprocessor uses the memory-mapping method, and thus add-ons are addressed in the same way as memory devices. Let us consider one of the control signals provided by the 68000, the AS (Address Strobe). It is set to a logic 0 state when the address line contains a valid address. A decoding circuit should therefore use this line when deriving device-selection signals (as in Fig. 2.5 and in Section 2.2.5). In addition to the AS line, the 68000 is provided with an input line, known as the DTACK (DaTa ACKnowledge) line, to accept an acknowledgement from external devices. Its function can be explained as follows. Since both the read and write signals of this microprocessor are supplied from a single line (R/W), the microprocessor requests information from an add-on by sending the add-on's address and setting the R/W line to a logic 1. The selected add-on will transfer the information to the microprocessor through the data bus and, at the same time, set the DTACK line. With this action, the device informs the microprocessor that the requested information is available on the data bus. A similar process is followed when the microprocessor sends information to an add-on. In this case, the microprocessor sets the R/W line to a logic 0, indicating a write state. When the add-on receives the information, it sets the DTACK line to a logic 0 state, indicating that it has received and accepted the data (see Example 2.3).

The following section gives examples to demonstrate the generation of device-control signals.

2.2.2 Examples of using control signals

Let us start by considering the generation of device-control signals from a 16-bit microprocessor, the Intel 8086. In addition to possessing a 16-line data bus, this microprocessor has 20 address lines, an active-low read line (RD), and an active-low write line (WR). It supports both memory mapping and input/output mapping by providing the M/IO signal described in the last section.

Example 2.1: **Design a circuit to generate four active-low control signals to allow an 8086 microprocessor to communicate with memory-mapped or input/output-mapped devices. The signals should permit the processor to read from or write to these devices individually.**

The requirement can be satisfied by using three control lines from the processor, namely the RD, WR and M/IO. The first two are active-low, whereas the M/IO line is low for input/output mapping and high for memory mapping. The required output signals are normally referred to as:

MEMR (MEMory Read)
MEMW (MEMory Write)
IOR (Input/Output Read)
IOW (Input/Output Write)

The MEMR and MEMW signals are used to identify the aim of the microprocessor to communicate (read or write) by using the memory-mapping method. On the other hand, the IOR and IOW signals indicate whether the microprocessor wants to read from or write to a device by using the input/output-mapping method.

The circuit of Fig. 2.4(a) can be used to produce the required four active-low signals. It uses four 2-input OR gates and a single inverting gate. The circuit therefore uses two ICs; the first is the 74LS32 (quad 2-input OR gate IC) and the second is the 74LS14 (hex schmitt-trigger inverting gates). Only one of the six inverting gates is used, leaving five gates available for other purposes, such as forming part of the circuit needed to generate the addressing signal (Section 2.2.3).

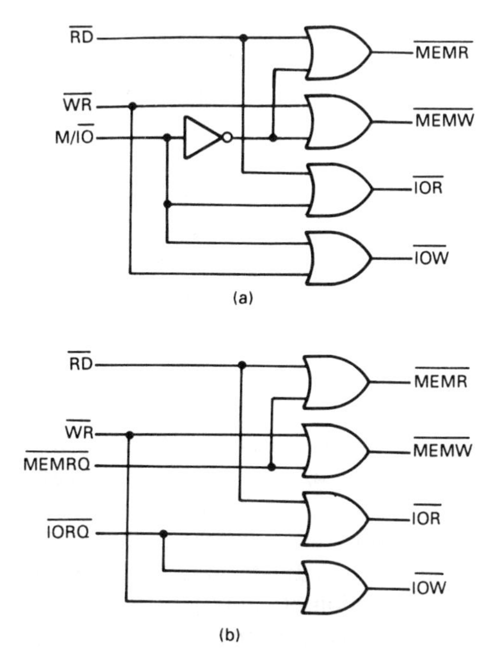

Fig. 2.4 Generating device-control signals: (a) using the 8086; (b) using the Z80.

Keep in mind that when the microprocessor activates one of these four output lines, it only identifies the type of mapping used and whether it wants to read or write. The selection is not complete without deriving a suitable addressing signal, such as signal CS in Fig. 2.3. This type of signal can be combined with any of the four outputs of Fig. 2.4(a) to produce a device-selection signal (Section 2.2.4).

Let us now consider deriving device-control signals from an 8-bit microprocessor, the Z80. It possesses an 8-bit data bus and a 16-bit address bus. Like the 8086, this microprocessor uses both input/output-mapping and memory-mapping methods.

Example 2.2: **Repeat Example 2.1, except this time derive the four signals from the Z80 microprocessor.**
The solution is very similar to that of Example 2.1, except that the Z80 provides one line for the input/output request (IORQ) and another for the memory request (MEMRQ). When the Z80 wants to receive information from an external source, it activates either of these lines and sets the read line (RD) to a logic 0 state. Similarly, writing to an external destination is performed by activating either of these lines as well as the write line (WR).

The required four signals can be produced by using these two lines with the read and write lines, as shown in Fig. 2.4(b), where all the signals are active-low.

Let us now consider generating device-control signals from the 68000 microprocessor. Since it uses memory mapping, reading from or writing to add-ons is treated as a memory read or a memory write respectively. Both the read and write signals are supplied on a single line and can be separated by using a single inverting gate (Fig. 2.1) to produce active-low read and write signals.

Example 2.3: **Design a circuit to allow a 68000 microprocessor to specify whether it wishes to read from or write to add-on devices.**
Figure 2.5(a) shows a simple circuit which can be used to allow the microprocessor to send information (write) to the add-on. Once the device receives the information, it activates the DTACK line, informing the microprocessor of the reception of data.

If the aim of the interface is to receive data from the add-on (read), then an inverting gate is required to invert the R/W signal before supplying it to the input of the OR gate.

In many applications, the add-on device is simple and has no means of setting the DTACK line. When interfacing such an add-on to the 68000, the device-selection signal can be fed back to the DTACK line, as shown in Fig. 2.5(b). Note that since the 68000 only uses memory mapping, the read and

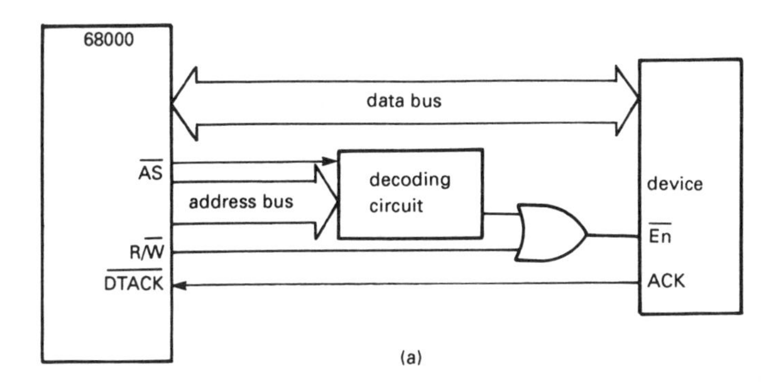

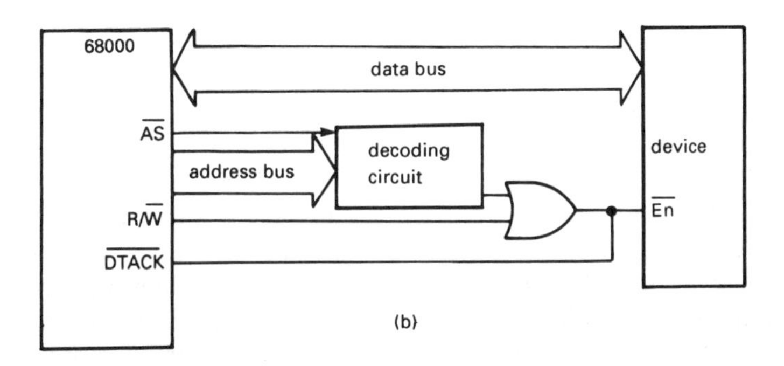

Fig. 2.5 Interfacing to the 68000 microprocessor: (a) acknowledgement from device; (b) acknowledgement from selection signal.

write signals are equivalent to the MEMR and MEMW signals of Example 2.1 respectively.

2.2.3 Deriving addressing signals

A microprocessor uses an addressing signal to identify the location of a device within the addressing area. When interfacing an add-on to a microcomputer, the designer needs to consult the memory map of the microcomputer (which is normally provided in the technical manual of the microcomputer) to find the area available for use in deriving addressing signals. The size and boundaries of the area vary from one microcomputer to another.

The addressing capability of one microprocessor can be different from that of another. For instance, the 8086 provides 20 address lines which are active during a memory map to produce the required addressing signal. During an input/output mapping, however, only the lower 16 address lines are active. This implies that the 8086 can address 1 megabyte of memory area and 64 kilobytes of input/output address area. The 68000 microprocessor, on the other hand, has 16 data lines and 23 address lines (A1 to A23) to address 8 megawords (each word consists of 16 bits). This means that the 68000 microprocessor can address 16 megabytes of memory. The number of address and data lines is lower in 8-bit microprocessors, such as the Z80. It possesses 16 address lines and can therefore address 64 kilobytes of memory. Its input/output addressing is achieved by providing an address on the lower eight address lines. It can therefore produce 256 addressing bytes.

A large number of add-ons can be addressed by a single unique location. In some cases, however (such as when an add-on board contains several individually addressable devices), a block of sequential addresses can be reserved for the interface. In such cases, one or more address locations are allocated to each device within the add-on. Ideally, a minimum number of addresses should be allocated to identify a device. Practically, however, a decoding circuit which reserves the minimum number of selection addresses per add-on may increase the complexity, cost and size of the interfacing circuit. A compromise can therefore be introduced in situations where the unused addressing areas exceed the requirements. In such a situation, a simpler decoding circuit can normally be designed at the expense of reserving a larger area for addressing an add-on. The designer, however, should not oversimplify the circuit, as the interface requirements may grow or change in future. This leads designers to leave an unused addressing area of what they consider sufficient size.

Note that the complexity of the decoding circuit is not always increased by selecting a smaller addressing area per device. This is because certain addresses can be produced by using a simple decoding circuit consisting of few gates, or a single decoding device, without sacrificing the addressing area of memory (as in Example 2.5). The design therefore depends on several factors, such as the size of the unused addressing area, the minimum number of addresses required per device, future requirements, and the knowledge of the designer of the availability of gates and/or decoding ICs which will simplify the decoding circuit.

The examples presented in the last section illustrate the generation of device-control signals. Let us now consider examples of deriving addressing signals. While doing so, we need to find a balance between the number of bytes that we reserve for the decoding and the complexity and cost of the required circuit.

Example 2.4: **Describe how to derive two active-low addressing signals from a microcomputer. Only one of the two signals needs to be activated at any**

instant of time and used to address an add-on device. Assume that the area between locations 8000H and 9FFFH is available for addressing by 16 address lines. The two addressing signals can be generated by decoding some or all of the 16 address lines. A circuit can therefore be designed to generate two signals, each identifying the area reserved for addressing one of the two devices.

Let us consider a simple solution based on using the circuit of Fig. 2.6. In this circuit, the two addressing signals are derived from decoding address lines A12 to A15 by using four 2-input OR gates (a 74LS32 IC) and two inverting gates (1/3 of a 74LS14 IC). Signal X1 will be low (logic 0 state) when A15 is high (a logic 1 state) while A14 and A13 are low. The circuit produces two active-low output signals X2 and X3. The logic state of these signals depends on the state of both X1 and A12. Both outputs will be high as long as X1 is high. If X1 is low, then the state of each output depends on the state of A12. When A12 is low, output X2 is low and X3 is high. On the other hand, when A12 is high, X2 is high and X3 is low. This means that X2 will be low (active) when the 16 address lines are defined by 1000 xxxx xxxx xxxx where x means don't care (any state). This identifies locations 8000H and 8FFFH. Similarly, signal X3 is low (active) when the address lines are represented by 1001 xxxx xxxx xxxx. This corresponds to locations 9000H to 9FFFH.

The circuit used in Fig. 2.6 is simple; however, it reserves the whole of the unused addressing area (8 kilobytes) for producing the two addressing signals. Any address with the lower 4 kilobytes identifies the first device, and any address within the upper 4 kilobytes identifies the second device. The reservation of the whole addressing area means that the microcomputer cannot be used to address other devices. This is therefore acceptable for applications where the two add-on devices form the only interface of the microcomputer. If the requirement for connecting other devices arises, then the circuit will need to be modified. This can be avoided if the circuit is designed to reserve less addressing areas, as shown in the following example.

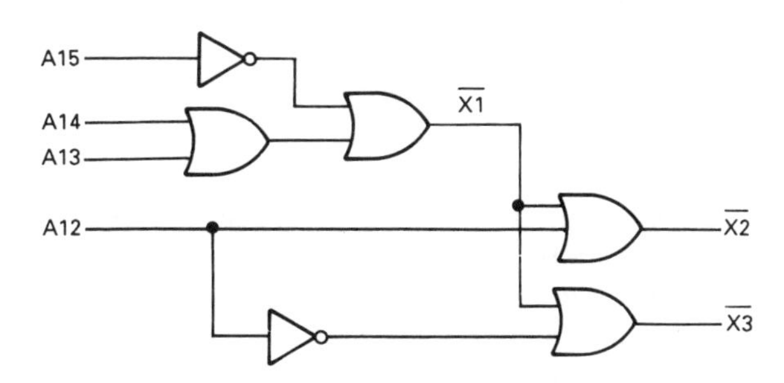

Fig. 2.6 Deriving addressing signals.

Example 2.5: **Modify the circuit presented in Example 2.4 so that a smaller addressing area is reserved for the two addressing signals.**
A smaller addressing area can be reserved if more address lines are included in the derivation of the addressing signals. Let us consider two solutions:

Solution 1

In Example 2.4, lines A12 to A15 are used to produce the addressing signals, where line A12 identifies which of the two devices is addressed. Let us now use address lines A2 to A15 to produce addressing signals with A2 identifying which device to select. Figure 2.7(a) shows a circuit which can be used to generate the two signals (an alternative circuit is given in Solution 2 below). The required signals are produced by employing 12 gates from four ICs. These are a 74LS32 (quad 2-input OR gates), a 74LS10 (triple 3-input NAND gates), and a 74LS14 (hex schmitt-trigger inverting gates). The circuit leaves five inverting gates available for use in other parts of the interfacing circuit. Two of these gates can be employed to invert and reinvert line A2 to reduce the effect of the propagation delay produced by decoding lines A3 to A15.

As in Example 2.4, the two output signals X2 and X3 are active-low. Their state depends on the state of both X1 and A2. Output X2 will be low only when both X1 and A2 are low, and output X3 will be low only when X1 is low and A2 is high. Therefore, as long as X1 is low, the state of line A2 defines which of the two outputs is active. The address needed to set X1 to a low state is represented by 1000 0111 1111 1xxx, which corresponds to addresses 87F8H and 87FFH.

Since output X2 is active only when both X1 and A2 are low, an addressing signal is produced through X2 when the address lines are represented by 1000 0111 1111 10xx, i.e. addresses 87F8H to 87FBH. Similarly, X3 is active only when A2 is high and X1 is low, i.e. when the address lines are 1000 0111 1111 11xx. This represents addresses 87FCH to 87FFH.

This demonstrates reducing the addressing area by the use of additional gates. The circuit of Fig. 2.7(a) only reserves 8 bytes for producing two addressing signals instead of the 8 kilobytes used by the circuit of Fig. 2.6. However, it employs several gates. The number of gates can be reduced by employing an alternative circuit, as shown in the following solution.

Solution 2

Instead of using the circuit of Fig. 2.7(a) to reduce the reserved addressing area, a simpler circuit can be constructed by connecting lines A15, A14, ..., A3 to the inputs of a 13-input NAND gate (the 74LS133 IC). This is shown in Fig. 2.7(b), where lines A14 and A13 are inverted before their connection to the input of the NAND gate. Thus, X1 will be low when both A13 and A14 are low and A3 to A12 as well as A15 are high. As long as X1 is low,

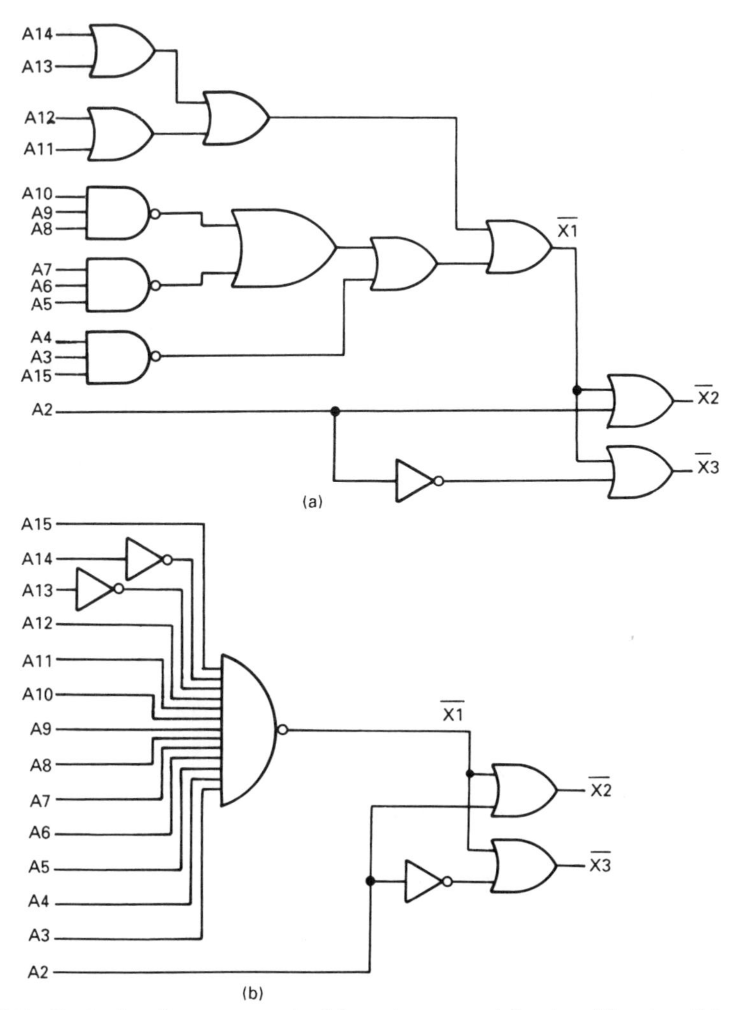

Fig. 2.7 Reducing the reserved addressing area: (a) using 12 gates; (b) using 6 gates.

the state of both outputs depends on the state of A2. When A2 is low, X2 is low and X3 is high, and when A2 is high X2 is high and X3 is low. This means that X2 is low when the 16-address lines are represented by 1001 1111 1111 10xx, which corresponds to addresses 9FF8H to 9FFBH. Similarly, X3 is low for the area corresponding to addresses 1001 1111 1111 11xx, which represents addresses 9FFCH to 9FFFH.

Comparison

Comparing this solution to Solution 1 reveals that this solution uses fewer gates to reserve the same size of addressing area as Solution 1. Note that the address of activating X2 and X3 in Solution 1 is different from that of Solution 2. Both, however, are within the area available for addressing.

2.2.4 Producing device-selection signals

Let us now consider examples of how to produce device-selection signals by combining addressing signals with device-control signals.

Example 2.6: **Design a circuit to provide an 8086 microprocessor with the ability to read from two add-ons using input/output mapping.**
The previous section showed how to combine address lines to produce addressing signals (such as X2 and X3 of Fig. 2.7). Since the 8086 uses 16 lines for addressing input and output devices, the circuit of Fig. 2.7 can be employed to provide the addressing signal. We also need to produce a device-control signal activated by a read instruction. This can be achieved as in Example 2.1, which shows how to produce an input/output read signal called the IOR.

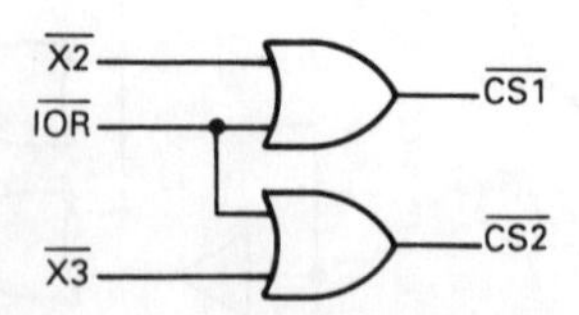

Fig. 2.8 Producing selection signals for input devices.

We therefore need to combine these signals as shown in Fig. 2.8 to provide the 8086 with the ability to read from each device using the input/output mapping method. The circuit provides two active-low outputs CS1 and CS2, each of which can be used to select a device. When IOR is low, the state of CS1 and CS2 depends on that of X2 and X3 respectively. Obviously, if an add-on requires an active-high selection signal, then any of the outputs of Fig. 2.8 can be inverted, or the circuit designed by using a NOR rather than an OR gate. The addressing area reserved by the circuit of Fig. 2.8 depends on the circuit used for deriving the X2 and X3 signals.

Clearly, if the requirement is to produce two device-selection signals for an output device, then the IOR line needs to be replaced by the IOW line in the circuit of Fig. 2.8.

Example 2.7: **What alterations are needed to the solution of Example 2.6 if the microprocessor is a Z80?**
The circuit of Fig. 2.4(b) is used in Example 2.2 to produce the IOR signal needed to identify the wish of the microprocessor to read from an input/output device. When activated, such a signal indicates that the lower eight address lines contain a valid input/output address. The addressing signal can therefore be produced by combining lines A0 to A7 of the microprocessor. They can, for example, be supplied to the eight input lines of an 8-input NAND gate (74LS30 IC). The output of the gate will be active (logic 0) when all of its inputs are at a logic 1 state, i.e. address FFH.

The required device-selection signals can be produced by using the circuit used in the last example (Fig. 2.8). Keep in mind that the IOR signal in the Z80 is produced by ORing the read and the IORQ signals, as in Fig. 2.4(b).

The above examples use the input/output-mapping method. The hardware needed to implement the memory-mapping approach can be applied by using similar circuits. Let us consider what alterations are needed to the circuit presented in Fig. 2.8 if the memory-mapping approach is adopted when using the 8086 microprocessor. The first alteration is the use of the MEMR signal instead of the IOR signal. Furthermore, and since the memory addressing in the 8086 is performed by employing 20 address lines, the device-selection signals need to be derived from the 20 address lines rather than the 16 lines used when applying the input/output-mapping method. If a device-selection signal is required at the same address as that of Example 2.5, then the circuit of Fig. 2.9 can be used.

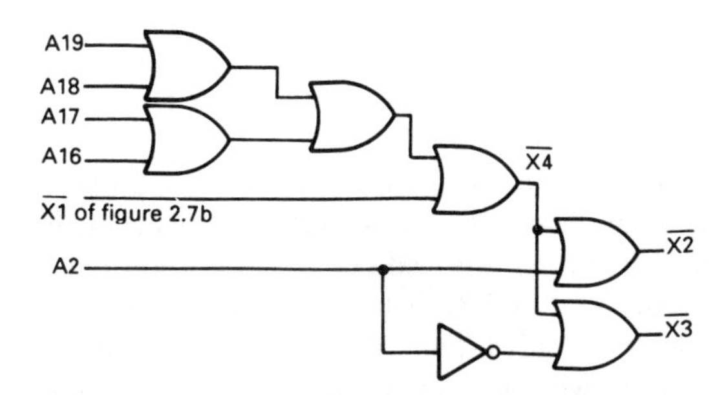

Fig. 2.9 Producing selection signals for memory-mapped devices.

If the microprocessor is a 68000, then the interface is achieved by using the memory-mapped approach, as in Example 2.3. It differs from that of using the 8086 in that the selection signal needs to be derived from 23 address lines and the read and write commands are on the same line.

2.2.5 Using an address decoder

Producing addressing signals normally requires the use of several logic gates. Instead of using such gates, one or more address decoder ICs can be employed. Figure 2.10(a) shows the use of the 74LS138 3-to-8-line decoder. It is one of the most widely used decoders. It has six input lines, three of which (A, B and C) are binary-weighted selection lines, and three of which (G1, G2 and G3) are enable lines. It provides eight output lines (Y0 to Y7), only one of which is activated (logic 0) at one time. The activated output is the one whose address is the binary combination of the three selection lines. This can only take place if the output is enabled by setting G1 = 1, G2 = 0, and G3 = 0. An example of another address decoder is the 74LS154, shown in Fig. 2.10(b). It is similar to the 74LS138 except that it is a 4-to-16-line decoder. It is a 24-pin IC providing four selection lines (A, B, C and D), two active-low enable lines (G1 and G2) and 16 output lines (Y0 to Y15).

In Fig. 2.10(a) one of the 74LS138 ICs is used to produce eight addressing signals Y0 to Y7. Address lines A13, A14 and A15 of the controlling microprocessor are connected to the enable lines G1, G2 and G3 respectively. The decoder is therefore enabled when lines A13, A14 and A15 are 1, 0 and 0 respectively. Lines A10, A11 and A12 are used as channel-selection lines by connecting them to inputs A, B and C of the IC. Consequently, the decoder can select eight devices with addresses ranging from 0010 0000 0000 0000 (i.e. 2000H) to 0011 1111 1111 1111 (i.e. 3FFFH). This represents an area of 8 kilobytes (8192 bytes) to select each of eight devices. This means that each device reserves 1 kilobyte.

The addressing area reserved for each device is further reduced by using a second 74LS138 decoder which is enabled when line A8 is high and both of address line A9 and output Y0 of the first decoder are low (line Y0 is activated when address lines A10, A11 and A12 are at a logic 0 state). This implies that the eight outputs of the second decoder cover the area 0010 0001 0000 0000 (i.e. 2100H) to 0010 0001 1111 1111 (i.e. 21FFH). This represents 256 bytes identifying eight addresses, i.e. 32 bytes per device.

The combination of the two decoders has therefore produced 15 addressing signals. The addressing area reserved per signal is 1 kilobyte for outputs Y1 to Y7 of the first decoder and 32 bytes for outputs Y0 to Y7 of the second decoder. This shows the advantage of using address decoders; they permit multi-device decoding without increasing the complexity, cost or size of the interfacing circuit and without sacrificing addressing area.

Note that the enable lines of the decoder can be supplied directly from control signals. For example, the AS line of the 68000 microprocessor can be supplied to an enable line (e.g. line G2) of the decoder whose other input lines are supplied from address lines. Similarly, lines IORQ and RD of the Z80 microprocessor can be used as inputs to the enable lines of the decoders of Fig. 2.10.

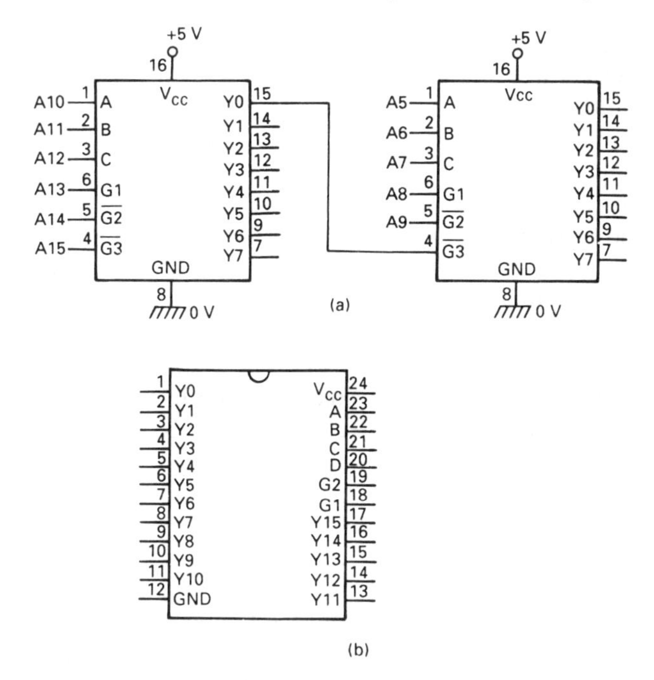

Fig. 2.10 Address decoders: (a) two 74LS138 ICs; (b) a 74LS154 IC.

Further examples of using address decoder ICs are given in Section 3.6.

2.3 Input ports

An input port is a 3-state buffer whose output is equal to its input only when enabled by a controlling device, such as a microprocessor. It normally consists of several channels (e.g. eight). An example of an input port is device A in Fig. 2.2, whose inputs can be supplied from the outputs of an add-on. When enabled, the port supplies information from the add-on to the microprocessor through the data bus.

An input port can be formed from using one of many ICs. One of the most widely used ICs is the 74LS244. It is an 8-line 3-state buffer which contains two active-low enable signals (1G and 2G). Each of these signals controls four buffer lines. The pinout of the IC is shown in Fig. 2.11(a) where line 1G controls the four lines whose inputs are A1, A2, A3 and A4 and whose outputs are C1, C2, C3 and C4 respectively. Similarly, line 2G controls the four lines B1, B2, B3 and B4 whose outputs are D1, D2, D3 and D4 respectively. An output line is at the same logic state as its corresponding input when its control line is activated (logic 0 state).

Therefore, by supplying the same command signal to both 1G and 2G lines, the IC will act as an 8-bit port. For instance, the selection signals CS1 in Fig. 2.8 can be used as a port-enable signal when connecting it to lines 1 and 19 of the 74LS244 IC. Note that the eight buffers of the IC are non-latching, which makes the IC suitable for use as input ports. That is because an enabled input port supplies information to a data bus which is time-shared by several devices and accepts signals from one device at a time. Thus, when using the 74LS244 as an input port, it allows each of the eight signals at its input to pass to the data bus only when instructed by the microprocessor to do so (when G1 and G2 set to a logic 0 state). Only one such IC is required for connection to an 8-bit data bus, whereas a 16-bit data bus requires two such ICs to receive 16-bit information.

Other non-latching 3-state buffer ICs can also be used as input ports to a microcomputer. The pinouts of two such ICs are shown in Fig. 2.11(b) and 2.11(c); they are the 74LS125 and 74LS126 quad 3-state buffer ICs. Each of these ICs contains four buffers, each of which can be enabled or disabled independently of the others. In Fig. 2.11(b) and 2.11(c), the individual buffers are numbered 1 to 4 with an input line (x), an output line (y), and an enable line (E). The difference between the two ICs is in the state of the enable line. It is active-low in the 74LS125 and active-high in the 74LS126. When using such ICs, a designer has a choice between supplying all the enable lines from a single source, or enabling each line individually. Clearly, the second approach requires the derivation of additional device-selection signals, but can provide more flexibility (single-line or multiple-line selection).

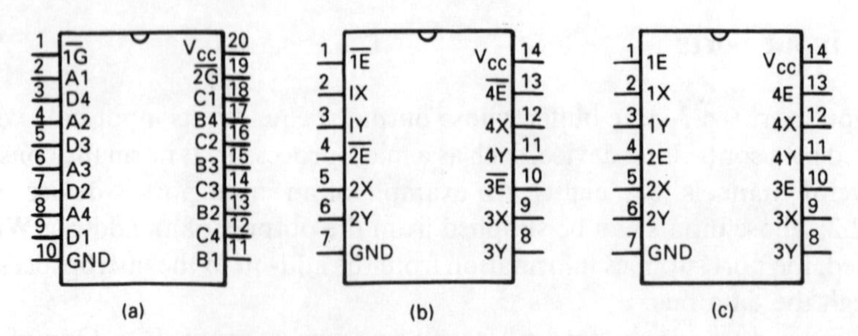

Fig. 2.11 Three-state non-latching buffers: (a) the 74LS244 octal buffer IC; (b) the 74LS125 quad buffer IC with active-low enable; (c) the 74LS126 quad buffer IC with active-high enable.

2.4 Output ports

The requirement imposed on an output port is different from that imposed on an input port. That is because the latter needs to put signals on the data bus only for a very short period of time. On the other hand, the former is supplied with signals from the data bus for a very short time (only when it is enabled), and is required to latch those signals so that the output will be available as long as the buffer is supplied with power. Consequently, an output port consists of 3-state latches.

A widely used IC is the 74LS374 octal D-type latch shown in Fig. 2.12. It contains eight 3-state edge-triggered flip-flops and provides two control lines; the first (pin 1) is the output-enable line, while the second (pin 11) is the clock line. The latter controls the input to the latches (in Fig. 2.12 lines D and Q represent the input and the output respectively). When the IC is utilized as an output port, the output-enable line is connected to the 0-V rails, so that the output is always enabled. When the clock line changes from a logic 0 to a logic 1 state, each of the outputs is latched to the same logic state as its corresponding input. In the latching mode, the IC will store the logic state of the eight lines. Therefore, the IC can be connected to the data bus. When the microprocessor wants to transfer information to the output, it delivers the information to the data bus and alters the state of the clock line to command the 74LS374 IC to accept and latch the information.

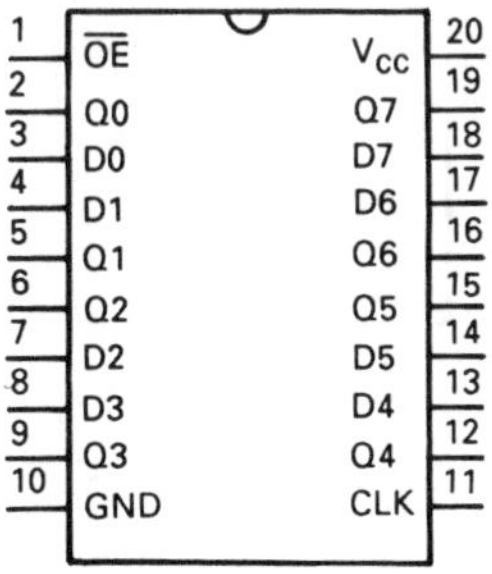

Fig. 2.12 The 74LS374 octal 3-state latch.

2.5 Address and data buffers

The address and data buses of microcomputers are capable of driving several external devices in addition to those already connected inside the microcomputer. If many external devices are required to be connected to these buses, then buffers can be introduced to increase the drive capability of the microcomputer. When designing add-on boards, a designer can, for example, employ additional ICs on each board to buffer the lines used for transferring data and commands between the microcomputer and the add-on.

The address bus is unidirectional; it passes signals from the microprocessor to the devices. It can be buffered by using the 74LS244 octal buffer IC described in Section 2.3 and shown in Fig. 2.11(a). It can buffer up to eight address lines supplied to its inputs. The IC is enabled by connecting its 1G and 2G lines to the 0-V rail. Control lines, such as the IOR and IOW, can also be buffered in the same way.

Since the data bus is bidirectional, it needs a different type of buffer. Use can be made of the 74LS245 octal bus transceiver IC shown connected to an 8-bit bus in Fig. 2.13. Pin 1 is the direction pin, and indicates to the IC whether the microprocessor wants to receive or transmit data. It can, for example, be connected to the IOR line. The IC has an enable pin (pin 19); if that pin is connected to the 0-V rail, the IC is always enabled. Instead, a device-selection signal can be used to enable the IC only when the microprocessor executes a read or a write to a particular address. Clearly, two 74LS245 ICs are required for a 16-bit bus, each connected to eight of the 16 data lines.

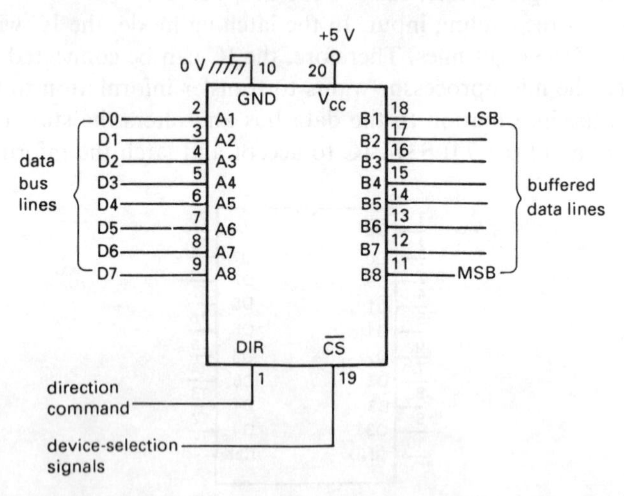

Fig. 2.13 Using the 74LS245 octal 3-state transceiver.

2.6 Introducing I/O ports to microcomputers

In order to provide a microcomputer with the ability to exchange data and/or control signals with add-ons, it is normally provided with input and output ports. These ports can either be incorporated within each add-on circuit, or they can be introduced to the microcomputer for general use.

The information presented in the previous sections can be employed to interface an 8-bit 3-state buffer, or an add-on containing such a buffer, to a microcomputer possessing an 8-bit data bus. A device-selection signal needs to be derived and utilized to activate a 3-state buffer. If the buffer is an input port, then it is non-latching (such as the 74LS244 1C). On the other hand, an output port is formed by employing a latching buffer (such as the 74LS374

IC). The interface of a 16-bit microcomputer to a 16-bit buffer or to an add-on containing such a buffer can be implemented in the same way, except that a 16-bit port can be formed by employing two buffers, each connected to eight lines. Both buffers are enabled from the same device-selection signal, as shown in Fig. 2.14.

Devices A and B in Fig. 2.14 are latching 3-state buffers (the 74LS374 ICs). The output-enable line of these ICs is connected to the 0-V rail so that the output is always enabled. The microcomputer supplies information through the data bus and activates the clock pin of each of the two ICs to make them receive and latch the information. Similarly, the figure utilizes two ICs to form a 16-bit port; they contain non-latching 3-state buffers (the 74LS244). The figure shows that the enable lines of these ICs are connected together and are supplied from a single device-selection signal from the microcomputer. In this way, the microcomputer reads all of the input lines simultaneously and updates all of the output lines simultaneously. Several of the examples presented in later chapters illustrate how to extract, set or clear some of the bits in an input or an output port (e.g. Examples 6.1 and 6.2).

Situations exist, however, where an 8-bit microcomputer needs to send 16-bit data to, or receive it from, an add-on. The sending requirement can be satisfied by using devices A, B and C shown in Fig. 2.15. These devices are 8-bit output ports (74LS374 ICs). The 16-bit data can be supplied as two successive bytes rather than one 16-bit word. The procedure consists of two steps. In the first step, the high byte of the result is supplied to device C, which is enabled by device-selection signal C3. This is followed by supplying the lower byte to an address which will result in activating both the C1 and C2. This commands devices A and B to receive and latch the signals available at their inputs, thereby updating the 16 output lines simultaneously.

Reception of 16-bit data by an 8-bit microcomputer can be accomplished by utilizing devices D, E and F of Fig. 2.15. Device D is a non-latching input port (74LS244 IC) and devices E and F are latching ports (74LS373 ICs, see Fig. 1.2) connected to receive signals from external devices. Note that the output-enable line of each of the two ICs is supplied from a device-selection line. Reading the 16-bit input can be accomplished in three steps. In the first step, the microcomputer supplies a device-selection signal which activates lines C6 and C7. This makes devices E and F capture the 16-bit input. This is followed by the second step, where the microcomputer enables device D (using line C4) and activates the output of device E (using line C5). With this action, the microcomputer reads the state of the first eight lines of the captured input. The third step is similar to the second step, except that the microcomputer activates the output of device F and enables device D to read the state of the other eight input lines.

Note that it is possible to supply the clock lines of devices A and B from a single selection signal, and to supply the latch-enable lines of devices E

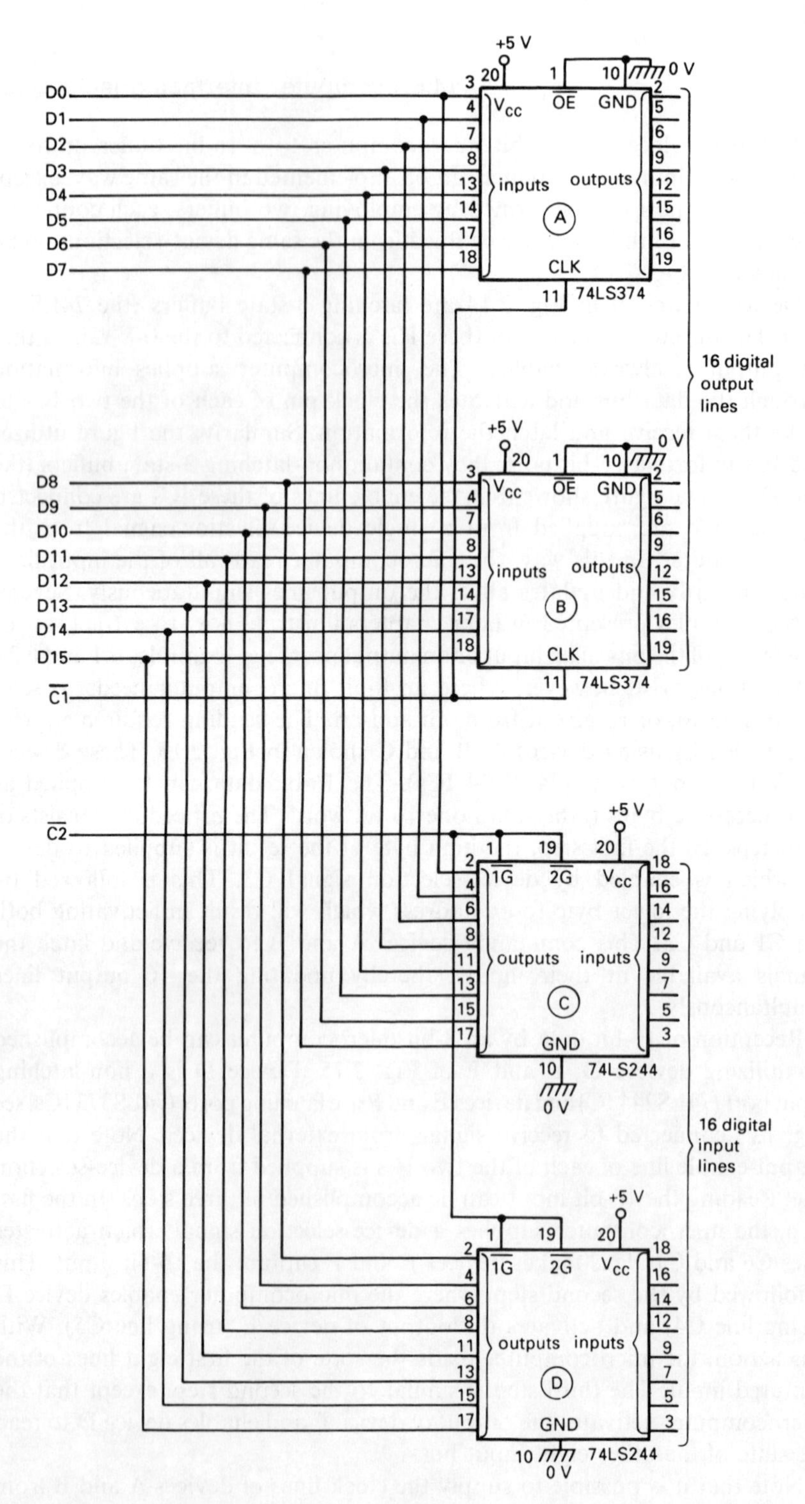

Fig. 2.14 Introducing input and output ports to a 16-bit microcomputer.

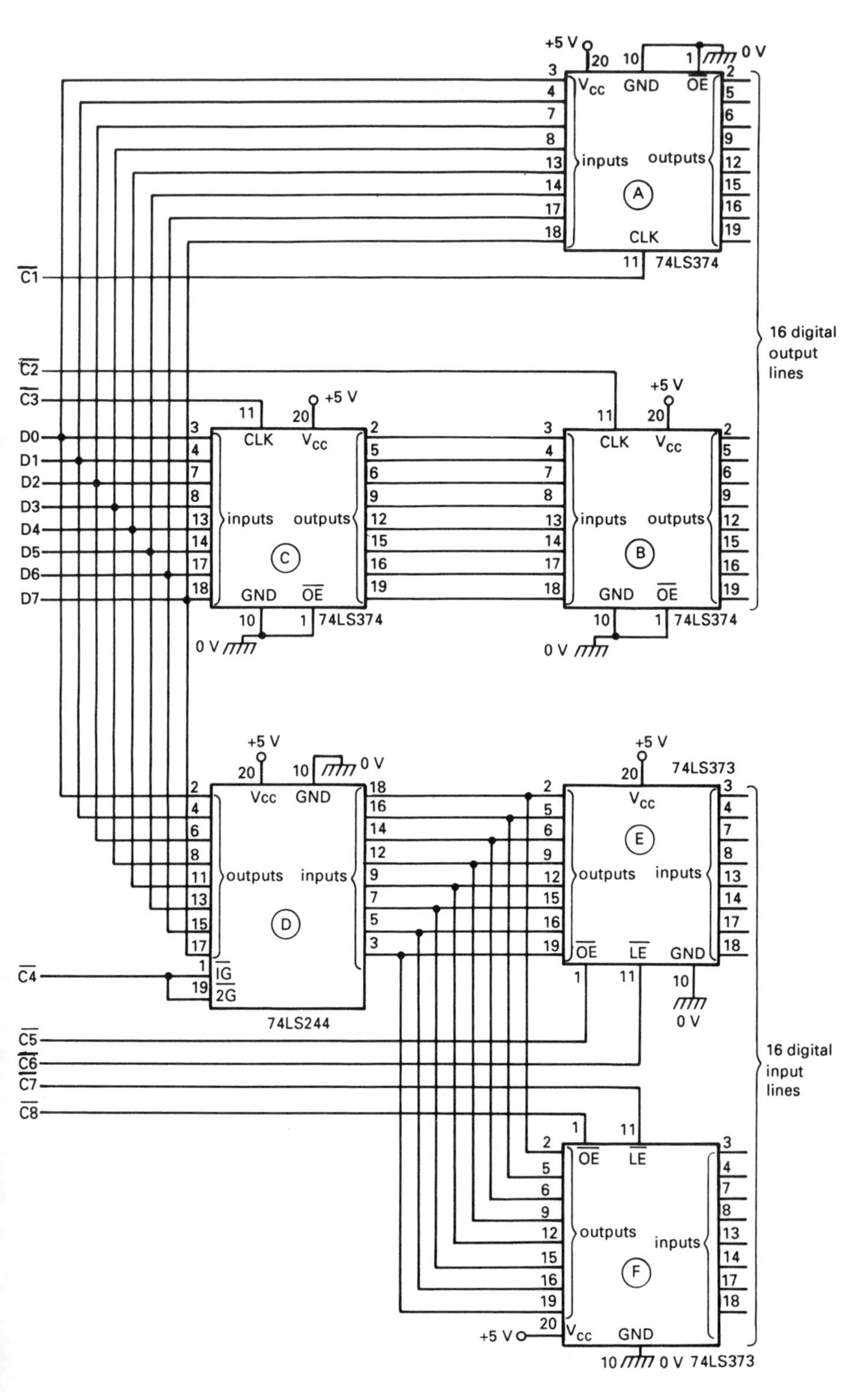

Fig. 2.15 Introducing input and output ports to an 8-bit microcomputer.

and F from another selection signal. The reason for using an individual selection signal per IC is to permit the use of the input and the output lines either as 8-bit or 16-bit individual ports.

The ports presented in Fig. 2.14 and 2.15 are used in many of the examples presented in this book. For the ease of reference, output devices A and B of Fig. 2.15 will be referred to as output ports 1 and 2 when used as two individual ports, and as the output port when used as a 16-bit port. Similarly, devices E and F will be referred to as input ports 1 and 2 when used as 8-bit ports, and as the input port when used as a single 16-bit port.

Chapter 3

Microcomputers and add-ons

3.1 Introduction

The addition of interfacing circuitry to a microcomputer provides a useful means of expanding capability and enhancing performance. It gives the microcomputer the ability to communicate with a variety of external devices, or other computer systems. It allows the integration of the microcomputer in a variety of applications ranging from a simple ON/OFF operation to intelligent data acquisition, signal processing and control in individual automation applications. The timing requirements of demanding applications are not necessarily best satisfied by the use of a high-price high-performance microcomputer. Instead, a simple microcomputer can be utilized with dedicated hardware. Therefore, if a microcomputer needs to be purchased for a specific application, the performance requirements should be specified and the designer should compare different approaches in order to obtain the optimum solution. Such a solution should satisfy the performance requirements and, at the same time, achieve the best combination of microcomputer, add-ons, software and cost.

In this chapter, the introduction and consequent effect of incorporating add-ons in microcomputers are considered. A typical situation is investigated where a choice exists between several ways to meet an interfacing requirement. It is explained that a microcomputer may not be capable of meeting the interfacing requirements of some applications efficiently without the inclusion of additional hardware. In this way, some tasks and decision-making operations are implemented by additional circuitry rather than by the microcomputer. This leaves more time for the microcomputer to attend to other tasks efficiently. In many situations, a choice exists between designing and purchasing such hardware. Add-on boards are available from various suppliers, especially for IBM and compatible microcomputers. The purchase of an add-on removes the need for designing the interface. It therefore forms the optimum solution to many interfacing problems. However, an add-on may not fully satisfy the requirements and the user may decide to design a board tailored to his or her needs. The cost of an add-on may prove to be too high for a computer owner aiming to interface his or her microcomputer to the outside world, or for a company requiring the incorporation of such add-ons for inclusion in its products.

In some situations, interface hardware already exists in a microcomputer, but it may not be capable of meeting the requirements of a new application. The user may be able to purchase replacement hardware, or design a new circuit, or it may be possible to modify the existing interface to suit the new requirements (the latter is explained in Section 16.5).

3.2 Interfacing microcomputers

This section considers personal and industrial microcomputers. More emphasis is given to the IBM and compatible types due to their wide use and the fact that IBM compatibility has become a standard. Users can select from a variety of off-the-shelf interface and expansion cards compatible with the IBM bus structure to satisfy their requirements. This gives them confidence that other add-ons will be available in future should they decide to expand their microcomputers. They can also design interface cards knowing that they can, if the need arises, transfer them to any of the many compatible machines, even if such machines are produced by a different manufacturer.

3.2.1 Personal microcomputers

Personal microcomputers are produced by many manufacturers and are used by millions of people worldwide for a variety of applications in offices, in laboratories, on shop floors and at home. They vary in performance, capability, price, weight, size and the availability of add-ons and software packages. Let us consider the IBM and compatible personal microcomputers whose success has led many manufacturers to produce a wide range of compatible software packages and add-ons. This has increased the range of applications of such microcomputers and enabled them to meet the requirements of very demanding applications. In the last few years, we have seen the production of compatible microcomputers by other manufacturers. In order to make such microcomputers attractive to the buyer, they need to offer a better performance, lower price, and/or to include additional features which are not included in the standard IBM microcomputers.

For instance, some personal microcomputers compatible with the IBM XT™ are based around the Intel 8086 microprocessor instead of the Intel 8088 microprocessor used in the IBM XT. Both processors have a 16-bit internal structure. The 8086, however, can use its 16-bit data bus for external communication rather than the 8-bit bus, as in the case of the 8088. The designers of such microcomputers need to produce interfacing signals compatible with those in the IBM microcomputer. In other words, a microcomputer based around the 8086 microprocessor needs to provide the same interfacing signals as the 8088 microcomputer. It therefore needs to communicate with add-ons as if it has an 8-bit bus, but uses its 16-bit bus for onboard operations, e.g. communicating with memory devices. Interfacing in this type of microcomputers is achieved through connectors within the computer casing. The connectors bring out a group of lines to form an expansion bus. In the XT-machines such a bus performs 8-bit interfacing and is referred to as the PC-bus or the XT-bus. When the AT™ version was produced by IBM, it was capable of achieving both 8-bit and 16-bit interfacing. It has the ability to operate software and add-ons designed for the XT type. However, communication between the microcomputer and an XT-type interface circuit is performed by the use of 8-bit interfacing. Consequently, a faster rate of exchange of information can be achieved if an interface circuit designed for an XT machine is redesigned to make use of the 16-bit data bus available in AT-type machines. Clearly, the improvement is possible only if the interface circuit needs information of higher resolution. Keep in mind that redesigning the circuit is not always the optimum solution, since a decision is needed on whether the aim is to improve performance, or to satisfy the requirements at a low cost and/or within the minimum development time.

This 16-bit bus is known as the ISA-bus (Industry Standard Architecture). It has a higher addressing capability and a faster transfer rate than the XT-bus. In order to make it downwardly compatible with the XT-bus, additional lines were added, making ISA-compatible expansion cards longer than those designed for the XT-bus.

The advance in technology has led to the design of much faster 32-bit processors, and with them have come much more advanced applications for microcomputers, requiring even more power and a much faster rate of transferring information. But in order to maintain compatibility, the ISA expansion bus stayed the same and started to form a bottleneck in some applications. The result was the design of new buses for communication with add-on boards. An example of a very widely used 32-bit bus is the EISA-bus (Extended Industry Standard Architecture), which is also downwardly compatible with the above-described buses. On the other hand, IBM has developed the MCA-bus (Micro-Channel Architecture), which is also a fast 32-bit bus, but is not compatible with the above-described architectures. This implies that it supports expansion cards specifically designed for this type of bus.

The continuous progress in the development of faster processors led to the design of the local bus. This represents a group of lines taken directly from the processor's bus and will therefore operate at a rate defined by the processor and not by the slower expansion bus. This is very attractive, since it provides faster processors with the ability to transmit information to high-performance peripherals at high speed. The bus is 32 bits wide for 32-bit processors, e.g. the Intel386DX and the Intel486 family. The local bus is introduced as an additional bus providing the user with the option of choosing which type of bus to purchase an interface card for within the computer. Even if not used at the time of purchase, the availability of local-bus connectors on a motherboard gives the user the ability to upgrade to such a bus at a later date. Interface cards requiring a very high rate of transfer of information can be connected directly to the local bus to achieve the highest possible rate of transfer (e.g. in graphics-based and data-storage and data-retrieval applications).

Suitable add-on cards are easily connected to this type of microcomputer through the onboard connectors (slots) and are limited by the specifications of the connection bus. This simplifies the interface task, because a user or a designer neither needs to modify the motherboard nor provide an expansion box to house the add-on circuitry. A motherboard with a higher number of expansion slots, or an expansion box, will be required when the user needs to include more interface boards than the available slots. Expansion boxes are particularly attractive for use in applications where the interface circuitry is not suitable for fitting inside the microcomputer.

There is a wide variation in the designed frequency of operation of microcomputers. For instance, the IBM PC/XT operates at a clock frequency of 4.77 MHz, and some XT-compatible machines operate at a higher frequency, e.g. 8 MHz. The requirement for a higher processing capability led to the design of faster processors, enabling microcomputers to operate at much higher frequencies, e.g. 66 MHz. For a given processor, operation at a higher frequency leads to a faster execution of programs. However, it creates the need for using faster components within the microcomputer, which increases its cost. One should, however, keep in mind that not all programs can be correctly run at different frequencies. For instance, wrong results will be produced when a faster (or a slower) compatible microcomputer executes programs containing software timing loops based on operation at a particular clock frequency. Such a difficulty can be overcome if the compatible microcomputer provides the user with the ability to change the frequency of the clock easily, e.g. from the keyboard or through a dedicated switch. Consequently, some personal microcomputers support operation at each of two or more clock frequencies.

Battery-powered portable personal microcomputers are widely used and are produced by many manufacturers. They combine high processing power and flexibility and are extremely attractive in a very wide range of applications. As far as interfacing applications are concerned, a portable microcomputer presents an ideal means of data collection, testing or programming on a factory shop floor or off site. The lack of internal space, however, can impose a restriction on the type of interface application based around such microcomputers.

3.2.2 Industrial microcomputers

There exist a variety of microcomputers dedicated for use in industry. They range from simple to sophisticated controllers for utilization in many applications, such as data acquisition and industrial control. Simple industrial microcomputers usually consist of a single PCB. In many applications, such a board is incorporated as a part of a larger controller to provide intelligence and/or to replace electronic circuitry, thereby reducing component numbers and controller size.

The cost of an industrial microcomputer depends on many factors, such as its performance, the features it offers, and whether it is designed for a specific application or for general-purpose use. The cost of such a microcomputer also depends on how easily it can be adapted to various application requirements and future expansions. This has led to the design and development of different types of industrial microcomputers. They provide various features which enable buyers to select the one that closely matches their requirements. Some industrial microcomputers can be configured by users. They can, for example, select one of a few operating modes by the use of onboard links or switches. They may also be able to set the magnitude of a reference signal or the gain of an amplifier by potentiometers or by rotary switches. Some microcomputers include built-in printers or plotters. Others provide means of interfacing to such equipment. Many industrial microcomputers are equipped with one or more disk drives to allow an operator to run one of many application programs stored on floppy disks or hard disk. The availability of such drives can also extend the storage capability of the microcomputer in data-collection applications.

A large number of industrial microcomputers do not need a VDU or a full-size keyboard. Instead they employ LEDs or a simple display to provide the required information. In order to enable their operation in certain environments, some industrial microcomputers include a sealed keyboard for protection against dust and damp. Others employ thumbwheel or on/off switches to replace the keyboard.

A microcomputer designed for use in an office or at home may not be suitable for operation in certain shop floor environments. It may, for instance, need a protective casing which increases the cost. Instead of using such a microcomputer, many manufacturers have designed microcomputers suitable for operation in industrial environments. A large number of such microcomputers are rack-mounted and the add-ons can be included within the rack. The utilization of an industrial microcomputer in a harsh environment often introduces the need for using an enclosure suitable for that particular environment. The enclosure may, for example, need to withstand vibration or excessive temperatures. When selecting an enclosure or a rack to house the boards of a microcomputer, it is normally advisable to select one which contains additional space for future expansion.

Industrial microcomputers are based around any of several processors, such as those mentioned in this book. Certain microcomputers include two or more processors: the 'master' controls the operation of the microcomputer, commanding and/or communicating with one or more 'slave' processors. Many industrial microcomputers are IBM-compatible, and a large number of these are offered in the form of a single board which can be integrated as a part of a system. Many such microcomputers possess a socket permitting the insertion of a math-coprocessor, or they accept replacement of the main processor with a higher-performance compatible processor. Flexibility will be increased if links are included to allow changing the frequency. This gives the user the option of including a faster processor in future.

Many industrial microcomputers provide digital and analogue input and output lines to which external devices can be interfaced. Others allow the expansion of the microcomputer by providing interfacing signals, which normally include data, address and control lines of the processor to allow the user to interface add-on circuits which he or she can purchase or design.

3.3 The optimum add-on

When attempting to expand a microcomputer, a choice exists between several solutions. The specification of the optimum add-on circuit depends on several factors, such as the requirements of the application and the allowable degree of involvement of the microcomputer in implementing the functions of the expansion. In a large number of applications, the expansion requirements are best satisfied by the design of a simple circuit consisting of few components, e.g. a general-purpose output port. On the other hand, there are many applications which require the introduction of complex multifunction hardware and software, e.g. a controller of a multiaxis robot arm. In between these two extremes, there exist a large number of applications with requirements that can be satisfied by a circuit of medium complexity, e.g. a multichannel A/D converter (Section 7.7).

3.3.1 Relation between software and hardware

Before purchasing or designing an add-on, consideration should be given to the amount of processing which needs to be carried out by the microcomputer to implement the interface. In most situations, there exists a choice between employing a circuit which meets the requirements with a minimum intervention from the microcomputer, or using simpler lower-cost hardware and make the microcomputer achieve the rest by software. The choice of an interface depends on its effect on the overall performance of the microcomputer. The use of a simple low-cost circuit is an attractive solution to many application problems. It reduces the cost of hardware, the size of the add-on, and the assembly time of the final product. The additional processing, however, reduces the time available for the microcomputer to attend to other tasks. This imposes a restriction on a large number of applications and it may be considered unacceptable. Furthermore, writing the software increases the development time of the project. Lengthy development time is normally justified when producing a large volume of the final product. In this case, the cost of software per add-on becomes small.

The designer should therefore find the best balance between software and hardware in order to design the optimum interface. We can therefore say that:

- Software can be used to replace hardware only if it is economical to do so.
- The optimum solution for many interfacing applications is the best combination between simplicity, low cost and low computation time.

Examples of choosing between software and hardware solutions are given in Section 4.4.

3.3.2 Optimization

One should always remember that the optimum interface should not only provide the most suitable input/output operation, but it should do so while maximizing or minimizing certain functions. For example, the design can be directed to allow the microcomputer to produce maximum performance. In this case, the performance is given the ultimate priority, whereas other functions, such as cost and development time, are given lower priority, but not ignored. Another application may assign the highest priority to the cost of the interface. In such an application, the optimum design is the one that implements the interface at a minimum cost. A decision is then needed on what to sacrifice in order to achieve minimum cost. In many situations, a very low-cost interface degrades the performance of the microcomputer. This is avoided in a large number of applications by specifying the ultimate goal to be the highest performance/price ratio. In this case, the designer concentrates on how to achieve a low-cost design which will allow the microcomputer to satisfy the interfacing requirements efficiently. Keep in mind that the resulting design may neither minimize cost nor maximize performance, but it maximizes the performance/price ratio. Other factors that are taken into account when aiming to design the optimum interface are minimizing the development time and ease of construction of the final product.

3.3.3 Important factors

In addition to defining the exact objectives of the expansion, consideration should be given to other factors such as whether or not the design should include the ability to adapt to future application needs. An example is given in Section 6.6.2, where the inclusion of more input channels than presently required is considered as an attractive solution. In that example, onboard jumpers are provided to inform the microcomputer which of the channels is in use. Another important factor which needs serious consideration when designing an interface circuit is the sampling rate of the interfaced signals. A decision is needed on whether the microcomputer can provide that sampling rate efficiently, or whether additional hardware should be introduced to provide the required signals. If the latter course is adopted, then what is the cost of that hardware and what restriction does it impose?

As the functions carried out by the add-on increase, the circuit becomes more complex and/or the microcomputer spends more time on operating the add-on. This can be avoided if some local intelligence is provided in the add-on circuit. For instance, an add-on can include a microprocessor which can control the operation of the expansion circuit and produce timing signals, thus removing the load from the microcomputer. A microprocessor is not expensive, but it requires additional components for its operation, such as memory ICs and interfacing circuitry. This may prove to be uneconomical in many applications and, in some situations, it can be replaced by the use of a microcontroller (an IC which includes a CPU, registers, memory, input/output ports and other features). Microcontrollers are widely used in a variety of applications, especially those requiring computing and decision-making power at a low cost.

3.3.4 Ready-made add-ons

A large number of ready-made add-ons are in the form of a PCB which can be housed in a box external to the microcomputer, or are specifically designed to plug directly into a slot within a microcomputer. Each add-on is normally a module providing a specific function, e.g. analogue-input channels. Many add-ons offer onboard options to increase their suitability for a variety of applications. Normally, the price of a ready-made add-on is directly proportional to its capabilities and the features it offers.

Let us consider an example of a ready-made add-on board designed for IBM PC/XT/AT-compatible machines. Let us assume that it offers an 8-channel A/D conversion. Each channel produces a 10-bit digital result at a rate of 10 kHz. The board can fit into one of the expansion slots and offers easy connection of input signals through screw terminals located on the edge of the board and appearing at the rear panel of the microcomputer. The board may be designed to interrupt the microcomputer at the end of each conversion, or it may be provided with a RAM area to store the collected data which can be read by the microcomputer when needed. The board may incorporate three sets of DIL switches or jumpers to enable users to customize the board for their applications. The first set permits application of any of several predetermined amplification factors (gains) to the analogue-input signals. The second set allows the user to select one of several input voltage ranges and to indicate whether the signals are unipolar or bipolar. The third set permits the user to activate individual or multichannel A/D conversion. The board manufacturer may specify that choosing to perform A/D conversion on a single specified channel speeds the rate of conversion and permits a more efficient sampling of an input signal. Configuring the card is simply a matter of setting the onboard switches or placing the jumpers according to the application requirements.

Even when deciding to purchase an add-on, a choice exists between a variety of options. One needs to consider all the options and select the one that will best suit the application. Instead of purchasing an add-on, use can be made of the examples given throughout this book to design a suitable add-on. For instance, the examples given in Chapter 7 can be used to design an add-on containing the features described above.

3.4 Operating add-ons by a microcomputer

When an add-on is interfaced to a microcomputer, the data bus is normally utilized for the transfer of information between them. The microcomputer normally attends to several tasks, one of which is operating or communicating with the add-on. The execution of tasks in the microcomputer can either be based on the time-sharing approach or performed on an interrupt basis. These two approaches are described in the following two sections, whereas four methods of transferring data between a microcomputer and an add-on are given in Section 16.3.

3.4.1 Program-controlled input/output

The term 'program-controlled input/output' is used to indicate that the instant of reading from or writing to an input or output device is decided by the software. The microcomputer either transfers information (commands and/or data) to the device, or receives information from it. In many situations, the microcomputer tests an input signal (normally digital) to deduce if a device needs servicing. The microcomputer performs the monitoring and the communication task as one of the tasks it is executing. This implies that use can be made of the information given in Section 1.4 to arrange the software so that the microcomputer interrogates the device in between attending to other tasks, or as a part of a software loop. It is shown in Example 1.2 that when the requirement of communicating with more than one device arises, the microcomputer interrogates each of the devices in sequence. One way of preventing, or minimizing, the possibility of losing valuable information is to prioritize the tasks. This means that the flag corresponding to the device which needs fast servicing is always tested first, or more often than the others. The designer should, however, ensure that this approach does not result in losing information from devices whose servicing is given a lower priority.

The program-controlled approach has the advantage of being simple, easy to implement and starting whenever the microcomputer decides to communicate. It therefore does not disturb the execution of other parts of the program. It is very suitable for a large number of applications, but it has two main limitations. The first is that an external device cannot inform the microcomputer that an immediate action is needed, but has to wait until the microcomputer asks for it. The other limitation is the time spent by the microcomputer interrogating all the devices. This, in some situations, consumes considerable computation time and, if not correctly planned, can degrade the overall performance of the microcomputer.

Several application examples are given in the following chapters illustrating how techniques can be introduced to reduce or to minimize the interrogation time which is associated with the implementation of the program-controlled approach. For instance, Example 7.4 explains that additional hardware can be used to store information generated by an add-on. This reduces the rate at which a microcomputer needs to interrogate the add-on. Section 4.3 gives an alternative solution in which the designer uses information on the behaviour of the system under consideration to reduce the time spent on the interrogation.

3.4.2 Interrupt-controlled input/output

Many applications employ devices requiring fast servicing. As explained in Section 1.9, such devices can have the ability to interrupt a microcomputer rather than wait for it to ask them. The microcomputer should therefore be able to sense the request and respond to it efficiently. The term 'interrupt-controlled input/output' is used to indicate that the interrupt-service routine results in reading from, or writing to, an input or an output device.

The interrupt request can either be generated from a device which needs servicing, or from a timing device to inform the microcomputer that it is time to execute the interrupt-service routine. Many microcomputers include a timing device which can be set by software to generate interrupts at the required time. Some timing devices can operate in the auto-reload mode, where the program sets that device once and it will be reloaded with the value and restarted after generating an interrupt. If, on the other hand, the interrupt request is generated by an external device, then that signal needs to be connected to a hardware interrupt line of the microcomputer. It is recommended that an externally originated interrupt request (or an input signal) is buffered by the use of a logic gate to reduce the possibility of damaging the internal circuits of the microcomputer. Further information on the use of interrupts is given in Section 1.9 and application examples are given in many parts of this book.

3.5 Alternative methods for implementing a task

In this section, the generation of a square wave is considered as an example to demonstrate the different options available to a designer in implementing a task. Each option has advantages and drawbacks. One way of generating a square wave is to complement the state of a digital-output line regularly. Let us assume that the waveform possesses an equal mark-to-space ratio, and each lasts for y μs. The frequency of this waveform depends on the time between successive complements of the output line and is equal to $1/(2 \times y)$. The resulting waveform can be observed on an oscilloscope, or it can be supplied to a speaker to produce sound effects using the simple circuit of Fig. 10.14. The tune can be varied by altering the frequency of the waveform.

3.5.1 Using the program-controlled output method

Example 3.1: **What is the procedure required to enable a microcomputer to supply a square wave to an external device through a digital-output line? Assume that this is the only task implemented by the microcomputer.**

Since the microcomputer is dedicated for the implementation of this task, then the program-controlled output method forms a simple and an attractive solution. The program can take the form of a continuous loop which does nothing except generating the waveform. It can be represented by the following procedure:

1 LET n = COUNT EQUIVALENT TO DURATION
2 REPEAT UNTIL EQUAL ZERO:

$$n = n - 1$$

3 COMPLEMENT THE DIGITAL OUTPUT LINE
4 GO TO STEP 1

The alteration of the frequency can therefore be achieved by altering the count. The method of sending signals through a single digital-output line without affecting other lines of the port is described in Chapter 4.

This example illustrates a cost-effective solution which satisfies the requirement by software. Only a single digital-output line is required. The microcomputer updates that line without the need to examine, or respond to, an input signal. If the value of n is high, the microcomputer will be idle for a long time. For instance, if the required frequency is 50 Hz, then the digital-output line will need to be complemented once every 10 ms. Instead of leaving the microcomputer waiting, it can execute thousands of instructions during that period, as shown in the following section.

3.5.2 Introducing additional hardware

Example 3.2: **How can hardware be utilized to minimize the processing time which the microcomputer spends on the generation of the square wave?**
A microcomputer-controlled square-wave generator can be produced by employing a simple circuit which supplies the waveform only when commanded by the microcomputer to do so. Let us start by considering the circuit of Fig. 3.1. It represents a square-wave generator based on the 555 timer IC. In the figure, the timer is connected for operation as a stand-alone square-wave generator which continuously produces a square wave. The frequency of the waveform is given by

$$f = \frac{1.44}{((R1 + 2R2)C1)}$$

which can be approximated to:

$$f = \frac{0.72}{(R2C1)}$$

when $R2 \gg R1$.

The mark (M) and the space (S) of the waveform are given by $M = 0.693(R1 + R2)C1$ and $S = 0.693(R2)C1$. This implies that when $R2 \gg R1$, the mark-to-space ratio is equal to 1. One method of giving an operator the ability to alter the frequency of the square wave linearly is to replace R2 by

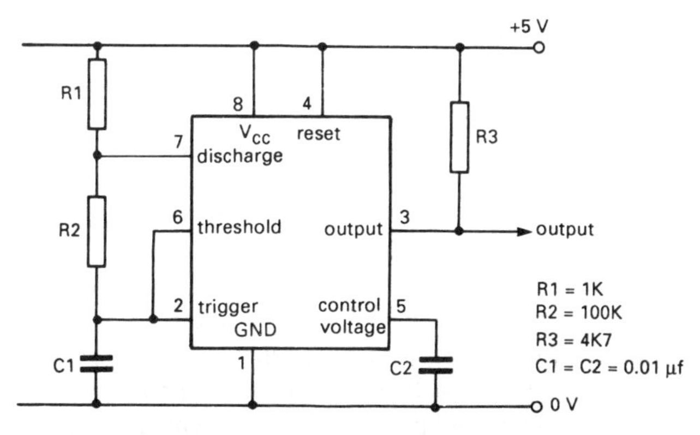

Fig. 3.1 Using the 555 timer as a square-wave generator.

a 100-kΩ potentiometer in series with a fixed resistor, e.g. 22 kΩ. Instead of using a potentiometer, the change of frequency can be accomplished by replacing capacitor C1 of Fig. 3.1 by a group of link-selectable capacitors. One end of each capacitor is connected to the 0-V line, whereas the other end is made available for connection to R2 via a link. The number of capacitors and their value depend on the required frequency stages.

Let us consider two methods which allow a microcomputer to control the instances of supplying such a waveform to a circuit. The first method is based on utilizing a single dual-input AND gate. One of the inputs of that gate is supplied from a digital-output line of the microcomputer and the other input is connected to the output of the 555 timer (pin 3). The square wave will be available at the output of the AND gate only when the microcomputer delivers a logic 1 to the input of that gate.

Instead of using the timer as a free-running square-wave generator, the circuit can be slightly modified to provide a microcomputer with the ability to control the signal-generation period directly. This can be achieved by changing the connection of pin 4 (the reset pin) where its direct connection to the +5-V supply is replaced by two connections. The first is a connection to the +5-V supply via a 10-kΩ resistor. The second connection provides a path for an active-low reset signal to be supplied to the timer from a digital-output line of a microcomputer.

Both methods employ a single latching digital-output line of the microcomputer. This implies that very little processing time is needed to control the square-wave generation process. All the microcomputer needs to do is to supply a logic 1 signal to a digital-output line. This signal will be latched and the square wave will be generated until the microcomputer issues a logic 0 signal. The software can be arranged so that at the start the microcomputer defaults to producing a logic 0 to prevent supplying the square wave until it is needed.

3.5.3 Using the interrupt-controlled output method

Example 3.3: **Repeat Example 3.1 and assume that the microcomputer is involved with the implementation of tasks besides generating the waveform. In this case, the square wave is produced by the use of an interrupt from an in-computer programmable timer.**

Before starting the timer, it needs to be supplied with a value (a count). This count corresponds to a time after which the timer interrupts the microcomputer. Obviously, the value depends on the frequency of the clock which operates the timer. The procedure consists of two parts; the first part is executed only at the start, whereas the second part is executed whenever an interrupt occurs. This implies that the second part is the generator of the waveform.

- Initialization
  1 ENABLE THE PROCESSOR TO ACCEPT AN INTERRUPT FROM TIMER
  2 LOAD TIMER WITH COUNT
  3 START TIMER
  4 OUTPUT A LOGIC 1 FROM THE OUTPUT TERMINAL
  5 CONTINUE WITH THE EXECUTION OF OTHER TASKS

- Interrupt routine
  1 LOAD TIMER WITH COUNT
  2 START TIMER
  3 COMPLEMENT THE OUTPUT TERMINAL
  4 RE-ENABLE INTERRUPT AND RETURN TO MAIN PROGRAM

The procedure shows that the interrupt routine is simple and requires a short processing time. The routine can even be simpler if the programmable timer can be operated in the auto-reload mode, where the timer is automatically reloaded with the count and restarts counting after generating the interrupt.

In many systems, considerable time improvement can be achieved by employing the interrupt-controlled approach rather than the program-controlled approach. Its use, however, imposes restrictions in some applications. For instance, in the above example, the microcomputer needs to execute the interrupt-service routine, which consists of four steps. It also needs to divert the program, save and restore certain registers and return to the original program. Let us assume that this takes 9 μs. If the rate of occurrence of the interrupt is not high, then the effect of its processing time can be ignored. If, on the other hand, an application imposes the requirement of updating the output at a high rate, say 100 kHz, then an interrupt needs to occur once per 10 μs. This does not leave sufficient time for the microcomputer to attend to other tasks. Consequently, any of the methods presented in the above two sections can be used for such an application.

3.5.4 Considerations for a multi-interrupt system

Let us now consider an example to illustrate a problem which may arise as a result of generating a square wave in a multi-interrupt system.

Example 3.4: **Assume that the generation of the square wave is achieved by the use of the interrupt described in Example 3.3. What restrictions are imposed if it is used as a low-priority interrupt which can be interrupted by another interrupt of a higher priority?**

The procedure of generating the waveform is the same as that of Example 3.3, except that the intialization part also defines the square-wave generating interrupt to be of a low priority. For simplicity, let us refer to the high-priority interrupt as L1 and the low-priority interrupt as L2. Consider the effect of the occurrence of L1 when the microcomputer is attending to the servicing routine of L2. Assume that L1 occurs at an instant before L2 updates the digital-output line. The occurrence of L1 diverts the path of the program from executing the interrupt-service routine of L2 to another path where it attends to the interrupt-service routine of L1. Such a diversion delays L2 from complementing the digital-output line and leads to extending the duration of one of the half-cycles of the waveform. The period of the delay is equal to the time taken to respond to L1, execute its interrupt-service routine, and return to L2.

Assume that attending to L1 consumes 50 μs. Its effect on the mark-to-space ratio of the square wave, and hence on its frequency, depends on the duration of one cycle of the waveform. If the waveform possesses a relatively low frequency, e.g. 50 Hz, then the duration of one cycle of the waveform is 20 ms. This implies that if the two interrupts coincide as described above, then the frequency of one cycle of the waveform will be altered by 0.25%. The rate of its occurrence depends on the rate at which the two interrupts concide. In many applications, such an effect is considered acceptable, especially when the probability of its occurrence is low. However, it may not be appropriate for many other applications, particularly if the waveform is generated at a very high fequency and/or if the rate at which the two interrupts coincide is high.

The optimum solution to the problem depends on the application. In many situations, it is possible to arrange the software in such a way that the high-priority interrupt consumes minimum time. In such situations, the resulting error is reduced and may become acceptable. Some multi-interrupt applications introduce the need to disable the high-priority interrupt during the execution of a specific low-priority interrupt, e.g. L2. If this is the case, then it is recommended to arrange the program so that the interrupt-service routine of L2 consumes a minimum time. This minimizes the interval of delaying the execution of L1 and forms a solution which is acceptable in a large number of applications. It is, however, not suitable for applications where L1 appears for a very short time and needs to be serviced immediately. The requirement of such applications can be satisfied by the adoption of an alternative method of generating the square wave, e.g. Example 3.2.

3.5.5 Discussion

The solutions presented above illustrate methods for solving a simple signal-generation problem. The designer can meet the requirements while minimizing, or maximizing, a certain function. For instance, the use of hardware provides a solution which introduces a very low software overhead and consumes minimum processing time. It allows the microcomputer to meet the requirements and be able to attend to other tasks efficiently and without degrading the performance of the system. The need for purchasing and constructing the hardware, however, imposes a restriction on its use. Also, the increase in the power consumption and physical size in some situations creates the need for utilizing an additional board.

An alternative solution is to meet the requirements by software. Example 3.1 employs the program-controlled method and presents a simple solution which is cost-effective and attractive to use, particularly in applications where the microcomputer has no requirement for attending to other tasks during the generation of the waveform. A third solution is presented in Example 3.3, where the generation of the waveform is done on an interrupt basis. This method is attractive to use in a large number of applications, especially in a multitasking real-time system. The method has many benefits, but it imposes some limitations as outlined in the previous two sections.

3.5.6 Conclusion

In order to be able to achieve the best solution to a design problem, the objectives of the task need to be correctly specified and the designer should be fully aware of the requirements and effects of performing other tasks by the microcomputer. The designer needs to ask several questions, such as:

- How many tasks do I need to run simultaneously?
- How often should each task be executed?
- Is there any time-critical task (or a part of a task)?
- What is the effect of the implementation of one task on others?

- What are the limitations on introducing alterations in future? The alterations may be in the form of adding, eliminating or modifying tasks.
- Which method is best suited to meeting the requirements?
- What is the development time?
- Are there sufficient resources for implementing the selected method?

If, for example, the interrupt-control method is considered, then the following questions should be asked:

- What is the interrupt-generating source and how can it be interfaced to the microcomputer?
- Is the microcomputer capable of accepting such an interrupt?
- How many interrupts are used in the system?
- Is there a need to prioritize interrupts? Is it allowed? How many priority levels are available in the microcomputer?
- Is there a need for additional hardware, and how easily can it be introduced?
- Is there sufficient area on the stack?
- Is there any part of the program that should not be interrupted? If so, then does the microcomputer allow the disabling of interrupts during the execution of this part of the program? Can the rest of the program tolerate the disabling of the interrupt?

3.6 Interface circuitry

It was explained in Chapter 2 that a microcomputer bus can drive a limited number of devices. When several devices are connected to a bus, its drive capability can be increased by employing buffers (see Section 2.5). These buffers are introduced to those bus lines which are employed for interfacing add-ons to the microcomputer. Since the price of such buffers is not high, they can be incorporated as a part of an add-on circuit (i.e. they can be placed on each add-on board). Besides increasing the drive capability, this approach also reduces the probability of damaging the microcomputer or other devices as a result of a fault on the introduced interface circuit. Address lines and control lines transfer commands in one direction. They therefore need unidirectional buffers, suh as the 74LS244 IC (Section 2.3). The use of the data bus to read from and write to add-ons creates the need for using bidirectional buffers, such as the 74LS235 IC (Section 2.5).

When introducing an interfacing device to a microcomputer, it is essential to consult the memory map, or the channel I/O assignment table, to select a suitable address for the add-ons. For example, in an IBM PC/XT/AT or compatible microcomputer, the input/output addressing area 300H to 31FH can be used for device addressing. Other addresses are available for interfacing certain add-ons, such as serial communication cards. If such cards are not employed, then their corresponding area can, if required, be used to address other types of add-ons. If, however, the need arises to utilize a serial communication card in future, then the addressing circuit of the add-on will need to be modified. Let us consider an example.

Example 3.5: **Derive a selection signal for an add-on which will be used in an XT- or an AT-type personal microcomputer.**

The addressing area available for input/output device selection is 300H to 31FH and is produced by lines A0 to A9. We therefore need to derive an addressing pulse at an unused location within that area. We can adopt a similar approach to those presented in Chapter 2 to generate a device-selection signal which implements the interface by the use of input/output mapping. This can be achieved in two stages. In the first stage, we derive an addressing signal for the available area (300H to 31FH), whereas in the second stage we introduce additional hardware to reduce the reserved addressing area.

Addressing signal

The addressing area 300H to 31FH can be represented in a binary form as 1100000000 to 1100011111, or simply 11000xxxxx where the x indicates a don't care value. This means that an addressing signal needs to be produced when A9 and A8 are at a logic 1 state, and A7, A6 and A5 at a logic 0 stage.

Special signals

By consulting the technical manual of this type of microcomputer, one realizes that the microcomputer produces an AEN (Address ENable signal). When this signal is at a logic 0 stage, it indicates that operation is in an addressing mode and not in a DMA (Direct Memory Access) mode. Therefore, this signal should also be taken into account when deriving the device-selection signal.

Simple addressing circuit

The configuration of Fig. 3.2 consists of a single 74LS138 address decoding IC which is enabled when line A9 is high and both AEN and A5 are low. The state of lines A6, A7 and A8 determines which of the output channels is activated (at a logic 0). The channel of interest here is the one which is activated when supplying any address within the area 300H to 31FH. It is the one at which A8 is high and both A7 and A6 are low, i.e. channel 1.

Reducing the reserved area

If line Y1 of Fig. 3.2 is employed to provide the addressing signal of the add-on, then only one selection signal will be produced for any of the addresses

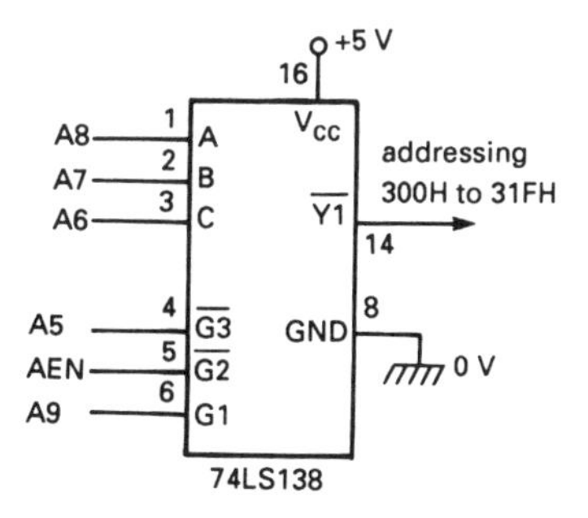

Fig. 3.2 Deriving an addressing signal.

within 300H to 31FH, i.e. 32 locations. We can therefore introduce an additional circuit to reserve less addressing area for the add-on. This will allow us to connect more add-ons in future. Figure 3.3(a) shows the use of two NAND gates to reduce the reserved area to 16 locations: 310H to 31FH. Similarly, the configuration of Fig. 3.3(b) uses an OR gate to reserve the addresses 300H to 30FH.

Further reduction of reserved area

An even smaller area can be reserved either by adding more gates, or by using another 74LS138 decoder IC instead of the gates. This is shown in Fig. 3.3(c), where the Y1 signal (from the circuit of Fig. 3.2) is used together with address line A4 to enable a second 74LS138 IC. Address lines A3, A2 and A1 are used to activate any of the eight output channels, Y0 to Y7 (the activated channel is the one whose number corresponds to the parallel combination of A3, A2 and A1). The outputs correspond to addresses 300H, 302H, 304H, 306H, 308H, 30AH, 30CH and 30EH.

Only one of these channels needs to be activated to define the selection address of the add-on. This leaves the additional addressing lines available for addressing other add-ons, or to implement other tasks. For instance, the example presented in the next section illustrates a use of such lines. An alternative example is employing such signals to select each of a few devices within a multidevice add-on. For instance, an add-on may include individually selected output channels.

Device-selection signal

Such a signal can be derived by combining any of the above-derived signals with suitable control lines provided by the microcomputer. This type of microcomputer provides the four signals IOR, IOW, MEMR and MEMW on its expansion connector. They can be combined with the addressing signals in a similar manner to that of Fig. 2.8.

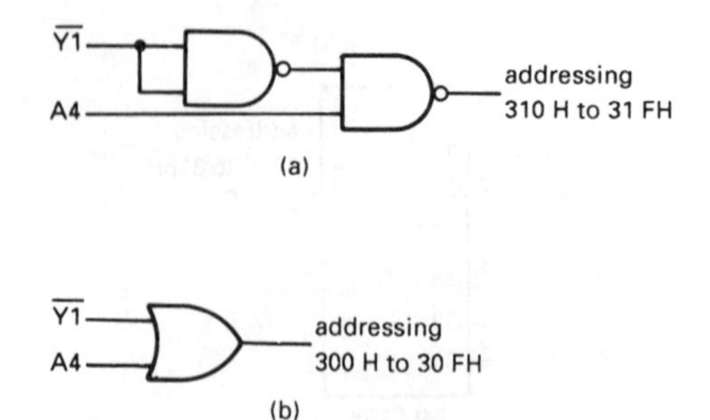

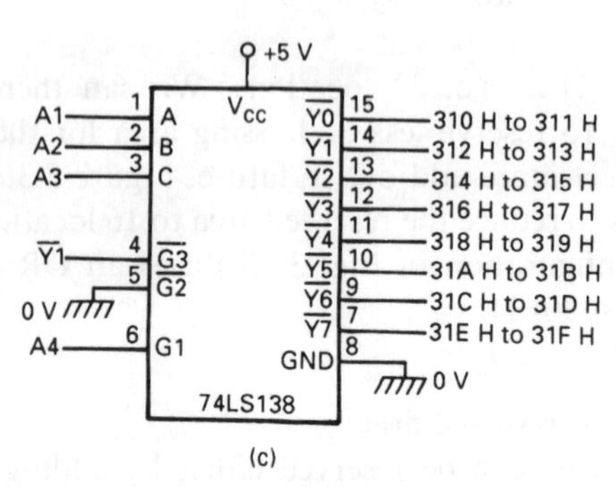

Fig. 3.3 Reducing the area reserved for addressing.

3.7 Flexibility of add-ons

An add-on can be designed to respond to a certain addressing signal which defines the location of the add-on within memory, or within the device-addressing area. If a second add-on is introduced which also responds to the same addressing signal, then the addressing circuit of one of the add-ons will need to be altered. This, however, involves a hardware change which in the case of using a PCB means cutting tracks and using wires to join certain points. Instead, the addressing circuit (which in many cases is located within the add-on board) can be provided with a flexible addressing feature. This is not difficult to implement and is shown in the following example.

Example 3.6: **How can the address-selection circuit of Example 3.5 be modified to include a flexible addressing feature by means of onboard links?**
The configuration of Fig. 3.3(c) provides eight individual outputs. They can be applied to select up to eight add-ons. The output lines can be connected as shown in Fig. 3.4 to allow an operator to choose any of the eight addresses to define thc addressing signal of an add-on. The choice is accomplished by

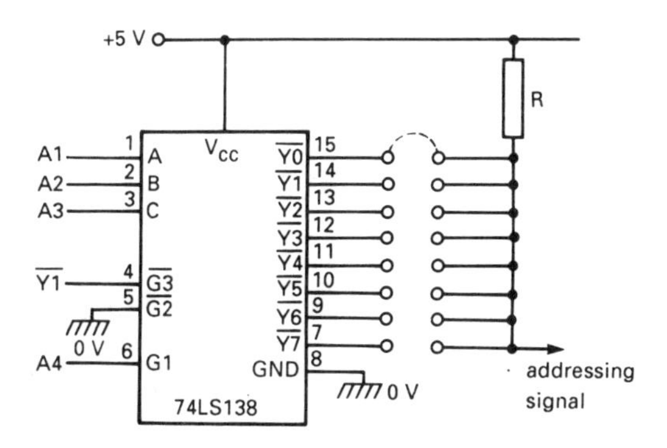

Fig. 3.4 Onboard address selection.

placing a link between an output line of the decoder and the common line. A resistor is connected to the +5-V supply so that when no link is placed, the line is at a logic 1 state, i.e. no selection. If a link is placed, then the addressing line will be activated only when the linked line is at a logic 0 state. With this approach the add-on can be given any of the following addresses: 310H, 312H, 314H, 316H, 318H, 31AH, 31CH and 31EH.

The above example illustrates a simple method which allows a user to locate an add-on in any of eight addresses. The method is not limited to the selections of an address; it can also be applied to provide an additional feature which allows the user to select the type of mapping (i.e. to select between memory mapping and input/output mapping). As an exercise, use the information presented in this and the previous chapters to design a circuit which will produce a device-selection signal and provide the selectable-mapping feature.

Chapter 4

Using digital-input and digital-output ports

4.1 Introduction

Computerized control equipment is used in various applications to improve performance and productivity. In such equipment, information is exchanged between microcomputers and external devices. This chapter presents examples of using input and output ports to interface microcomputers to external circuits and considers a few types of load. It explains that the requirement of an interface can normally be satisfied in any of several ways. The designer should therefore consider all possible solutions and choose the one which best fulfils the requirements. The introduction of input and output ports to microcomputers is presented in Chapter 2, where 8-bit and 16-bit interfacing is illustrated. It is explained that input ports are non-latching buffers and output ports are of the latching type. Section 4.4 compares the software and the hardware approaches as ways of satisfying interfacing requirements. It presents examples and illustrates how to satisfy the requirement by either approach.

One should remember that interface lines should be well protected, since short-circuiting an output line, or supplying an input line with a wrong signal, could permanently damage the internal circuitry of the microcomputer. The degree of protection depends on the type of application and the level of signals handled by the circuit. Chapter 10 gives examples of the use of diodes and zener diodes for protection. Further protection can be achieved by using fuses and placing current-limiting resistors at interface lines. Isolation is strongly recommended, particularly in noisy or high-power applications. The use of isolation devices (e.g. transformers and optocouplers) is presented in later chapters.

Section 4.5 deals with the inclusion of synchronization devices in interface circuits, explaining the advantages of their use and giving a few examples.

The behaviour of much microcomputer-based equipment depends on the type of load it is subjected to. Section 4.6 considers different types of load and explains how they affect the relationship between current, voltage and power.

4.2 Examples of receiving and sending digital signals

This section presents examples on supplying signals to microcomputers through input ports and generating signals from microcomputers through output ports. Further examples of using both types of ports are given throughout the book.

4.2.1 Using input ports

Through an N-bit input port, a microcomputer can receive N-bit information from an external device. Alternatively, such a port can be incorporated to read N individual signals. The port will then be considered as a multichannel digital-input device. The microcomputer reads all the channels simultaneously by executing a single instruction. After reading the port, the microcomputer can examine all the channels either simultaneously or individually. The former case is adopted in situations where the microcomputer aims to detect if there has been a change of state in any of the N input channels since they were last examined. It therefore compares the new reading with the previous one and only takes an action if the two values are not the same. The action can be in the form of examining the bits individually to detect the ones whose state have been altered. The following example illustrates how a microcomputer can examine the state of a single digital-input line.

Example 4.1: **How can a microcomputer use its input port to detect the state of a SPST (Single-Pole Single-Throw) switch?**
The configuration of Fig. 4.1(a) generates a digital signal (Y) with a logic state dependent on the position of the contacts of a SPST switch. The signal can be supplied to a microcomputer through a digital-input line, e.g. line 0 of an input port. When the switch is open, line 0 is at a logic-low state. Closing the switch changes the state of that line to a logic high. By monitoring the state of line 0, the microcomputer can detect any change in the setting of the switch. The connection of Fig. 4.1(b) allows the generation of a logic-high signal when the switch is open and a logic-low signal when the switch is closed. Both of these connections will be referred to later in the book.

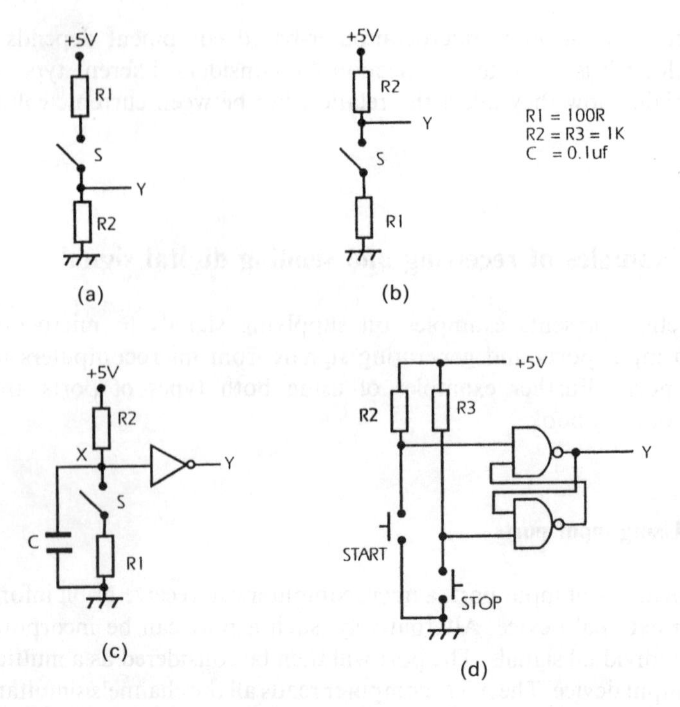

Fig. 4.1 Interfacing an SPST switch to a microcomputer: (a) low state when switch is open; (b) high state when switch is open; (c) debouncing with a capacitor; (d) debouncing with a flip-flop.

Debouncing of a switch

When a switch is opened or closed, its output bounces between logic-low and logic-high states. This happens because switches do not have ideal contacts. In order to prevent misrepresentation of the state of a switch, a software or a hardware solution can be adopted. The software solution is based on introducing a short interval (several milliseconds) between detecting a change in the state of a switch and re-examining it. The hardware solution, on the other hand, can be based on the use of a capacitor, as shown in Fig. 4.1(c). When the switch is open, the capacitor is fully charged and point X is at a logic-high state. Closing the switch forms a discharge path for the capacitor through resistor R1. This results in reducing the voltage at point X to indicate a logic-low state. When the switch is opened again, the capacitor will start to charge to the supply voltage through resistor R2. This changes the state of point X to a logic high. Rather than supplying the input line of the microcomputer with a slowly changing voltage, a schmitt-trigger inverting gate is incorporated in Fig. 14.1(c) to interface the switch to the input line of the microcomputer. The 74LS14 IC contains six such gates and has the same pinout as the 74LS04 IC shown in the Appendix. One should keep in mind, however, that the inverting gate will deliver a signal of the opposite logic state to the input line of the microcomputer.

Using an 8-bit input port

Let us use the configuration of Fig. 4.1(c) to illustrate how the microcomputer detects the state of the input line. We start by considering an interface through an 8-bit input port of a microcomputer possessing an 8-bit data bus. The procedure of detecting the state of the switch can be presented by the following two steps, where 'y' is a byte in RAM and the '&' character represents a software AND

```
1  READ INPUT PORT AND STORE READING IN 'y'
2  IF (y & 0000 0001) = 1
        THEN SWITCH IS CLOSED
   ELSE   SWITCH IS OPEN
```

The procedure reads the input port as 1 byte. It then uses a software AND to extract the logic state of 1 out of the 8 bits (bit 0).

Using a 16-bit input port

If a 16-bit port is utilized with a microcomputer possessing a 16-bit data bus, then the procedure is very similar to that of an 8-bit port, except that 'y' is a 2-byte word and the first line of step 2 becomes:

IF (y & 0000 0000 0000 0001) = 1

Example 4.2: **How can a microcomputer-based industrial controller be provided with a user interface through two PTM (Push-to-Make) switches? One requests the start of the controller and the other commands the termination of the operation.**

The requirement here is to detect the activation of two momentarily closed switches and take an action according to which of the two was last pressed. Rather than using two individual connections, a flip-flop can be employed to perform a few functions. It detects and indicates the closure of either of the two switches, it performs the debouncing and it latches the signal to eliminate the need of monitoring the switches at a high rate. The connection is shown in Fig. 4.1(d) where pressing the START switch sends output line Y to a logic-high state. That state will be latched and not affected by the bounce in the switch, or by its repeated activation. The state of the output line will change to a logic low only if the STOP switch is pressed. Again, bounce and reactivation of that switch will not have an effect on the state of the output line, whose state can only be changed to a logic high again by pressing the START switch.

The above examples illustrate reading digital signals through an input port of a microcomputer. Interfacing analogue signals to microcomputers is described in Chapter 7. Examples of both types of interface are given throughout the book. In many applications, however, the microcomputer does not need to read the actual value of an analogue signal, but it only needs to know whether that value is above or below a certain level. In such applications, an analogue comparator can be utilized. It provides the microcomputer with a logic level whose state depends on the relative values of two analogue signals at the input terminals of the microcomputer.

Example 4.3: **How can a microcomputer detect when the voltage of a test point drops below a fixed level?**

The analogue voltage of the test point can be examined by the use of an analogue comparator. The circuit of Fig. 4.2 illustrates the introduction of a fixed voltage level by two resistors (R1 and R2) to form a voltage divider. The value of the supply voltage V_{cc} is higher than that of test point b. The ratio of the value of the two resistors is selected according to the required voltage level of the test point. The voltage at point a is given by:

$$V_a = V_{cc} \times R2/(R1 + R2)$$

Point a is connected to the positive terminal of the comparator and point b to the negative terminal. Consequently, output line Y will be at a logic-low state as long as the voltage at the test point exceeds the voltage at point a. The microcomputer examines the state of line Y through a digital-input line and only takes an action when its state changes to a logic high.

Comparators are very fast devices; the switching time for the LM311 used in Fig. 4.2 is 100 ns. The LM311 can be operated from the +5-V power supply of the microcomputer and logic circuit, but it is shown connected to V_{cc} (up to +36 V). Rather than operating from a single power supply, the IC can operate from a split supply (up to ±18 V) with an absolute maximum differential input voltage of ±30 V (without exceeding the supply voltage). Resistor R3 is employed as a pull-up resistor. It is required because the comparator possesses an open-collector output terminal. This allows setting the level of the output voltage according to the requirement of the circuit connected to its output line. The LM311 is capable of operation up to 40 V, but in the case of interfacing to a digital port of a microcomputer, the pull-up resistor is connected to the +5-V line.

Note that if V_{cc} is obtained from an external power supply, then its 0-V line should be connected to that of the +5-V supply. Care should be taken if an external power supply is used, since a faulty connection or an excessive voltage level can damage the interface circuit as well as the internal parts of the microcomputer.

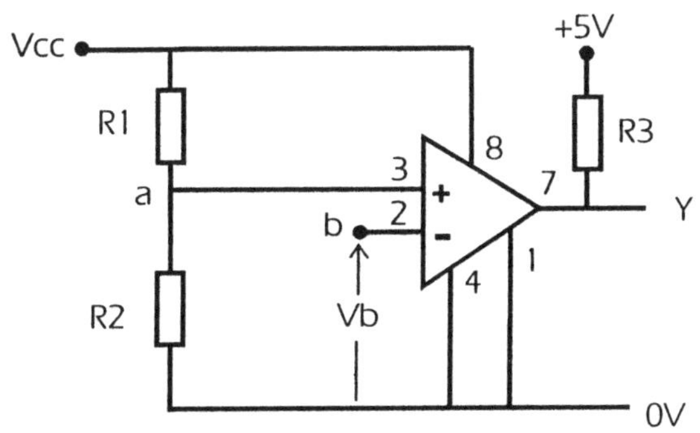

Fig. 4.2 Using a comparator.

4.2.2 Using output ports

A microcomputer with an N-bit processor can write an N-bit number to an N-bit output port in one step. Depending on the application requirements, all the lines of an output port may be interfaced to a single device, or, alternatively, individual port lines may be interfaced to different devices. Sending information from a port to a single device is a simple task, since all the microcomputer needs to do is to write the information to the port. More work is needed if different parts of the program write to different lines within a single output port, as shown in the following example.

Example 4.4: **How can the software be arranged to permit the use of a single 8-bit output port to supply information to a model-train controller? The controller accepts a 6-bit number identifying the required speed, a single bit to define the direction of motion, and another bit to start/stop the train.**
In this application, the microcomputer sends three different commands through a single output port. Let us assume that each command is set by a different part of the program. This implies that the software should be able to update some of the output lines without affecting the others. Since writing to the port updates all the bits of that port in one step, then it is best to reserve a single byte in memory to store the contents of the output port. In this way, each of the three parts of the program can update their corresponding bits without affecting the others. Changing the speed can be represented by the following two instructions where byte 'x' contains a 6-bit number representing the required speed (the upper 2 bits are 00). Byte 'y' represents the memory byte containing the port information. Symbols '|' and '&' represent a software OR and AND respectively.

```
1  y = (y & 11000000) | x
2  TRANSFER y TO THE OUTPUT PORT
```

The first instruction performs two actions. It starts with the one enclosed by the brackets to clear the least significant six bits from the memory location by ANDing them with 000000. The second step within the instruction is to write the new speed into the memory location by using a software OR. Note that the upper two bits of 'x' have to be 00 to prevent altering the upper two bits of 'y'. Updating each of the other two bits of the output port can follow a similar procedure where clearing a bit is performed by a software AND with a 0, and setting it is achieved by a software OR with a 1.

The above procedure updates the output port immediately after updating the memory location. In some applications, it is preferred to synchronize the operations. Sending information to the controller can, for instance, only be permitted through a regularly-occurring interrupt. In this way, the three different parts of the program will alter their corresponding bits in byte 'y' and let the interrupt send the information to the port at the appropriate instant.

Rather than supplying the signals directly from a microcomputer, there are many applications where a microcomputer is employed to control the flow of signals between devices. In other applications, time-consuming tasks are performed by hardware, leaving the microcomputer free to handle other activities. The control that a microcomputer is given over the hardware-generated signals may be limited to enabling or disabling the flow of these signals. Several examples using this method of control are presented in later chapters.

Control with logic gates

Logic gates form a simple, low-cost method of microcomputer-based control of external digital signals.

AND gates

An AND gate sets its output line to a high state only when all of its input lines are at a high logic level. If a digital-input signal is connected to one input of a 2-input AND gate (point X in Fig. 4.3(a)), and its other input is connected to an output port of a microcomputer, the flow of the digital signal to the output of the gate depends on the state of the output line from the microcomputer. A low-state signal from the microcomputer forces the output of the gate to a logic-low state regardless of the state of the digital signal at point X. On the other hand, a high-state logic signal from the microcomputer makes the output of the gate at the same state as the input signal at point X. This implies that the gate is incorporated as a microcomputer-controlled digital switch. A single 74LS08 IC contains four 2-input AND gates and can therefore be combined with the output lines of

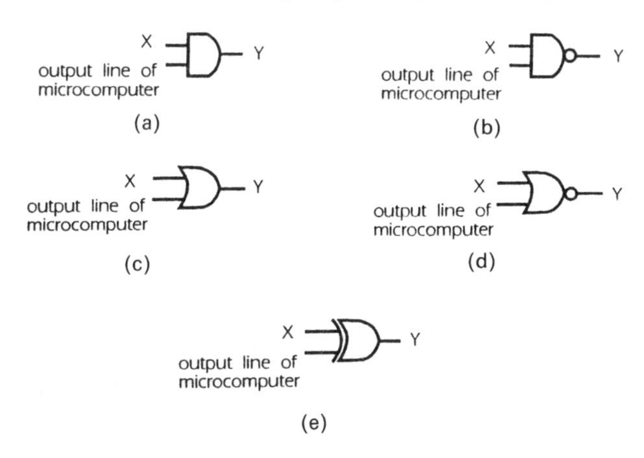

Fig. 4.3 Signal manipulation: '(a) using an AND gate; (b) using a NAND gate; (c) using an OR gate; (d) using a NOR gate; (e) using an Ex-OR gate.

a microcomputer to form four digital switches. The pinout of this IC is given in the Appendix.

NAND gates

A 2-input NAND gate is used in Fig. 4.3(b) to form a different type of microcomputer-controlled digital switch (four such gates are available in a single 74LS00 IC, see Appendix). The operation of this type of gate is similar to that of an AND gate, except that its output is inverted. Consequently, the use of such a gate as a controlled digital switch will invert the digital signal supplied to point X. Using a NAND gate in this configuration can therefore provide an additional inversion function which can serve to reduce the number of components in applications where both of these functions are required.

OR gates

Figure 4.3(c) shows a 2-input OR gate. The output of such a gate is at a logic-low state only when both of its inputs are at a logic-low state. Using this statement, an output line of a microcomputer can be set to a logic-high state to force the output of such a gate to a logic-high state regardless of the state of the other input line. The transfer of a digital signal through the gate from point X to Y will take place only when the microcomputer sets its output line to a logic-low state. The 74LS32 IC contains four 2-input OR gates (see Appendix).

NOR gates

Figure 4.3(d) employs a 2-input NOR gate (four such gates are contained in a 74LS02 IC; see Appendix). The output of this type of gate is at a logic-high state only when both of its inputs are at a logic-low state. This means that the microcomputer can block the flow of signal X through the gate by supplying a logic-high signal from an output port. No matter what state signal X takes, the output will be at a logic-low state. In this respect, the output of the gate is the same as that of an AND gate when conduction is blocked by a microcomputer. The state of the output, however, is different when conduction is permitted, since the NOR gate will invert the digital signal applied to point X when the microcomputer sets its output line to a logic-low state.

Ex-OR gate

Rather than employing a gate to form a switch, Fig. 4.3(e) illustrates how an Ex-OR gate can be incorporated as a controlled digital inverter. The output of such a gate is at a logic-high state when its input signals are at different logic states. Otherwise, its output is at a logic-low state. This relationship can be used to form a digital inversion when an output line from a microcomputer is connected to an input of this gate and is set to a logic-high state. This changes the output to an opposite logic state to that at point X. If the microcomputer alters the logic state of its output line to a logic low, then the output of the gate will be at the same logic state as the digital signal at point X. Four such gates are available within a single 74LS86 IC, as shown in the Appendix.

Conclusion and application

The above shows that a single logic gate can be combined with a single digital output line from a microcomputer, to form either a digital switch or a digital inverter operating under the control of the microcomputer. The relationship between the input and output signals depends on the type of gate used in the application.

Let us consider an application where a microcomputer is employed to select any of three clock signals to drive a digital counter. Following the above discussion, AND gates can be employed to provide the microcomputer with the ability to select any of the three signals. A logic-low signal from a microcomputer blocks the transmission of any signal through the gate, leaving the state of its output line at a logic low. The output of the three AND gates can be connected to an OR gate, as shown in Fig. 4.4(a). With this simple circuit, the microcomputer can select the source of the clock signal to the digital counter by setting a single output line high and keeping the other two lines at a logic-low state.

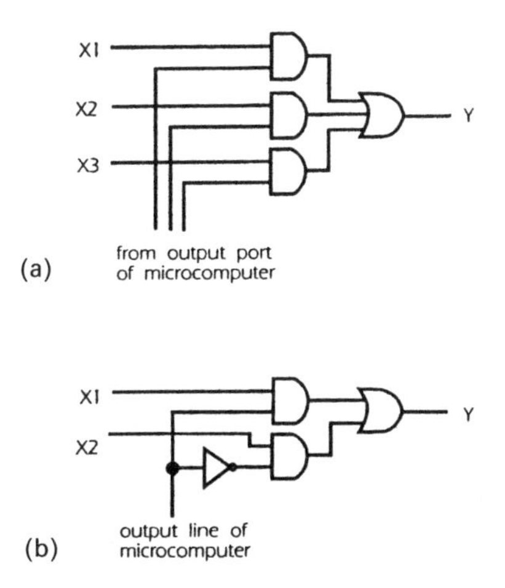

Fig. 4.4 Path selection by a microcomputer: (a) one out of three; (b) one out of two.

In some configurations, it is possible to reduce the number of output lines from a microcomputer by employing additional circuit components. The circuit of Fig. 4.4(b) is an example. It employs an inverting gate to allow a microcomputer to use a single output line in the selection of any of two digital sources. A logic-high signal from the microcomputer permits the transfer of signal X1 to the output line (Y). Changing the state of that line to a logic low blocks signal X1 and permits the conduction of signal X2 to the output. However, the advantage of using two lines from the microcomputer is in providing the microcomputer with the additional ability to block both signals.

Rather than using gates, the selection and diversion of digital signals is performed in many applications by the use of multiplexers. These are explained in Chapter 5. They become more attractive to use than gates when the number of signals to be selected is higher.

4.3 Alternative solutions to a monitoring problem

This section illustrates how any of several solutions can be used to satisfy the requirements of an interfacing problem. Each solution is best suited for a certain type of application. Let us consider an application where an 8-bit microcomputer is utilized to control an industrial process. The period of operation is controlled by two external devices. The first provides the microcomputer with an 8-bit digital value commanding the start and specifying the type of operation. The second device supplies the microcomputer with a command to stop the process. This command is supplied through a single active-low digital-input line. Assume that the activation time of each signal is 2 ms and an inactive start command is represented by 00000000.

4.3.1 Continuous monitoring

One way of satisfying the requirement is to interface the 8-bit start command to an 8-bit input port of the microcomputer and connect the line carrying the stop command to one of the lines of another input port. The microcomputer can be programmed to monitor both ports, seeking the detection of a change of state. If the operation requires a fast detection of a change, then continuous monitoring forms one solution. This is based on the use of the program-controlled method where the state of the ports is examined at the highest possible rate in between the implementation of other tasks. The speed of response depends on the duration between successive readings of the ports.

This approach provides a simple and an easy-to-apply solution. However, it does not make the best use of the capabilities of the microcomputer. It may form a suitable solution when the microcomputer is utilized as an on/off controller which synchronizes the operation of a multidevice control process.

The following sections present alternative solutions.

4.3.2 Optimizing the monitoring time

Simple techniques can be employed to simplify the implementation of the task. One can use his/her knowledge of the system operation to perform the monitoring goal more efficiently. If, for example, it is not possible to commence implementing a new type of operation before terminating another, then after receiving a start command, the microcomputer only needs to test the stop signal. Similarly, only the start command needs to be examined after detecting the stop signal.

Let us now present a procedure to enable the microcomputer to monitor the state of the inputs, perform the control process, and execute another task (t). The latter is executed only while waiting for the activation of the start command. Once this command is received, the microcomputer will be devoted to implementing the process which consists of several tasks (nm).

As in the previous solution, the program-controlled method can be adopted. The microcomputer examines the input port of interest after executing each task and takes a corresponding action only if a change is detected. The procedure uses two 8-bit variables. The first (n) stores the number of the task being executed, and the second variable (y) contains the previous start command. Assume that, in the beginning, the process is at a stop state:

```
 1  LET n = 0 AND y = 0
 2  READ PORT 1 AND STORE VALUE IN x
 3  If x = 0
        THEN GO TO STEP 4
    ELSE
        LET y = x,
        START THE REQUESTED PROCESS,
        GO TO STEP 6
 4  EXECUTE TASK t
 5  GO TO STEP 2
 6  READ PORT 2
 7  IF (READING & 0000 0001) = 0
        THEN STOP THE PROCESS AND GO TO STEP 2
 8  n = n + 1
 9  If n > nm
        THEN LET n = 1
10  EXECUTE TASK n
11  GO TO STEP 6
```

The procedure imposes the restriction that the execution of tasks must be structured so that the time spent on the execution of a single task is less than 1 ms. This is because each of the start and stop commands can be active for only 1 ms. Although 1 ms is considered a long time in terms of microcomputer execution, the imposed restriction may limit the suitability of this method for use in certain microcomputer-controlled equipment. A designer must determine the processing time needed to execute long modules and possibly divide long tasks into smaller ones.

The following section presents an example of how to overcome this restriction.

4.3.3 Using an interrupt

In order to remove the restriction of completing the tasks within a 1-ms interval, an interrupt request can be arranged to occur regularly to command the microcomputer to read its input ports (an alternative use of interrupts is given in the next section). In this way, the microcomputer can execute tasks continuously, leaving the interrupt-service routine to read the start and stop commands. An interrupt can, for example, be generated once every millisecond by using a timer. The interrupt-service routine reads the two inputs, seeking a change of state. If a change is detected, then it stores the reading, sets one of two flags, and returns to resume the execution of the main program. The flags can be examined by the main program, which detects a change, takes a corresponding action, and resets the flag. This solution is acceptable as long as the execution time of the interrupt-service routine will not introduce unacceptable delays to other parts of the program (see Section 1.5). After completion of the interrupt-service routine, the execution of the main program resumes from where it was left before the occurrence of the interrupt.

4.3.4 Introducing additional hardware

We have so far seen that each of the solutions presented in the previous sections has certain limitations. Let us therefore consider an alternative solution which involves the use of additional hardware. Figure 4.5 shows a configuration where logic gates are utilized to produce an interrupt-request signal only when a command is introduced by either of the two external devices. The configuration employs two ICs: a 7425 Dual 4-input NOR gate and a 7400 Quad 2-input NAND gate. The former possesses a strobe line per gate. It is not needed in this configuration and is therefore connected to the +5-V line. Let us assume that the interrupt is activated by a transition from a logic 0 to a logic 1; therefore, the active-low stop signal is inverted by a NAND gate and supplied to one of the interrupt-request lines of the microcomputer. The other interrupt-request signal is generated by the second part of the interface circuit. When the start lines are inactive (the external device supplies 00000000), the interrupt-request line is supplied with a logic 0 signal. When the external device supplies a value other than 00000000, the state of the interrupt-request changes to a logic 1, thereby requesting an interrupt. As can be seen in Fig. 4.5, the start lines are still supplied to an input port. This is because the microcomputer needs to read the value in order to recognize which type of operation is requested.

4.4 Software and hardware solutions

In the majority of microcomputer-controlled applications, there exists a choice between satisfying the requirements by software, hardware, or a combination of both. The optimum solution relies on obtaining the best balance between software and hardware, taking into consideration other factors such as cost and development time. In addition to reducing the number of components in an interfacing circuit, software solutions ease the task of introducing future modifications. Such modifications can be achieved by changing a program (in a disk or an EPROM) rather than having to redesign the circuit. The use of software solutions is, however, limited by such factors as the cost of writing the software. Many industrial controllers are produced in large quantities, thus making the cost of software per

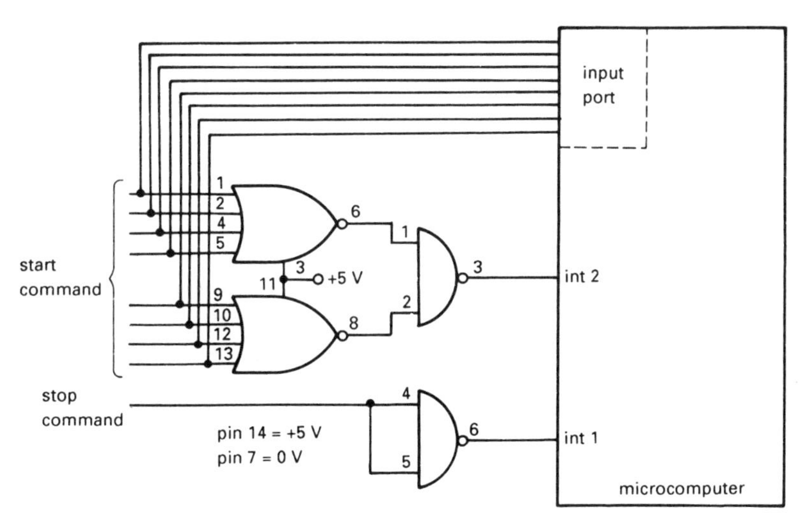

Fig. 4.5 Satisfying interfacing requirements with hardware.

controller very small. Another restriction on the use of a software solution is the time needed for its execution. This limits the number of functions that can be implemented by software economically.

Let us consider a typical example of using software to replace the need for a 2-input AND gate. In order to simulate this gate by software, the two digital-input signals need to be supplied to the microcomputer through an input port, say to lines 0 and 1 of an 8-bit input port. An AND gate produces a logic 1 signal only when all of its inputs are at a logic 1 state. Otherwise, its output is at a logic 0 state. The function of an AND gate can therefore be simulated by the following sequential steps, where '&' represents a software AND:

```
1  READ PORT AND STORE READING IN x
2  x = (x & 0000 0011)
3  IF x = 0000 0011
        THEN OUTPUT = 1
   ELSE
        OUTPUT = 0
```

The function of an OR gate can be simulated in the same manner, the difference being that the output of an OR gate will be at a logic 0 state only when all the inputs are at a logic 0 state. Otherwise it is at a logic 1. The procedure is therefore very similar to that of an AND gate, except that step 3 tests for x not equal to zero.

Example 4.5: **A project introduces the requirement of utilizing the circuit given in Fig. 4.6. A designer can adopt the hardware approach and construct such a circuit, or can use the software approach to perform the function of the circuit. Compare the two approaches.**

Solution 1 The hardware solution

The hardware solution is based on the use of the circuit of Fig. 4.6. Three ICs are needed: two 74LS32 (quad 2-input OR-gate) ICs and one 74LS30 (8-input NAND gate) IC. Such ICs are not expensive and the interface to the microcomputer is implemented through a single digital-input line.

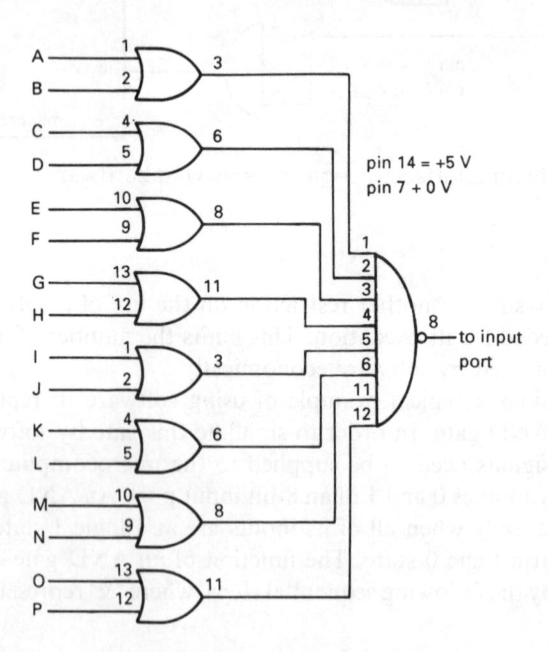

Fig. 4.6 Interface circuit for Example 4.5.

Solution 2 The software approach

The software solution is based on supplying the 16 digital-input signals to the microcomputer through two 8-bit input ports and writing software to simulate the functions of the gates. A microcomputer with an 8-bit bus will therefore read the values by two steps (eight lines at a time), whereas a microcomputer with a 16-bit bus can read the 16 lines in one go. After reading the signals, the program can simulate the function of each individual gate, by ORing individual bits and ANDing the result. This, however, is a time-consuming process which can be replaced by executing two instructions, as shown below.

Reducing the execution time
The execution time can be reduced by connecting signals A, C, E, G, I, K, M and O to one input port, and signals B, D, F, H, J, L, N and P to the other port. In this case, the microcomputer reads the two ports and obtains two bytes which can be ORed together by executing a single instruction. This produces a byte whose first bit corresponds to ORing signal A and B, whose second bit corresponds to ORing signal C with D, ... and so on. The contents of the resulting byte are therefore equivalent to the state of the signals transferred to the inputs of the NAND gate in Fig. 4.6. The next step is to perform the function of the NAND gate. Its output is at a logic 0 state only when all the input lines are at a logic 1 state. The program should therefore test the result of executing the OR instruction which is stored in a byte. If the contents of that byte are equal to 11111111, then the output is a logic 0. Otherwise, it is a logic 1.

This solution is easy to apply and can be represented by the following steps, where both 'w' and 'y' are bytes. The '|' represents a software OR.

```
1  READ PORT 1
2  READ PORT 2
3  LET y = (PORT 1)|(PORT 2)
4  If y = 11111111
         THEN w = 0
   ELSE
         w = 1.
```

Comparing the two solutions

The hardware solution is easy to apply and requires the use of three low-cost ICs. However, it requires board area and increases component cost as well as assembly time. The software solution, on the other hand, has been simplified by correctly designing the procedure and connecting the input signals to those digital lines which result in minimum software operations. Since a microcomputer can execute such a procedure in a few microseconds, the software overhead is small as long as the rate of monitoring the inputs is not extremely high. This makes the software approach attractive in many interfacing applications. Its adoption replaces the need for the three ICs required for the hardware solution. However, it needs 16 input lines to the microcomputer instead of the single line needed by the hardware solution.

A decision of which method to adopt depends on the availability of the 16 digital-input lines within a given system. The decision is also affected by other factors such as the availability of physical space for the hardware, or the lack of spare memory area to contain the software.

Example 4.6: **Consider an application where a microcomputer commences the implementation of a control task after obtaining a digital value equivalent**

to the peak of a sine wave. Such a value can be obtained by interfacing an A/D converter to the microcomputer (Chapter 7). Let us assume that the converter is already interfaced to the microcomputer. It delivers the digital value through an 8-bit input port. We therefore need to define how the microcomputer can detect the peak and read its value.

(1) The software approach

Since the microcomputer commences the implementation of the control task after the detection of the peak, the program-controlled method can be employed to detect the peak. The microcomputer begins its operation by continuously reading the amplitude of the waveform (through the A/D converter). Each reading provides a digital value which is compared with the previous reading to detect if the peak is reached or not. The accuracy of the result depends on the rate of monitoring the amplitude. When using this method, the requirement of high accuracy may prevent the microcomputer from attending to other tasks until the peak is detected.

(2) Techniques for minimizing noise effects

When using a microcomputer to detect the peak of a time-varying signal, one should keep in mind that noise effects may lead to the detection of a false peak. One way of reducing the noise effect is to average out the values received from the A/D converter. Furthermore, the designer can inform the microcomputer of the condition in which a detected peak value represents the sought-after result. One way of achieving this task is to program the microcomputer so that it will not end the procedure after detecting an increase followed by a decrease in magnitude. The search for the 'real' peak should continue for a certain length of time, e.g. 200 μs, in which no averaged value higher than the detected peak is obtained. One can introduce other conditions, such as a 'real' peak of a sine wave will not occur as a result of a sharp change in amplitude and the overall slope increases towards the peak. Can you suggest other conditions?

(3) A modified solution

The application may impose the requirement that the microcomputer should perform other tasks besides searching for the peak value. These tasks may be in the form of displaying a message, initializing devices, establishing communication with other equipment, etc. Instead of diverting the microcomputer from the continuous monitoring of the signal a user may use his or her knowledge of the characteristics of the system. If, for example, it is known that the peak will not occur within the first millisecond of the start-up procedure, then there is no requirement for the microcomputer to monitor the signal during that period. Instead, it can be employed to perform the required tasks as long as their execution period does not consume more than 1 ms.

(4) The hardware approach

Instead of spending execution time waiting for the peak to occur, hardware can be utilized to detect the peak, thereby allowing the microcomputer to attend to other tasks. The required hardware is shown in Fig. 4.7. At the beginning, the microcomputer resets the circuit by commanding the closure of the controlled switch (e.g. an analogue switch, see Section 5.2) to discharge the capacitor. As the amplitude of the input signal increases, the capacitor gets charged until the peak is reached. Any decrease in the amplitude of the waveform will not affect the charge stored on the capacitor (because the diode stops the discharge). The input device can be a 741 IC, whereas it is preferable that the output device possesses a very high input impedance, e.g. a 3140E IC, to minimize the discharge of the capacitor. The output voltage is equal to the voltage stored on the capacitor and can therefore be read by the microcomputer at a later instant.

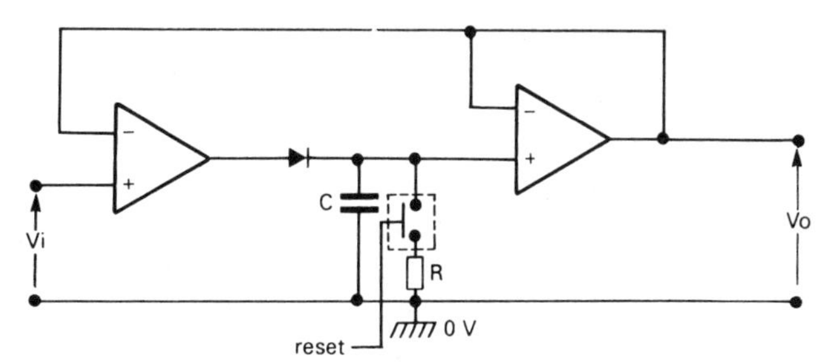

Fig. 4.7 Peak detector.

(5) Discussion

If, during start-up, there is no requirement for the microcomputer to do anything except monitor the input signal, then the program-controlled software approach forms an attractive solution. This approach may become less attractive in applications where the microcomputer needs to attend to other tasks. In such situations, the requirements may be met by introducing simple modifications based on the knowledge of the particular application. Although the hardware approach requires several components, it is easy to apply and does not impose on the microcomputer the requirement of reading the amplitude at the instant of occurrence of the peak.

4.5 Synchronized operations

Synchronized devices are 'clocked' with a series of pulses. Actions are taken when the proper clock transition appears, producing results which depend on the logical state of the inputs at that instant. Using D-type flip-flops to form a shift register is an example of a synchronized operation, where shifting the digital information takes place when a clock pulse is received.

In synchronized operations, digital signals will stay at the same logic state for at least one complete clock cycle. In other words, only complete pulses are transferred through a synchronized system. This can eliminate problems that may appear when a signal occurs halfway through an operation, or when a signal is not there long enough to be recognized. Not taking an action until a clock pulse occurs causes a delay in sensing the asynchronous signal which, in the worst case, equals the duration of a single clock pulse. This implies that clocking a device with pulses of a higher frequency reduces the time needed to sense an asynchronous action. There are, however, many factors which set a limit to the clock frequency, as highlighted by the following example.

Example 4.7: **How many shift stages are needed if one part of a synchronized circuit is a shift register employed to shift a digital signal by 40 μs?**
Let us start by assuming that eight shift stages are sufficient to fulfil the requirement. This is convenvient, because there are many single ICs containing 8-stage shift registers, or two 4-stage registers which can be cascaded to form an 8-stage register. To get a shift of 40 μs in eight stages, one needs to clock the shift register with a frequency of 200 kHz (5-μs period of each clock pulse).

If the 5-μs interval is considered long, then a higher clock frequency is required, but this creates the need for using a higher number of shift stages. Let us assume that we want to reduce the clock period to 1 μs. This corresponds to a clock frequency of 1 MHz. With this frequency, a 40-μs shift will need 40 shift stages. This will increase the number of ICs from one to five. The result is, obviously, an increase in both the cost and the physical size of the circuit board. On the other hand, it provides a faster detection of changes in the input signal, since the duration of each stage is $\frac{1}{5}$ of that of a 200-kHz clock signal.

Clocked digital latches are very attractive devices for use in designing sychronized circuits.

- A 'digital latch' is a device whose output follows its input as long as it is enabled. Once it is disabled, the state of its output will be held at the last logical state it took when the device was enabled. The device is therefore a memory unit. Writing to this unit is performed by enabling the latch.

- A 'clocked-digital latch' is one with a dedicated input line for a clock signal. The input to the latch is examined when a clock pulse is received. The signal will be stored and the output changes accordingly.

The 74LS74 IC (used in a few examples within this book) is a dual clocked D-type logical latch. When clocked, this type of latch transfers the logic level present at its D input line (pin 2 in Fig. 4.8) to its Q output line (pin 5). Note that pin 6 delivers an output signal of an opposite logic state to that of pin 5. It is provided to eliminate the need for an additional inverting gate in digital circuits. The latched information will be held until the arrival of the next clock pulse. Each of the two D-type latches within the 74LS74 IC possesses active-low set and reset command lines. When not needed, these lines should be connected to the +5-V supply. This is shown in Fig. 4.8(a), where one of the D-type latches is connected to synchronize signal X with a microcomputer-controlled circuit. Actions in the circuit take place when a clock pulse is supplied. A change in the state of signal X will be transferred to the Q output line of the latch and hence to the rest of the circuit only when a clock pulse is applied.

In some situations, however, signal X occurs for a very short time interval and may not be recognized if the clock was not fast enough. A D-type latch can be connected to capture the change in X and store it for the microcomputer to read. The connection is shown in Fig. 4.8(b), where signal X is supplied to the CLK input rather than to the D input line. In this way, pulsing line X will change the state of output line Q to a high-logic state. Once the microcomputer, or its associated circuit, senses the change in the state of line Q, a logic-low pulse can be delivered to clear the latch through its reset line.

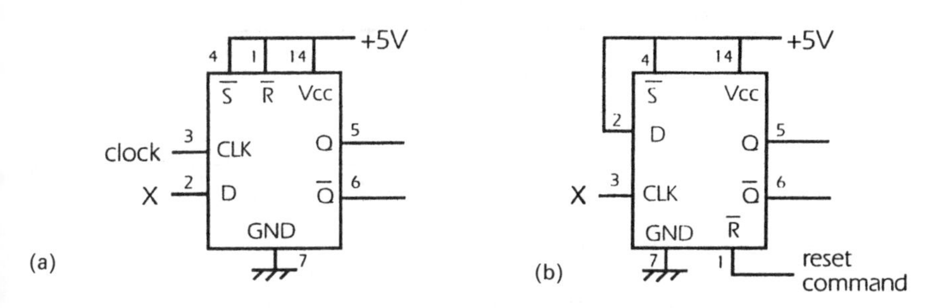

Fig. 4.8 Synchronized operations: (a) latching signal X; (b) capturing fast signals.

Example 4.8: **Design a circuit to allow a microcomputer to respond to any of two interrupts. Assume that the microcomputer has a single external interrupt line and that each of the interrupts appears for a very short time.**

The two interrupt sources need to be multiplexed and delivered to the single interrupt line of the microcomputer. Example 1.3 presents a method of multiplexing interrupts, but it may not be suitable if the duration of the interrupt signal was too low for the microcomputer to recognize which of the two sources requested attention. An alternative solution can be based on the use of latches. The circuit of Fig. 4.9 employs the 74LS74 IC, which contains two D-type latches. Each interrupt signal is connected to the CLK input line of a latch whose D input line is connected to the +5-V line. A logic-high signal will then be stored when an interrupt-request signal is received. The outputs of the latches are ORed together to request an interrupt as soon as either of the two interruption sources is activated. The microcomputer, however, may be executing an instruction of a longer duration than the period of the interrupting pulse. As soon as the microcomputer responds to the interrupt request, it reads the states of the two digital-input lines to detect which of the interrupts has occurred (which could be both). The microcomputer will then pulse its digital-output line to reset the latches before executing the interrupt-service routine.

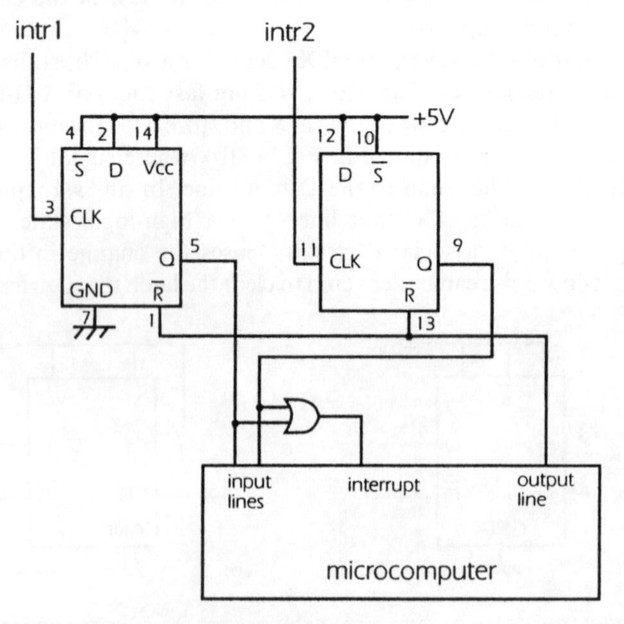

Fig. 4.9 Multiplexing fast interrupt signals.

Example 4.9: **In an industrial application, a microcomputer is employed to monitor various activities. The state of operation depends on the sequence of signals (events) delivered by four local controllers. Rather than interrupting the microcomputer with every event, design a circuit which will only store the event that occurred last.**

Assume that the occurrence of an event will be reported in the form of a high logic level. This signal can be stored in a D-type latch and examined by the microcomputer. The requirement of storing only the last of a few events means that the interface circuit should use the event to clear any previously stored indications as a new indication is being stored. With this approach, all the microcomputer needs to do is to read the output of the latches through an input port. The event whose corresponding output line is high is the one that has occurred last.

The requirement can be satisfied by the circuit of Fig. 4.10, where four events are assumed. When an event indication is received, it will be stored in a D-type latch clocked by high-frequency clock pulses (two 74LS74 ICs are employed to provide four latches). The storage can only take place when a clock pulse is received. This changes the state of output line Q of the triggered latch to a logic high. The latch will maintain its output at that state until reset by a low logic level applied to its reset line. A 3-input OR gate is connected to reset each latch. The input lines of each OR gate are connected to the output lines of all the latches other than the one it is employed to reset. The output of an OR gate is at a logic-low state as long as all of its input lines are supplied with logic-low signals. When an indication of an event is received and stored, the output line of the corresponding latch will change state to a logic high. Since this line is connected to three OR gates, then a change of its state alters the state of the output lines of these

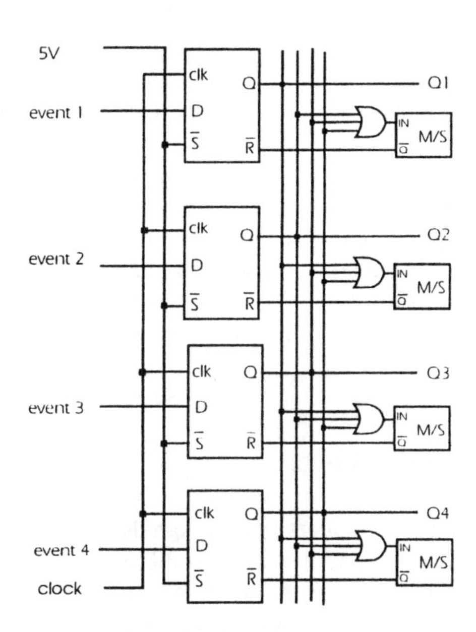

Fig. 4.10 Last-event detection circuit.

gates to a logic high. This cannot be used directly to reset the the other latches. This is because the reset line should be pulsed rather than supplied with a logic level. The formation of the pulse is accomplished by the use of a monostable is described in Section 9.5.1). The inverted output of the monostable is used so that a logic-low pulse will be supplied to reset the latches.

All these actions are taken by the logic circuit without any interference from the microcomputer. The latter, however, can read the latches regularly through four digital-input lines. The higher the rate of reading the latches, the sooner the microcomputer will realize the occurrence of an event.

Example 4.10: **Repeat the previous example, but this time let the interface circuit store the first event.**

The requirement can be satisfied by using a circuit similar to that of the previous example. The main difference is to stop responding to changes in the inputs once an event is detected. This can be accomplished by using the circuit of Fig. 4.11, where the output lines of all the latches are supplied to 2-input OR gates as well as to an input port of a microcomputer. When any of the latches receives a logic-high signal corresponding to the occurrence of an event, its Q output line changes state to a logic high. The use of OR gates means that the state of the signal at point X1 becomes high when any of the Q output lines changes state to a logic high. This forces the signal at point X2 to a logic-high state regardless of the state of the clock pulses delivered to the other input line of the OR gate. This stops the clock pulses from reaching the clock input lines of the latches, thereby stopping the latches from responding to signals at their inputs. The circuit will therefore store the first event only.

Note that the reset lines of the latches are connected to a digital-output line of the microcomputer. This gives the latter the ability to pulse these lines in order to clear the latches. This takes place at start-up and after the microcomputer completes reading the stored event through its four digital-input lines.

4.6 Loads and their effects

Microcomputer-based controllers supply signals to different types of loads, e.g. resistive, inductive, capacitive, or a combination of all. The load may be single phase, or multiphase. A resistive load dissipates energy, a pure inductive load stores energy in a magnetic field, and a pure capacitive load stores energy in an electric field. The instantaneous relationship between the

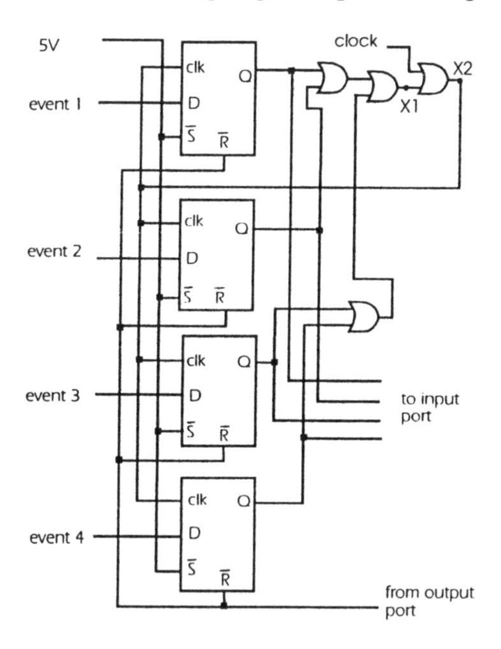

Fig. 4.11 First-event detection circuit.

voltage and current varies according to the type of load. This can affect microcomputer-based measurement and control.

4.6.1 Inductive loads

An a.c. induction motor is an inductive load. It seems to a power supply as a resistor (R) in series with an inductance (L), as shown in Fig. 4.12(a). The inductance stores energy and will use it to maintain the flow of current when a change of state occurs. The relationship between the voltage and current in an inductance is given by:

$$V = L(dI/dt) \tag{4.1}$$

where L is the inductance, and dI/dt is the rate of change of current per unit time.

Effect on d.c. signals

Let us consider supplying an inductive load from a d.c. supply through a switching device. Powering the device applies voltage to the inductive circuit and tries to pass current. The inductance opposes the introduced change, leading to a gradual increase of current to a steady-state value defined by:

$$i = V/R \tag{4.2}$$

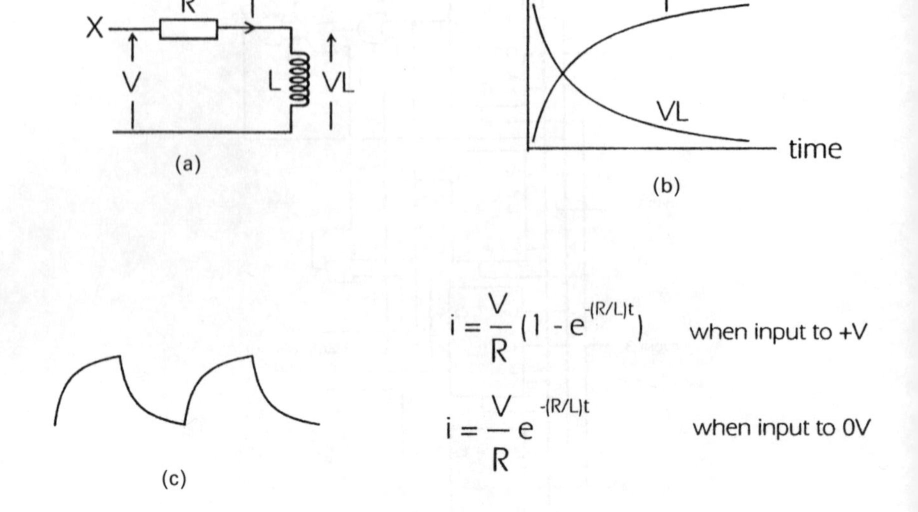

Fig. 4.12 Inductive loads: (a) circuit; (b) timing signals; (c) response.

The instantaneous change of current is shown in Fig. 4.12(b), which is represented by an exponential relationship defined by the equations within the same figure. The gradual change of the voltage across the inductor is also shown in the same figure. If the state of operation is altered by switching the input of the circuit to the 0-V line rather than to the d.c. volts, then the result is a gradual reduction of current. That is because the inductance will use its stored energy to maintain the flow of current. The instantaneous change of signals applied to an inductive load is best observed when energizing such a load from a repetitive waveform. Let us assume that a microcomputer energizes an inductive circuit with a square wave. The waveform of the generated current is a distorted square wave whose sharp edges are 'smoothed' by the effect of the inductance, as shown in Fig. 4.12(c). The time taken for the output signal to reach steady state depends on the value of the inductance as well as the resistance, as given by the equations of Fig. 4.12. Attention should therefore be paid to these values and the amount of distortion introduced to the waveforms. For example, a limit may be imposed on the maximum frequency that can be applied to an inductive circuit. This limit depends on the maximum acceptable distortion in the waveform. At very high frequencies, the waveform may never reach the required steady-state value.

Effect on a.c. signals

When a.c. power is applied to an inductive circuit, the current will lag the voltage by a phase angle K, defined by:

$$\tan K = wL/R \tag{4.3}$$

where K is presented in radians and is referred to as the 'power-factor' angle, and $w = 2\pi f$, where f is the frequency, and $\pi = 3.14159$.

Let us now consider the effect of applying a sine wave to a resistive load and then to an inductive load. Figure 4.13(a) shows the relationship between the voltage and current in a resistive load. The value of the current is given by the ratio V/R and there is no phase shift between the voltage and the current waveforms. Figure 4.13(b), on the other hand, shows the waveforms in an inductive circuit, such as the one given in Fig. 4.12(a). The waveforms show a clear phase shift where the current lags the voltage by a phase angle K which depends on the value of both R and L. This effect is illustrated in Fig. 4.13(c) by the use of a vector representation. The circuit components are shown to consist of a real part and an imaginary part. They can be represented by R + jwL where the j identifies the imaginary part. The impedance Z is the vector resulting from combining the real and the imaginary parts. The effect of the frequency and the value of each component on the angle and magnitude of that vector can be seen from Equation 4.3 and the equation given in Fig. 4.13.

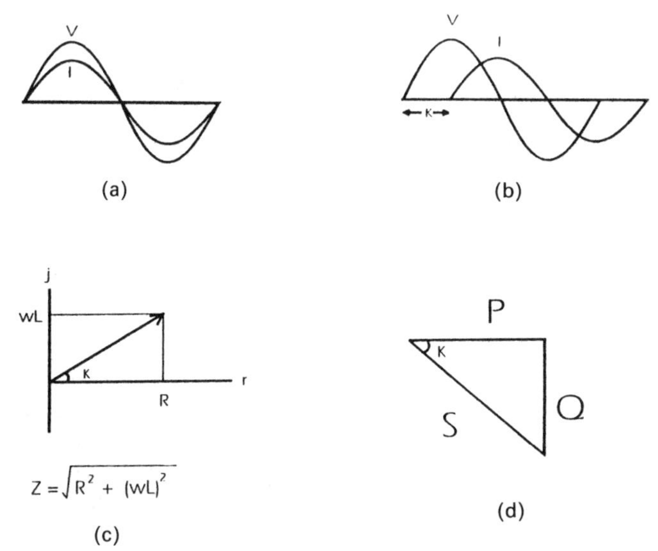

Fig. 4.13 Applying an a.c. signal to resistive and inductive loads: (a) resistive load; (b) inductive load; (c) effect of circuit components; (d) power components.

Power consideration

The phase angle also affects the power reaching the load resistor. This can be explained with the aid of Fig. 4.13(d), where the power is represented by the following three components:

- The 'apparent power' (S) is measured in volt-amperes (VA), or kilovolt-amperes (kVA), and is given by:

$$S = VI \tag{4.4}$$

  It is represented in Fig. 4.13(d) as the vector sum of the two other components (P and Q).
- The 'average power' (P) is what is dissipated in the resistive part and is measured in watts (W) or kilowatts (kW). It is related to the applied voltage and current by:

$$P = VI \cos K \tag{4.5}$$

  where K is given in Equation 4.3 and cos K is referred to as the 'power factor'.
- The 'reactive power' (Q) is the power in the inductive part of the load. It is measured in vars or kilovars and is defined by:

$$Q = VI \sin K \tag{4.6}$$

If R is the actual load, then the power supply needs to deliver voltage and current represented by S, which is higher than the required power P. The proportion of P is represented by the 'power factor', which is given by the ratio P/S where the optimum value is 1.

There are ways of reducing the power factor angle, as will be explained later in this chapter.

4.6.2 Capacitive loads

There are many examples of capacitive loads, e.g. power supplies within electronics equipment. Such supplies rectify the a.c. signal of the mains before converting it to a d.c. voltage to drive computers and electronics circuits. The converted (rectified) voltage is smoothed by the use of capacitors, as shown in Section 16.6. This type of power supply is a capacitive load receiving current to charge its internal capacitors and then using that charge to maintain the level of voltage, as explained in Section 16.6.

Effect on d.c. signals

When voltage is applied to a capacitive load, such as that of Fig. 4.14(a), current starts to flow to charge the capacitor. The increase in the capacitor's charge reduces the current gradually to a steady-state value of zero when the capacitor is fully charged. The initial value of the current is:

$$I = V/R \tag{4.7}$$

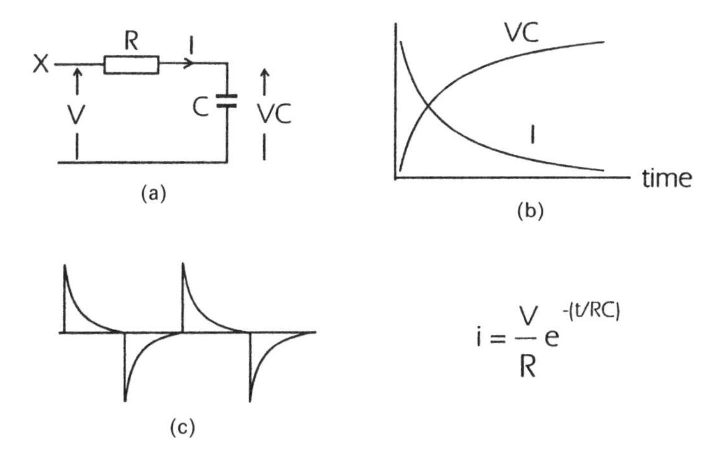

Fig. 4.14 Capacitive load: (a) circuit; (b) timing signals; (c) response.

and the gradual change of the value of current is represented by the equation of Fig. 4.14. Removing the applied voltage and connecting the input to the 0-V line discharges the capacitor through the resistor. The current will start with a value of (V/R) and will flow in the opposite direction to the charging current. The instantaneous change of current is given by the equation of Fig. 4.14, where the rate of decay depends on the capacitance.

Applying and removing power by the use of a square wave will distort the shape of the current waveform, where the current represents charging and discharging the capacitor, as shown in Fig. 4.14(c).

Varying the rate of charge and discharge

Since the rate of charging and discharging the capacitor depends on the value of the resistor (R), it is possible to have a rate of charge different from that of discharge. Figure 4.15 presents a circuit incorporating two diodes in order to define the direction of current through each resistor (each diode permits conduction in one direction and not the other, as described in Section 10.2). The result is a charge time set by R1 and C and a discharge time defined by the value of R2 and C.

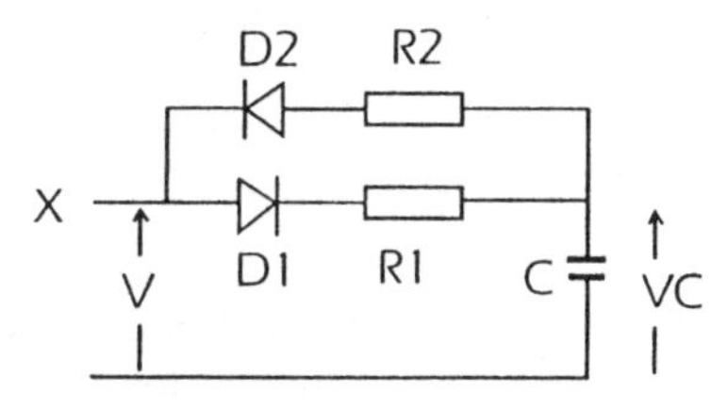

Fig. 4.15 Altering the rate of charge and discharge.

Effect on a.c. signals

When operating from an a.c. signal, a capacitive load gives a leading phase shift between the voltage and current, as shown in Fig. 4.16(a). The phase shift angle is given by:

$$\tan H = 1/(\ wCR) \tag{4.8}$$

The relationship between the circuit components can be represented by the relationship $R - j(1/wC)$. The imaginary part of the equation is negative, which means a leading phase angle. This implies that the current will lead the voltage, as shown in Fig. 4.16(a). A vector representation is given in Fig. 4.16(b). Examining it reveals that the magnitude and the phase angle of the vector depend on the value of each of the circuit components. The figure also gives an equation representing the impedance *Z* of the circuit. Compare this equation with that of Fig. 4.13 to see the effect of the frequency on the impedance of different types of loads. The effect of the frequency on the phase angle can be seen by comparing Equations 4.3 and 4.8.

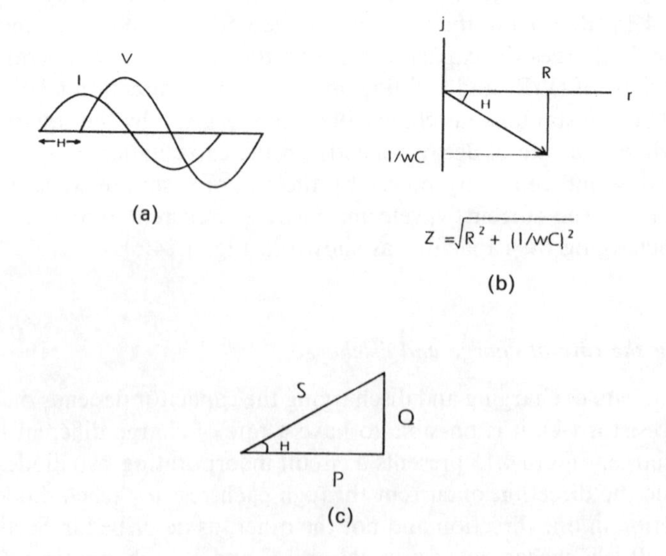

Fig. 4.16 Applying an a.c. signal to a capacitive load: (a) phase shift; (b) effect of circuit components; (c) power components.

Power consideration

The relationship between the three components of power is shown in Fig. 4.16(c) and is similar to that of the inductive load, except that the phase angle is leading, rather than lagging.

Integration and differentiation

The charging and discharging behaviour of a capacitive load is used in the construction of integrators and differentiators. Figure 4.17(a) shows an integrator. The relationship between the input and output voltages is represented by the equations given in the figure. The equations illustrate that the rate of change in the value of the output voltage depends on the value of both R and C. Choosing the value of C to be 1 μF and both R1 and R2 to be 100 kΩ, the time constant $T = \text{R1C}$ becomes 100 ms. The circuit can be modified to provide a path to discharge the capacitor before starting the integration process. An example of such a circuit is given in Fig. 10.10(a). Swapping the position of the resistor and the capacitor will alter the function of the circuit of Fig. 4.17(a) from an integrator to a differentiator. The resulting circuit is shown in Fig. 4.17(b), where equations are also presented illustrating the relationship between the input and output voltages and showing the effect of the value of both R and C.

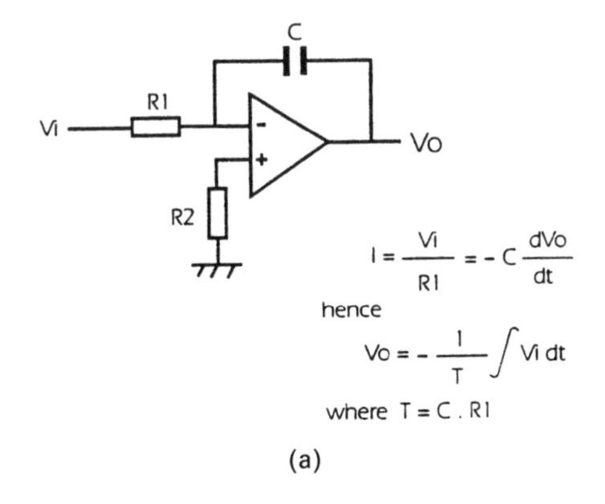

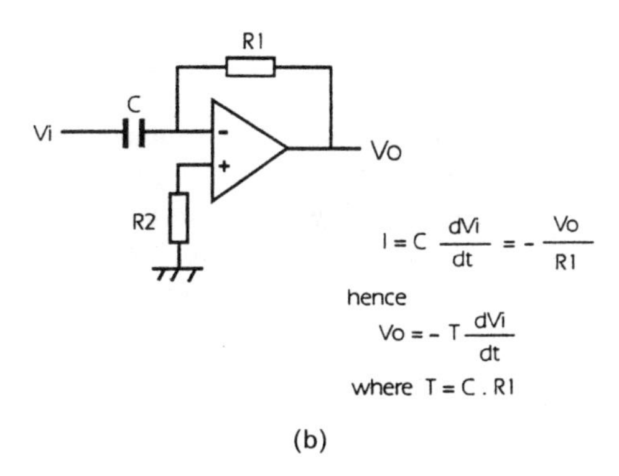

Fig. 4.17 Resistor–capacitor circuits: (a) integrator; (b) differentiator.

Reducing the phase angle

In many industrial sites the load is inductive and (as mentioned in Section 4.6.1) it is preferred to reduce the power-factor angle. One way of reducing it is to connect capacitors in parallel with the point where the load connects to the power supply. The effect of the additional capacitors is the introduction of a leading phase angle. The resulting phase shift is the sum of the leading and the lagging angles taking into consideration that they oppose each other. Assuming that the leading angle is smaller than the lagging, then the result is a smaller lagging power-factor angle and hence a smaller apparent power (S) for a certain value of average power (P). The effect of the inductor is shown in Fig. 4.18(a) as vector Q representing the reactive power. The inclusion of a capacitor will reduce the power factor angle leading to a smaller value of Q represented by vector Q1 in Fig. 4.18(b) which is the effective value. Consequentely, the inclusion of additional capacitors reduced the apparent power from S to S1 for the same value of P.

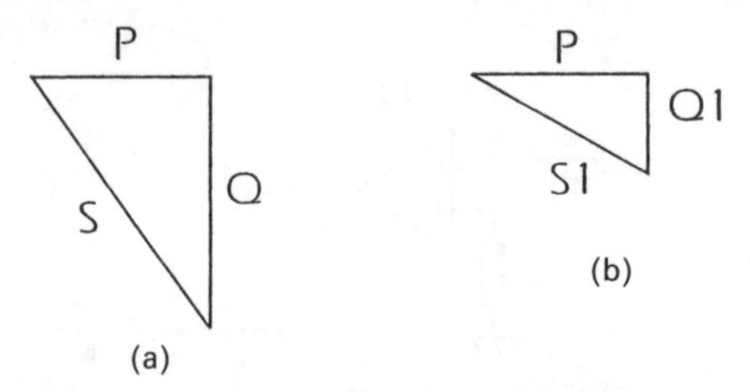

Fig. 4.18 Reducing the phase angle: (a) without capacitors; (b) with capacitors.

Capacitors can be grouped together to form capacitor banks which can be switched by a controlling microcomputer to improve the power factor. Relays can be incorporated as the switching devices and the capacitor value is increased by switching more banks in parallel, as illustrated in the following example.

***Example 4.11:* A work area contains inductive loads, some or all of which can be operated at any instant of time. How can a microcomputer be used to maintain the power factor at a level not less than 0.95?**

The power factor is represented by cos K. This implies that a 0.95 power factor corresponds to a phase angle of 18°. To get this angle, the microcomputer must first measure the power-factor angle, which corresponds to the interval between the zero-crossing points of the voltage and current. The measurement can be performed with the aid of two zero-cross detectors. Such detectors are explained in Section 10.9, where circuits are presented. If a pulse is obtained from each zero-cross detector, then the duration between the two pulses can be measured by a circuit similar to that of Fig. 9.26. Such a circuit supplies the microcomputer with a digital number proportional to the delay between the two pulses. The microcomputer converts this number to a phase angle by using a conversion factor which depends on both the frequency of the clock supplied to the counter and the frequency of the a.c. signal (for example, for a 50-Hz supply, the duration of a single a.c. cycle is 20 ms which corresponds to 360 electrical degrees).

If the a.c. signal is obtained from the mains, then one has a choice of simplifying the solution and assuming that the frequency is constant, or measuring it on line. The decision depends on how stable the supply frequency is and on how accurate the required control is. If it is assumed constant, then one can program the duration of a full cycle into the microcomputer. If, on the other hand, one decides to measure the frequency, then the following is required.

Since the circuit produces sequential pulses corresponding to the zero-crossing instants, the duration of one cycle is the time between two successive zero-crossing pulses of the same signal. This can be counted by a second counter and supplied to the microcomputer (remember that the microcomputer obtains a count representing the phase angle from the first counter). The reading of the second counter represents 360°. The phase angle can then be calculated from the proportion of the two counts (the use of counters is explained in Chapter 9).

If the measured power-factor angle is more than 18 electrical degrees, then the microcomputer issues a command to close an additional relay, thereby increasing the capacitance and correcting (reducing) the power-factor angle.

4.6.3 Three-phase loads

Figure 4.19 shows three sine waves of a 3-phase system. The phase angle between the voltage of one phase and another is 120 electrical degrees (one-third of a cycle regardless of the frequency). Generating 3-phase signals under the control of a microcomputer can be performed by controlling the

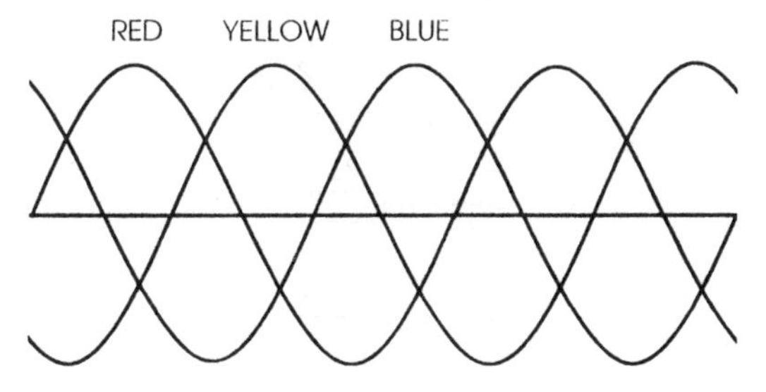

Fig. 4.19 Three-phase voltages.

operation of switching devices connected to form a 3-phase inverter (as shown in Fig. 15.14). The sequence of turning on the switching devices can be obtained from examining the three waveforms of Fig. 4.19, starting from the left-hand side. Let us refer to the waveforms as R, Y and B. One can clearly see that the sequence of crossing the 0-V line is:

$$R+, B-, Y+, R-, B+, Y-$$

where the '+' and '−' signs represent crossing to the positive half-cycle and the negative half-cycle respectively.

A 3-phase load can be connected as either star or delta. A star load is shown in Fig. 4.20(a), which includes a neutral line N in addition to the three phase lines. The red-phase current passes through load Z1, the yellow-phase current passes through Z2 and the blue-phase current passes through Z3. The voltage can be considered either line-to-line or line-to-neutral, where the former is greater than the latter by a factor equal to the square root of 3. The voltage across each of the three loads is the line-to-neutral voltage and the three currents can be represented by:

$$I_{RN} = \frac{V1}{Z1}, \; I_{YN} = \frac{V2}{Z2}, \; I_{BN} = \frac{V3}{Z3} \tag{4.9}$$

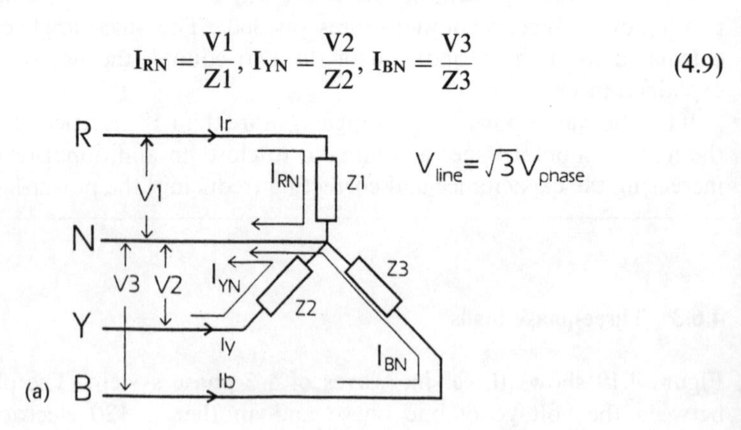

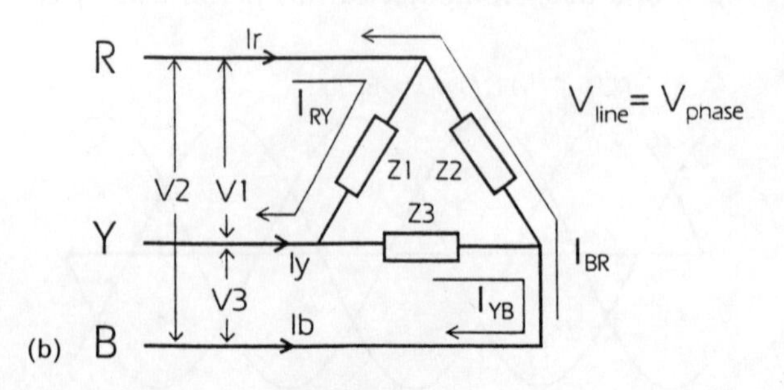

Fig. 4.20 Connections to a 3-phase load: (a) star; (b) delta.

The neutral current can be represented by:

$$I_N = -(I_{RN} + I_{YN} + I_{BN}) \tag{4.10}$$

If the load is balanced, then there will be no current through the neutral line. The load is not always balanced and one should pay attention to the magnitude of the current flowing through the neutral line. This is not always easy to define exactly, since a 3-phase supply may be distributed in a building where different loads are connected between each of the three phases and the neutral.

In a delta load, the line-to-line voltage is applied across each of the three loads Z1, Z2 and Z3, as shown in Fig. 4.20(b). The currents flowing through the three loads can be represented by:

$$I_{RY} = \frac{V1}{Z1}, \ I_{BR} = \frac{V2}{Z2}, \ I_{YB} = \frac{V3}{Z3} \tag{4.11}$$

The current delivered by the red phase is:

$$I_r = I_{RY} + I_{RB} = I_{RY} - I_{BR} \tag{4.12}$$

The current delivered by the yellow phase is:

$$I_y = I_{YB} + I_{YR} = I_{YB} - I_{RY} \tag{4.13}$$

and the current delivered by the blue phase is:

$$I_b = I_{BR} + I_{BY} = I_{BR} - I_{YB} \tag{4.14}$$

When using a microcomputer to control the operation of a 3-phase system, it is important to know the number of signals to be measured and controlled by the microcomputer. The microcomputer may, for example, need to sample three line voltages, three line currents, neutral current and other signals, e.g. temperature, speed, and direction of motion. More is given on 3-phase systems in Section 15.3.

Chapter 5

Multiplexing, storing and data-conversion techniques

5.1 Introduction

This chapter begins by explaining how analogue and digital signals can be multiplexed and demultiplexed. It gives examples of popular multichannel multiplexing ICs and single-channel analogue switches. Section 5.3 considers storing analogue signals by the use of sample-and-hold devices. It explains how a microcomputer can control the operation of such a device (the latching devices presented in the previous chapters are examples of storing digital signals). Sections 5.4 and 5.5 are devoted to explaining popular data-conversion methods. They highlight the advantages and drawbacks of these methods. Section 5.6 explains how multichannel data converters can be constructed by time-sharing a single-channel converter. The last section of this chapter describes causes of errors in practical data converters and how they contribute to deviations in the behaviour of a converter from the ideal one.

Many of the techniques presented in this chapter are used in later chapters.

5.2 Multiplexers and demultiplexers

Multiplexers are digitally controlled channel selectors. They are utilized for transmission of information from one of several sources to a single destination. The source channel is selected by means of an address supplied to the multiplexer. Demultiplexers, on the other hand, are employed for transmitting information from one source to one of several selected destinations. Therefore, demultiplexers perform the inverse process of multiplexers and in many cases the same device can be used either as a multiplexer or as a demultiplexer.

Popular multiplexing ICs possess 4, 8 or 16 channels with an on-chip binary channel decoder. The decoder reduces the number of digital lines needed to permit a controlling device to perform the channel-selection task. For example, an 8-channel multiplexer IC provides three channel-selection lines. The activated channel is the one represented by the binary combination of the signal carried by the three lines. When switching from one channel to another, there is a small time delay in the device between disconnecting one channel and connecting the next; this is known as 'break before make'. It ensures that no input channels are ever momentarily shorted together. Multiplexing devices possessing an enable line provide a controlling device with the ability to disable the multiplexer. In the disable mode, the output is not affected by any of the input signals. The advantage of using the enable line is highlighted in Example 6.7.

5.2.1 Analogue multiplexers

Analogue multiplexing ICs are available from several manufacturers. Figure 5.1 shows three digitally controlled bidirectional analogue multiplexing/demultiplexing ICs. These ICs offer fast switching and a low ON resistance, but differ in the number of input and output channels. They contain two voltage supply lines, V_{cc} (pin 16) and V_{ee} (pin 7), representing the upper and lower voltage limits, where the difference between them should not exceed 10 V. An analogue signal of a value between these two voltages can be conducted through the multiplexed channels. An active-low enable line (pin 6) is provided within each IC to permit a controlling device to turn the output on and off. Each of the three ICs shown in Fig. 5.1 is continuously enabled by joining its enable line to the 0-V line. Line V_{ee} of each of the three ICs is also connected to the 0-V line to limit the conduction of positive voltages through the channels.

Figure 5.1(a) shows the 74HCT4051 IC. It provides eight independent channels on one side and a single common line on the other side. Any of the eight channels can be connected to the common line according to a binary number delivered by a controlling device through digital lines S2, S1 and S0 (line S2 represents the most significant bit of the binary number). For instance, if the controlling device supplies the binary number 000, then line A0 (pin 13) is connected to the COM line (pin 3). Changing the number to 111 connects line A7 to the COM line.

Many applications impose the requirement of using two common lines. Rather than using two 74HCT4051 ICs, use can be made of the 74HCT4052 IC, which contains two 4-channel multiplexers/demultiplexers. The channels are drawn in Fig. 5.1(b) in two groups. One channel will be selected from each group at one time according to a binary number delivered to lines S1 and S0, where line S1 represents the most significant bit of the binary number. This implies that lines A0 and B0 will be connected to lines COM A and COM B respectively, when both lines S1 and S0 are supplied with a logic-low

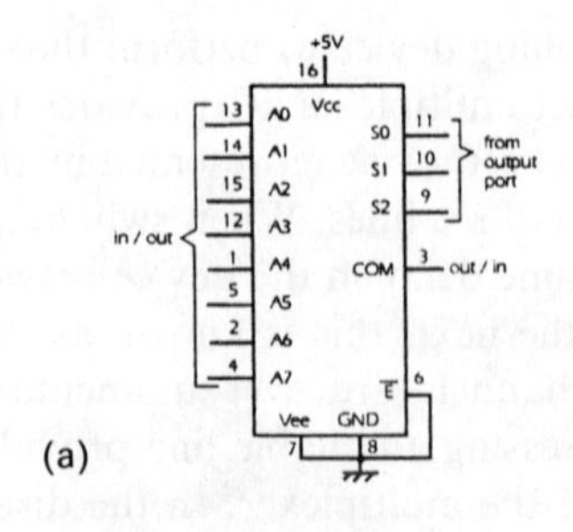

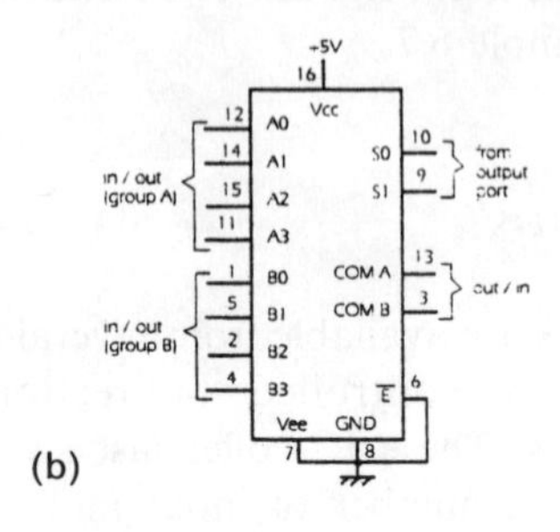

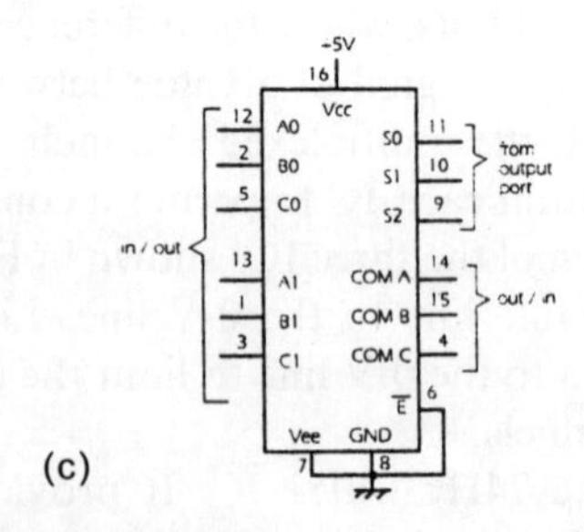

Fig. 5.1 Analogue multiplexer/demultiplexers: (a) 74HCT4051; (b) 74HCT4052; (c) 74HCT4053.

level. Changing the number to 10 will lead to connecting lines A2 and B2 being connected to lines COM A and COM B respectively.

Another IC is shown in Fig. 5.1(c). It is the 74HCT4053 IC triple 2-channel bidirectional multiplexer/demultiplexer. The IC possesses three common lines on one side and lines Ax, By and Cz on the other side, where *x*, *y* and *z* are either 0 or 1. The 3-bit binary number delivered to lines S2, S1 and S0 results in connecting three channels to the three common lines, where *x*, *y* and *z* correspond to lines S2, S1 and S0 respectively. For example, if a controlling device supplies lines S2, S1 and S0 with 000, then lines A0, B0 and C0 are connected to common lines COM A, COM B and COM C respectively. Changing the number to 101 connects lines A1, B0 and C1 to the three common lines.

Attention should therefore be paid to the internal resistance of analogue multiplexers, as this causes a voltage drop when current flows through the device. This problem can be overcome by the use of an FET input amplifier to buffer the output.

5.2.2 Digital multiplexers

A digital multiplexer operates on the same principle as the analogue multiplexer, except, as the name implies, it multiplexes digital signals. An example of a digital multiplexer is the 74LS151 (8-channel) and 74LS153 (dual 4-channel) ICs. Channel selection is similar to that of the analogue multiplexers. Figure 5.2 shows the configuration needed to achieve channel selection by a microcomputer through an output port. The 74LS151 IC is used and the selected channel is the one whose number is equal to the binary combination of the three channel-selection lines A, B and C (the most significant bit is supplied through line C). The IC has two outputs, Y and W, where the latter delivers an inverted logic state signal. The enable line, which is referred to as the strobe (S), should be at a logic 0 state during multiplexing. When this line is driven high, outputs Y and W are set to a logic 0 and a logic 1 respectively, independent of the state of the inputs.

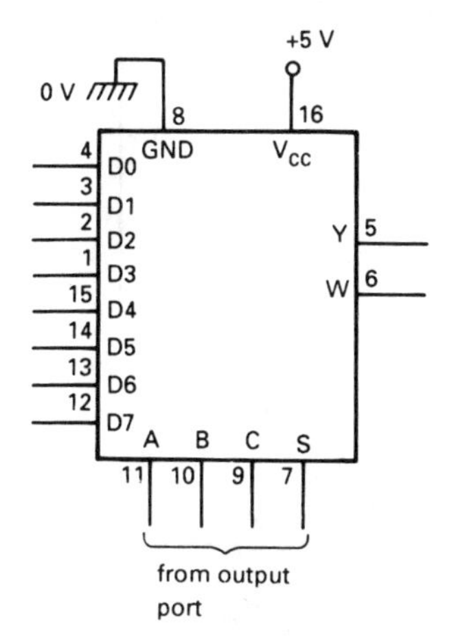

Fig. 5.2 The 74LS151 digital multiplexer.

5.2.3 Analogue switches

Multiplexing can also be accomplished by using analogue switches. These are available from several manufacturers and can be easily interfaced to TTL and CMOS devices. A switch will be closed when its associated digital control line is enabled.

Figure 5.3(a) shows the 74HCT4066 quad-bilateral-analogue switching IC supplied from a +5-V power supply. It is a 14-pin IC which is widely used and contains four independent analogue switches. These switches are bilateral, i.e. they allow conduction in either direction. A switch will be closed when its digital control line is supplied with a logic 1, and will be opened when supplied with a logic 0.

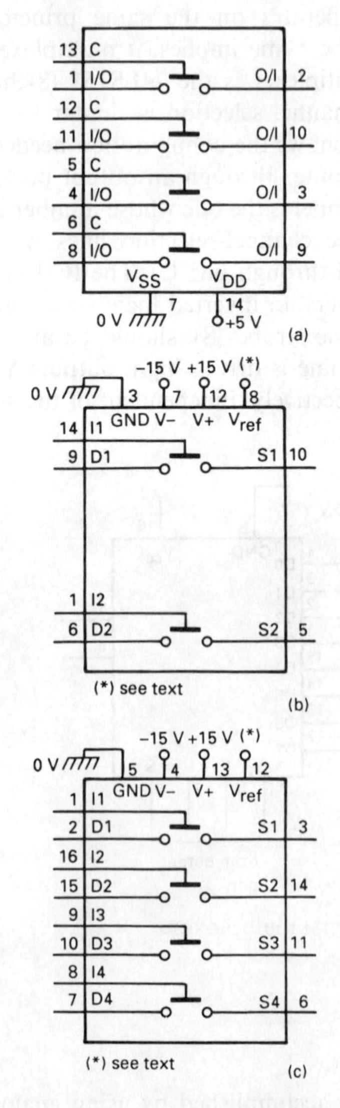

Fig. 5.3 Analogue switches: (a) 74HCT4066; (b) DG200; (c) DG201.

Figures 5.3(b) and 5.3(c) present the DG200 (dual-analogue switches) and DG201 (quad-analogue switches) ICs. In these ICs, a switch will be closed when its digital control line is supplied with a logic 0, and will be opened when supplied with a logic 1. Both of these ICs possess an internal reference voltage derived from the positive supply voltage (V+) by the use of an on-chip voltage divider. It produces a +2.4-V reference voltage at pin 12 when the IC is operated from V+ of +15 V. For operation from a lower power-supply voltage, an external resistor is needed between pins 12 and 13. It is equal to 100 Ω for operation from a +12-V power supply with the control pin driven from a TTL signal. The variation of the ON resistance of the analogue switch with the change of the input voltage becomes smaller when the range between V+ and V− is larger.

Examples of using analogue switches are given in later chapters, e.g. Example 7.8.

5.2.4 Comparison

Multiplexing with an N-channel multiplexer is different from using N analogue switches. Each analogue switch can be enabled or disabled independent of the other switches, whereas in a multiplexer all the channels share the same output path to which any one of the channels can be connected. It is therefore possible to connect N analogue switches to emulate the operation of an N-channel multiplexer. Although this is not usually economical, it forms an attractive solution to some interfacing problems, as in Example 6.8.

Examples of using multiplexing devices under the control of a microcomputer are presented in Chapters 6 and 7.

5.3 Sample-and-hold (S&H)

This is a device which can be instructed to capture an analogue voltage signal and store it on a capacitor. It has two modes of operation: the first is the *sample* mode, in which the output of the device tracks the input signal; the second is the *hold* mode, in which a sample of the input signal is stored. During the hold mode, the output of the device is a relatively constant voltage representing the magnitude of the input signal at the instant of obtaining the last sample. The value will be held until the hold mode is disabled. The device can then be instructed to resume the continuous sampling.

S&H circuits are widely used in many applications; for example, a peak detector relies on the use of an S&H for its operation (Example 4.6). Another example is the utilization of an S&H as part of an A/D conversion circuit

(Chapter 7). In such a circuit, the S&H holds the value of the analogue signal at a stable level during the conversion period. S&H circuits are also employed in applications requiring simultaneous A/D conversion of several signals (Example 7.10). Other applications include high-speed sampling of an analogue signal in a system employing a slow A/D converter (Example 7.11).

The simple circuit shown in Fig. 5.4(a) represents an S&H. It consists of a switch in series with a capacitor. When the switch is closed, the capacitor will charge to the value of the input voltage. When it is opened, the capacitor

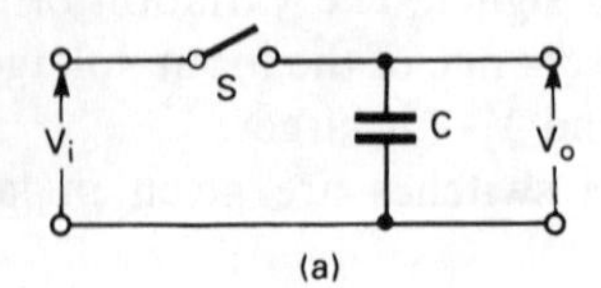

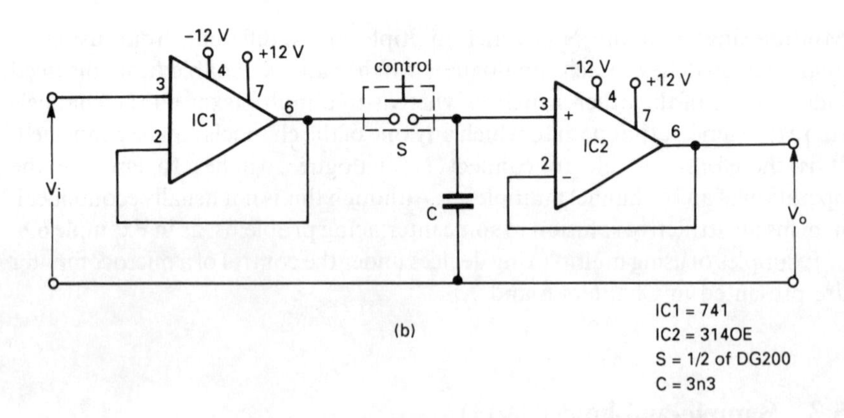

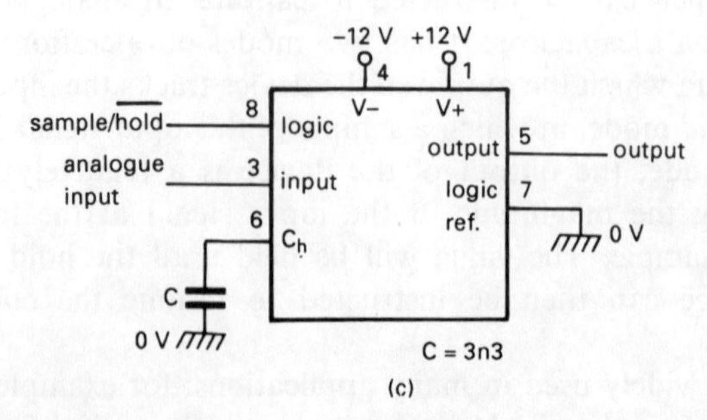

Fig. 5.4 Sample-and-hold circuits: (a) typical; (b) practical; (c) commercial (The National Semiconductor® LF398).

will hold the value until the switch is closed again. The switch can be a bipolar transistor, a MOSFET, or an analogue switch.

In order to prevent the S&H from loading the source of the input signal, a high input impedance buffer amplifier is normally used at the input of the S&H. This buffer should be capable of supplying enough current to charge the capacitor rapidly. In order to reduce undesirable decay in the captured value, an S&H circuit employs a high-quality low-leakage capacitor such as polystyrene, polypropylene or Teflon. The value of the capacitor depends on many factors. A low value reduces the charging time, but increases the rate of decay in the held value. On the other hand, a high value increases the time needed to charge the capacitor, but reduces the decay. The discharge of the capacitor is minimized by using an FET input amplifier with a very low input bias current to buffer the sampled signal. For example, an S&H circuit is presented in Fig. 5.4(b). It includes a 741 IC as the input buffer and a 3140E as the output buffer. The selection of the operation mode is achieved by the use of an analogue switch, such as the ones described in the previous section. Rather than constructing an S&H circuit, a commercial S&H can be used, e.g. the National Semiconductor® LF398 shown in Fig. 5.4(c). It includes the sample/hold selection switch as well as the input and output buffers.

A microcomputer controls the operation of an S&H by setting or clearing a single digital-output line connected to the enable input of the switching device within the S&H. A logic 1 instructs the LF398 of Fig. 5.4(c) to operate in continuous sampling mode, whereas a logic 0 commands it to hold the value. In the circuit of Fig. 5.4(b), however, the logic state required to close or open an analogue switch depends on which analogue switching device is used (refer to the previous section).

Either of two approaches can be employed to operate the S&H. In the first, the device is normally operated in the sample mode and is only switched to the hold mode at a specified instant. The second approach keeps the device in the hold mode and only switches to the sample mode when a new value is required. The choice between these approaches depends on the application. When the S&H is switched from the sample to the hold mode, a small spike (known as the 'sample-to-hold transient') appears on the output; time should therefore be allowed for that spike to settle before reading the output. Time should also be allowed for the capacitor to charge and the buffer to settle. It should be remembered that the larger the value of the hold capacitor, the longer is the settling time.

Samples from continuous signals are normally captured by an S&H at regular intervals. If, for example, the triangular wave is sampled at regular intervals, then the sampled signal will be in the form of steps whose number depends on the rate of capturing the signal. The process therefore converts a continuous analogue signal into a sampled analogue signal with an amplitude equal to that of the actual signal during the sampling instants.

5.4 Digital-to-analogue (D/A) conversion methods

A D/A converter enables a digital computer to communicate with the analogue world. It decodes a digital-input signal to produce an output in the form of an analogue voltage or current signal. The use of D/A converters provides the digital computer with the ability to control analogue devices. Such converters are utilized in supplying analogue signals to control robotic mechanisms, generate waveforms and many other applications including their use in achieving A/D conversion, as will be explained. Two of the most popular D/A conversion methods are:

- binary-weighted method
- ladder method

In each of these methods a group of resistors is connected to a fixed reference voltage and driven to produce an analogue output at the required magnitude.

5.4.1 Binary-weighted method

This method is based on generating an analogue voltage by the use of a stable reference voltage (V_r) and a group of resistors with values weighted in a binary fashion as shown in Fig. 5.5(a). The closure of a switch connects a resistor to the reference voltage V_r, which results in producing a current whose value depends on the magnitude of the resistor. Since these resistors are binary-weighted, the closure of a switch results in supplying a binary-weighted current. The currents produced by the closure of a given combination of switches are added together and supplied to an operational amplifier to form the required analogue signal.

At any instant of time, the connection of some or all of the resistors is controlled by the selection logic and is achieved by using analogue switches. The selection logic can be an output port of a microcomputer through which digital values are supplied for conversion to the analogue form. The selection switches are digitally controlled for connection to V_r or to 0 V according to the setting of the digital input (whether it is a logic 1 or a logic 0). The resulting analogue-output voltage (V_o) is proportional to the digital input supplied by the selection logic circuit.

The stability and accuracy of this type of D/A converter depends on the stability of V_r and on the accuracy of the resistors and their ability to track each other over a range of temperature. The resolution is determined by the number of resistor branches used.

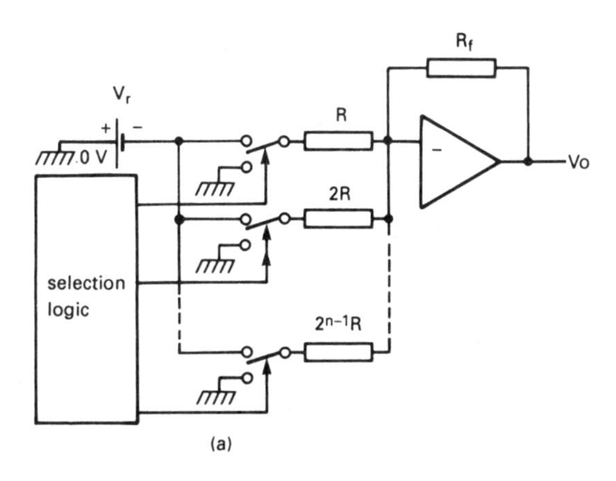

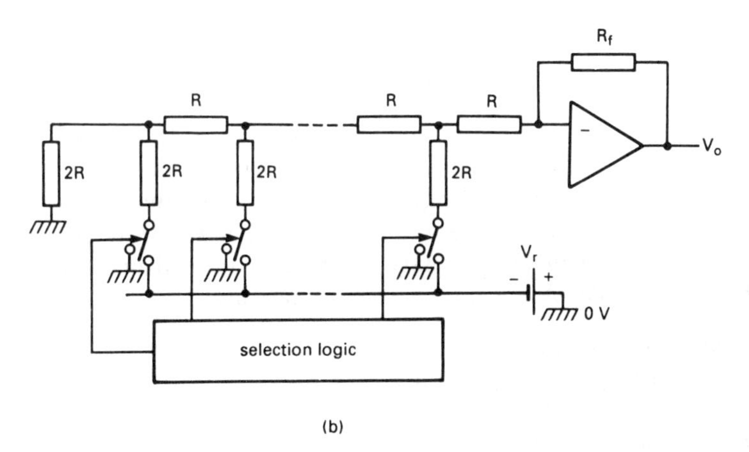

Fig. 5.5 D/A conversion methods: (a) binary-weighted method; (b) ladder method.

5.4.2 Ladder method

This method of D/A conversion is similar to that of the binary-weighted method described above, except that it only uses two values of resistors as shown in Fig. 5.5(b). The series resistors possess a value of R, while the shunt resistors possess a value of 2R. The reference voltage produces currents which flow in each of these resistors. The current path is determined by the setting of each of the switches. This results in supplying the current either to the input of the amplifier or to the 0-V point. The combination of the R–2R resistors results in dividing the current into two equal parts at each junction, producing binary-weighted currents.

When compared to the binary-weighted method, the ladder method uses only two values of resistors. This makes the selection and the matching of the resistors easier.

5.5 Analogue-to-digital (A/D) conversion methods

An A/D converter encodes an analogue-input signal producing an output in the form of a digital number which can be read by a digital device, such as a microcomputer. The use of an A/D converter therefore enables a microcomputer to receive information from analogue devices. There are various types of A/D converters, and they vary in their conversion speed, accuracy, resolution, price, etc. For example, A/D converters used for video applications are normally ultrafast, while a slow A/D converter can be utilized to measure the level of water in a tank. Several A/D conversion techniques have been developed; they can be grouped in two parts. The first is the *open-loop* type, and the second is the *closed-loop* type. An open-loop converter generates a digital code directly upon the application of an input voltage. On the other hand, a closed-loop A/D converter generates a sequence of digital codes, reconverts each one in turn back to an analogue value and compares it to the input. The resulting digital output will be the closest value of the reconstructed analogue voltage compared to the real analogue voltage.

The most popular A/D conversion techniques are:

1 Open-loop techniques:
- voltage-to-frequency method
- analogue-to-pulsewidth method
- simultaneous-conversion method

2 Closed-loop techniques:
- ramp method
- dual-slope method
- successive-approximation method
- tracking method

5.5.1 Open-loop techniques

(1) Voltage-to-frequency (V/F) method

This method is based on producing pulses whose frequency is proportional to the input voltage. These pulses are produced by supplying the analogue input voltage to a voltage-controlled oscillator (VCO). The resulting pulses are supplied to a counter which will in turn generate a count proportional to the input voltage when read at regular time intervals by a controlling device, such as a microcomputer. Therefore, the implementation of this method requires the involvement of the microcomputer for a very short time, since all that it needs to do is to read the counter at constant intervals. This can be achieved by the use of the interrupt-controlled method, where an interrupt is generated at regular intervals. It is recommended that the interrupt is given a high priority to ensure a constant time-base signal.

This type of converter is represented in block diagram form in Fig. 5.6. The main drawback of this approach is the dependency of the accuracy of the result on the linearity of the VCO. The accuracy of the resulting count also depends on the time between successive readings of the counter by the microcomputer. At low voltages, the number of pulses produced per unit time is low. A more accurate result can be obtained if the interval between successive readings is high, which is not always desirable. Therefore, a compromise is required between the conversion time and accuracy.

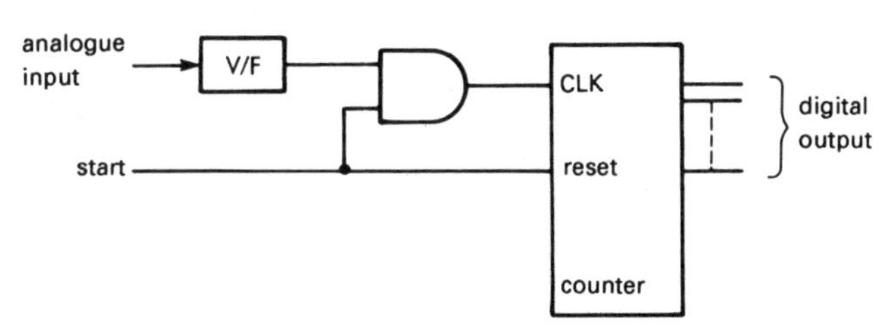

Fig. 5.6 V/F conversion method.

(2) Analogue-to-pulsewidth method

In this method of conversion, an integrator is utilized to generate a ramp; it starts from zero and its output rises at a constant slope. At the same time, a counter starts counting from zero. The ramp is compared with the analogue-input signal, and when the magnitude of the ramp exceeds the analogue-input signal the output of the comparator will change sign and it stops the counter. The contents of the counter are equivalent to the magnitude of the analogue-input signal. Figure 5.7 shows a block diagram of the circuit required to achieve the conversion.

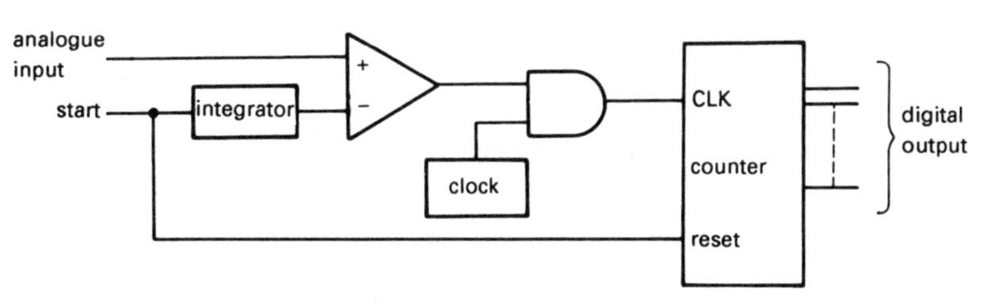

Fig. 5.7 Analogue-to-pulsewidth conversion.

There may be situations where a noise spike appears and alters the output of the comparator. In order to avoid an error in such a situation, the counter can be designed to continue its counting as long as the input is higher than the ramp (even after the output of the comparator changes sign). This minimizes the effect of the spike. The counter will therefore be reset only at the beginning of a new ramp. If the result of conversion is to be read as soon as possible, then the designer will need to introduce a special technique to prevent false readings. The designer may therefore choose to read the counter after detecting a change of sign and allow a small delay before reading it again to see if the sign change was due to a noise spike, or if it is the actual result. This conversion method is therefore not recommended in noisy environments.

The conversion time of this method is proportional to the duration of the ramp and is normally preferred to be low. On the other hand (and similar to the V/F conversion method), better accuracy can be achieved if the duration is longer. The time required to reach the result of the conversion depends on the amplitude of the signal. It also depends on the number of bits of the required count (the resolution), the frequency of the clock, and the rate of rise of the ramp. The method relies on the accuracy of the ramp and on the stability of the clock frequency.

(3) Simultaneous-conversion method

This method is also known as the *parallel* or the *flash* method and is based on using a parallel bank of voltage comparators. Each comparator responds to a different level of input voltage and all the comparators change state simultaneously. The voltage levels are produced by using a chain of series

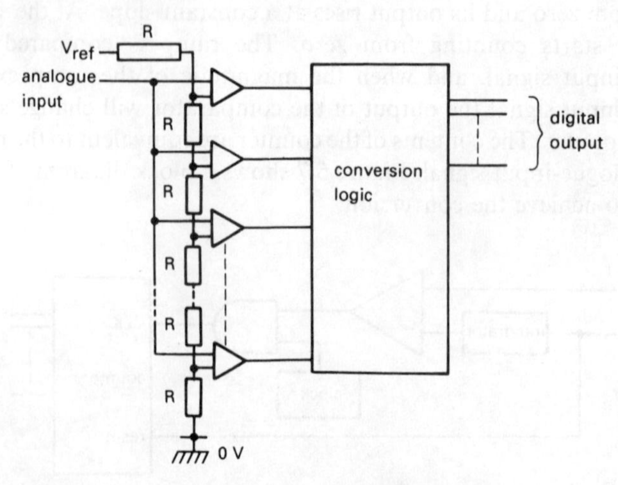

Fig. 5.8 Simultaneous-conversion method.

resistors and a voltage reference, as shown in Fig. 5.8. Each comparator produces a logic 1 output only if the input voltage is higher than the corresponding reference. Otherwise, the output is at a logic 0 state. The conversion process can therefore be considered as a one-step operation.

The number of the required comparators depends on the required resolution, as each comparator identifies a voltage level. This implies that 255 comparators are needed to obtain an 8-bit result. The advantage of this approach is the extremely high speed of conversion; this also ensures that the input signal will not change during conversion. The disadvantages are the high number of comparators used in such a converter and the need to encode the digital outputs to an *n*-bit binary number.

5.5.2 Closed-loop methods

(1) Ramp method

In this method a comparator is employed to compare two signals, the first being the analogue-input signal and the second being the output of a D/A converter, as shown in Fig. 5.9. The conversion is performed by starting a counter to count from 0 and supplying its output to the digital inputs of the D/A converter. The analogue output of the converter is compared with the analogue-input signal and the process continues until the output of the D/A converter becomes equal to the input signal. This causes the output of the comparator to change state and stops the counter. The resulting count is proportional to the amplitude of the analogue-input signal. An *n*-bit conversion requires the use of an *n*-bit counter and an *n*-bit D/A converter.

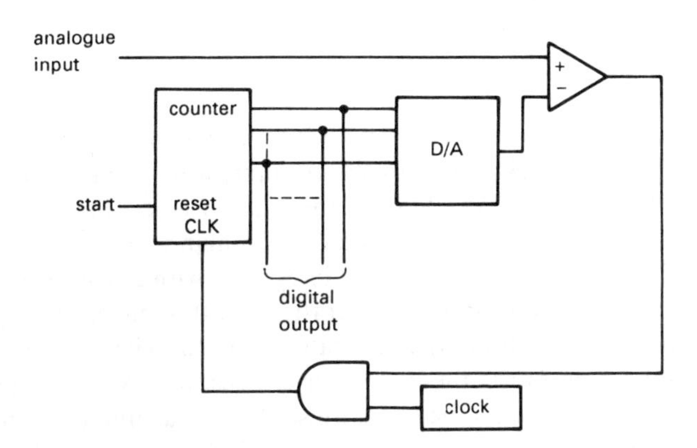

Fig. 5.9 Ramp method.

This method is simple to apply, but the conversion time is relatively long and is proportional to both the amplitude of the input voltage and the frequency of the clock. Care should be taken when selecting the frequency of the clock; the resulting count for any allowable amplitude of the input signal should not result in overflowing the counter before reaching the end of conversion.

After obtaining the digital result, any decrease in the amplitude of the input signal will not affect the output of the comparator, and such a decrease will therefore not be recognized unless the conversion is repeated. If, on the other hand, the magnitude of the input signal increases, then the output of the D/A converter becomes less than the input signal and will therefore result in increasing the count to produce a digital result proportional to the new value. This can be overcome by using an up/down counter as in the tracking method described at the end of this section.

(2) Dual-slope method

This method is based on using an integrator and a negative reference voltage as shown in Fig. 5.10. The conversion is in two stages. In the first stage

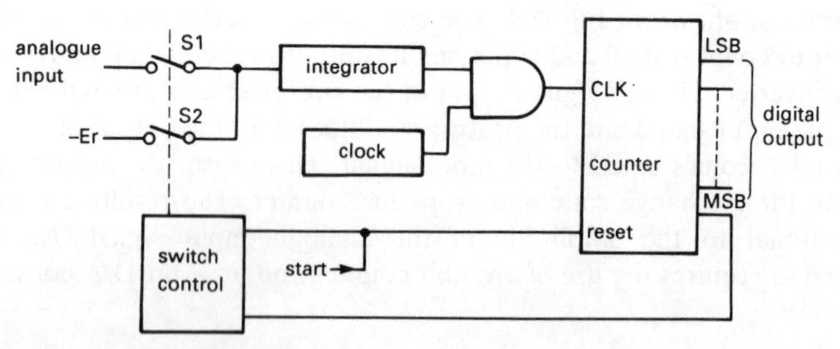

Fig. 5.10 Dual-slope method.

switch S1 is closed to supply the input voltage to the integrator. This enables the transmission of clock pulses to the counter, which starts and continues to count these pulses until the MSB of the counter is set. This bit is supplied to a switch-control circuit which issues a command to open S1 and close S2. This action supplies the input of the integrator with the negative reference (E_r) and starts the second stage of operation. In this stage, the capacitor of the integrator starts to discharge at a rate set by E_r. The counter continues to count until the output of the integrator crosses the 0-V level; this disables the AND gate and prevents the clock pulses from reaching the counter. The output of the integrator can therefore be used to indicate the end of conversion.

The duration from the start of the count until the setting of the MSB is fixed and is independent of the magnitude of the input voltage; this is because the counter counts the clock pulses, which are of a fixed rate. The time required to decrease the voltage of the capacitor to zero at a fixed rate (defined by E_r) is proportional to the magnitude of the input voltage signal. Therefore, the count achieved during the discharge period (which is the output of the counter without the MSB) is proportional to the magnitude of the input signal. The conversion time of this method is directly proportional to the magnitude of the input voltage. Accurate results can be obtained by using an accurate reference and a linear integrator.

(3) Successive-approximation method

This is one of the most popular methods of A/D conversion due to its relatively short conversion time and low cost. Its operation can be explained with the aid of Fig. 5.11. At the start of conversion, a digital value is supplied

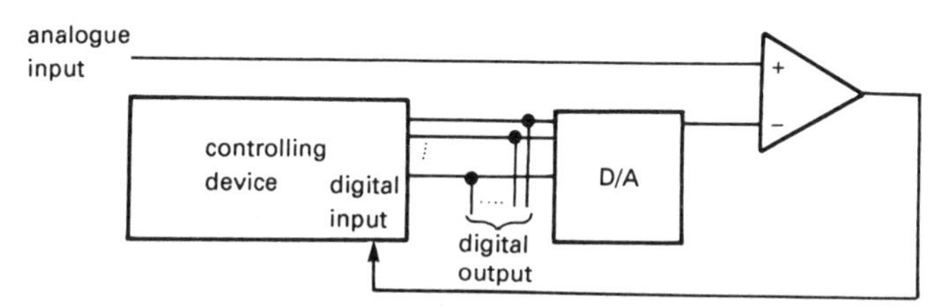

Fig. 5.11 Successive-approximation method.

by a controlling device to a D/A converter. Only the MSB of this value is set to a logic 1 and all other bits are set to 0 (this corresponds to half of full-scale). The output of the D/A converter is compared with the analogue-input signal. If the output of the comparator is high, then the output of the D/A is smaller than the input signal. Thus the MSB is left set, otherwise it will be reset to 0. This is followed by setting the next MSB and the process is repeated for all the bits of the D/A.

The conversion time is therefore dependent on the number of bits of the D/A converter (the resolution) and is relatively short, as only n trials are needed for n-bit resolution. It also depends on the time needed by the controlling device to turn on a bit, wait for the comparator to settle, examine the results of the comparison and, if required, turn off that bit. The conversion time is constant and unaffected by the amplitude of the input signal. It is important to keep the input voltage constant during the conversion; this can be achieved by the use of a S&H circuit (Section 5.3). Examples of the application of the successive-approximation method are given in Chapters 7 and 8.

(4) Tracking method

In this method an up/down counter is utilized to produce a count proportional to the analogue-input signal with the aid of a D/A converter and a comparator

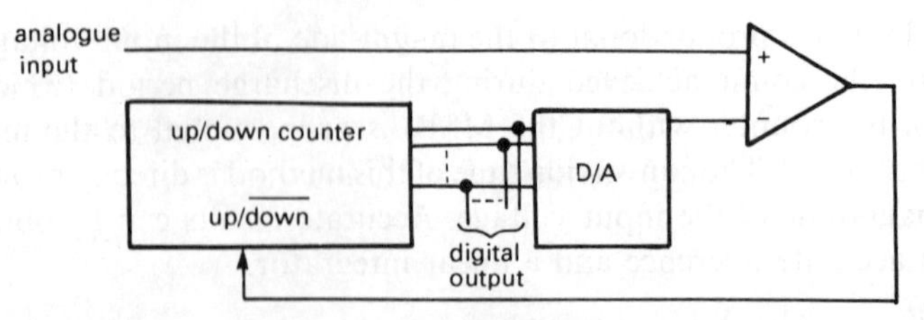

Fig. 5.12 Tracking method.

as shown in Fig. 5.12. Whenever there is a difference between the analogue-input signal and the output of the D/A, the counter will count either up or down according to the output of the comparator. If the input signal is greater than the output of the D/A, then the output of the comparator is high and the counter will count up. If, on the other hand, the input signal is smaller, then the counter will count down. Once the count is settled, it will be proportional to the magnitude of the input signal. Any variation in the signal will therefore change the count, i.e. the digital output of the converter will track the input signal and is continuously available at the output. This method of conversion does not impose the requirement for using a S&H. It is suitable for many applications, especially for digitizing a slowly changing analogue signal.

5.6 Multichannel data conversion

The purpose of this section is to present an application which employs devices given in this chapter. Multichannel D/A and A/D conversion are taken as examples. Such converters can be designed to include S&Hs as storage devices and multiplexers or demultiplexers as digitally controlled analogue-channel selectors.

5.6.1 Multichannel D/A conversion

A digital system can generate several analogue voltages simultaneously by employing several D/A converters (this is explained in Section 6.6). Each of these converters needs to be interfaced to the digital system so that it can be supplied with a latched digital value. Alternatively, a single D/A converter can be employed together with several S&Hs through a demultiplexer, as shown in Fig. 6.7. The number of S&Hs is equal to the number of required channels. This forms a multichannel D/A converter where the analogue value of each channel is stored on the capacitor of its corresponding S&H. Due to leakage, the magnitude of the analogue voltage across the capacitor of the S&H will reduce slowly. The output of the D/A converter is therefore connected sequentially to each S&H to refresh its value. The refreshing rate depends on the leakage within the S&H. The operation of the configuratation of Fig. 6.7 is described in Example 6.7. Further information on multichannel D/A conversion is given in Chapters 6 and 8.

5.6.2 Multichannel A/D conversion

A/D converters are utilized to enable a digital system, such as a microcomputer, to measure analogue signals such as speed and temperature (Chapter 7). Many applications introduce the requirement of several A/D conversion channels.

The use of multiplexers allows the time-sharing of a single A/D among a number of analogue channels. It provides a low-cost approach to multichannel A/D conversion as it replaces the need for using an A/D converter per channel. Since the price of an A/D converter is proportional to its resolution, the time-sharing approach is particularly attractive when a high-resolution A/D converter is used. However, when the A/D converter is shared, its maximum rate of conversion per channel drops. Therefore, the use of the time-sharing approach may not be acceptable for applications requiring fast sampling.

Figure 7.9 gives an example of a typical multichannel A/D converter. The microcomputer decides on which channel to connect to the input of the A/D converter and does so by supplying the address of that channel to the multiplexer. After selecting a channel, the microcomputer commands the closure of the switching device within the S&H circuit. The charge on the capacitor will change according to the amplitude of the newly connected signal. Before starting the A/D conversion process, the microcomputer instructs the S&H to open its switch, thereby storing a sample of the signal. Multichannel A/D conversion is described in Section 7.7 and application examples are given in Section 7.8. Chapter 8 gives an example of using a multichannel A/D conversion IC based on the time-sharing approach.

5.7 Practical data converters

The relation between analogue and digital values is considered in Section 1.10, where an ideal conversion device is considered. In practical converters, however, errors occur, resulting in a departure from the ideal transfer function. Let us consider the most common types of error:

- *Offset error*
  This represents the analogue value by which the transfer function deviates from zero as shown in Fig. 5.13(a).

- *Gain error*
  This defines the difference in slope between the actual and the ideal transfer function as shown in Fig. 5.13(a).
- *Linearity error*
  This represents the maximum deviation of any point within the actual transfer function from the ideal transfer function. The deviation between the two functions appears as a curvature from the straight line; see Fig. 5.13(b).
- *Differential non-linearity error*
  The maximum difference between the ideal and the actual size of 1 LSB (least significant bit).

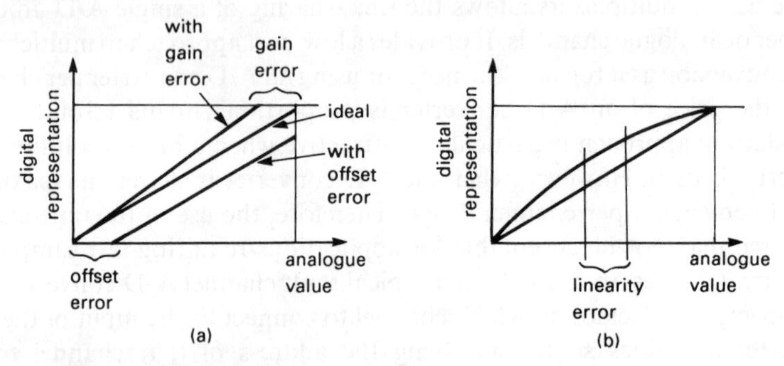

Fig. 5.13 Errors in converter transfer function: (a) offset and gain errors; (b) linearity error.

Quantization error and resolution are defined in Section 1.10. Another useful term is the *monotonocity* of a D/A converter. If the output of a D/A converter increases or remains constant as a result of increasing the value of the digital input, then the converter is called *monotonic*.

Figure 5.14(a) shows an ideal transfer function of a 3-bit D/A converter where each digital value produces an equivalent analogue value. It provides staircase characteristics with an equal step size. Figure 5.14(b) shows a transfer function with an unequal step size. A converter possessing this type of transfer function is considered non-monotonic, because of the decrease in the value of the analogue output when the converter is supplied with the digital value 110. Note that the small increase in amplitude resulting from supplying the value 011 satisfies the monotonicity condition, since it represents an increase in the analogue value as a result of an increase in the digital input. In Fig. 5.14(b), the differential non-linearity is represented by the difference between the transition from 100 to 101.

The output of a non-ideal D/A converter will not change immediately after a change in the value of the digital input. A short finite time is needed; it is

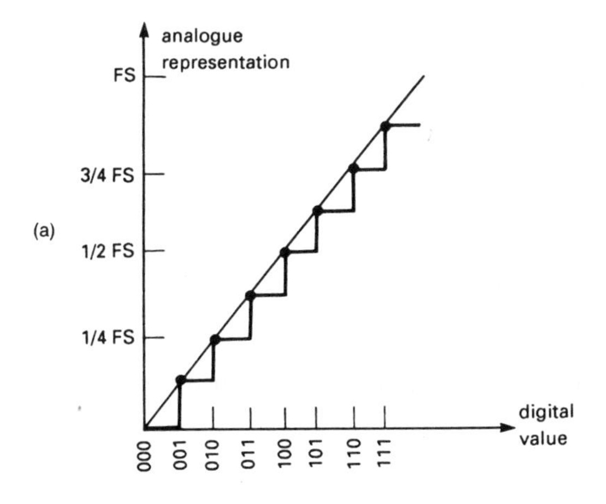

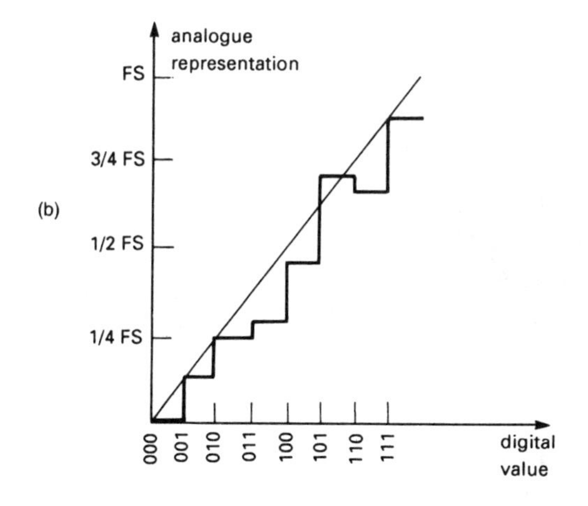

Fig. 5.14 D/A transfer function: (a) ideal converter; (b) non-monotonic converter.

known as the *settling time*. It represents the time taken for the output to settle within $\pm\frac{1}{2}$ LSB. The settling time is short when the change in the input is small, e.g. 1 LSB. It increases as the size of the change in the input value increases.

A transient spike may occur at the output of a non-ideal D/A converter as a result of a change in a few of the input digital bits simultaneously, e.g. 011 to 100. It is known as a *glitch* and occurs because the switches within the D/A converter are non-ideal, i.e. it occurs due to the time needed to change the state of these switches. One way of removing glitches is to place an S&H device at the output of the D/A converter in a similar way to that presented in Fig. 6.6(b) (although that figure uses the S&H for another purpose; see Example 6.6).

Let us now consider the transfer function given in Fig. 5.15(a). It shows that an A/D converter quantizes an input signal to the nearest level; that is, the maximum difference between an analogue signal and its digital representation is $\pm\frac{1}{2}$ LSB. Figure 5.15(b) presents the transfer function which deviates from the ideal, It shows a missing code in between 001 and 011, i.e. a converter with this transfer function will not produce the digital value 010. A missing code can exist as a result of using a non-monotonic D/A converter as a part of the A/D conversion circuit.

More information on data converters and their use is given in the following chapters.

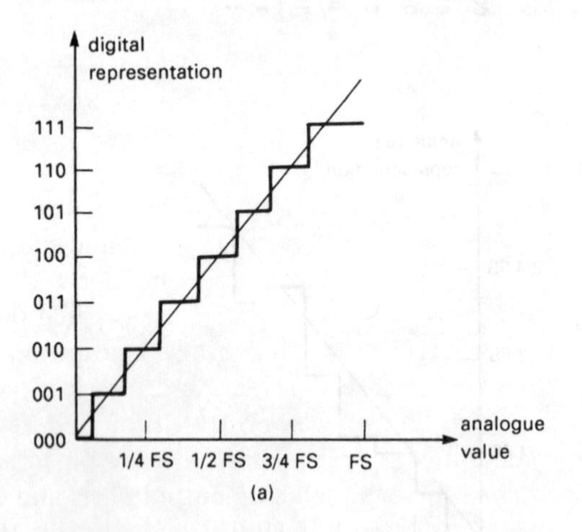

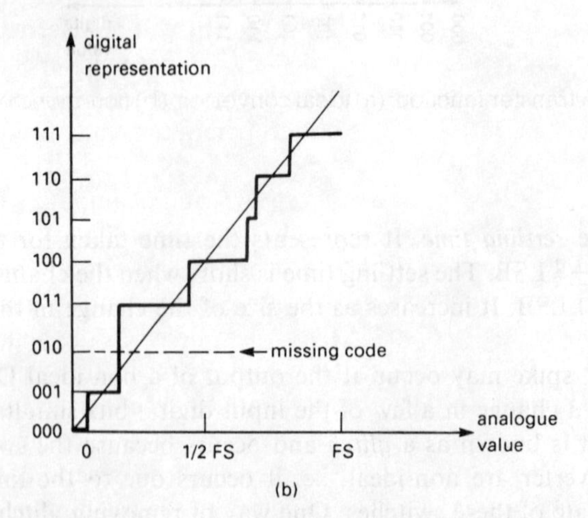

Fig. 5.15 A/D transfer function: (a) ideal; (b) including missing code.

Chapter 6

Interfacing D/A converters to microcomputers

6.1 Introduction

It has been shown in the previous chapters how a microcomputer can communicate with the outside world through input and output ports. Such ports are available in some but not all microcomputers. The large number of microcomputers that lack these ports normally provide the user with an access to the address bus, data bus, control lines and power supplies. This makes it possible to interface such ports, as described in Chapters 2 and 3. The examples given in this chapter demonstrate 8-bit and 16-bit interfacing by using two microcomputers. The first is presented in Fig. 6.1(a). It provides two 8-bit input ports and two 8-bit latching output ports and will be referred to in the rest of the book as the one which uses 8-bit interfacing. On the other hand, the microcomputer of Fig. 6.1(b) will be referred to as the one using 16-bit interfacing. It provides the same number of lines as the microcomputer of Fig. 6.1(a), except it employs a 16-bit input port and a 16-bit latching output port. Each microcomputer reads from, or writes to, one port at a time.

The above indicates that each of the two microcomputers of Fig. 6.1 can communicate with the outside world through its digital ports. There are numerous applications where a microcomputer needs to control analogue devices, or operate as a part of a control process using analogue signals. Each of the microcomputers of Fig. 6.1 can be given the ability to generate such signals by interfacing it to D/A converters. In this chapter, examples are presented demonstrating such an interface. The converters are presented in a block diagram form at this stage. Once you develop an understanding of the interface, you can refer to Chapter 8, where several commercial D/A converters are presented. Further examples of the use of such converters are presented in later chapters.

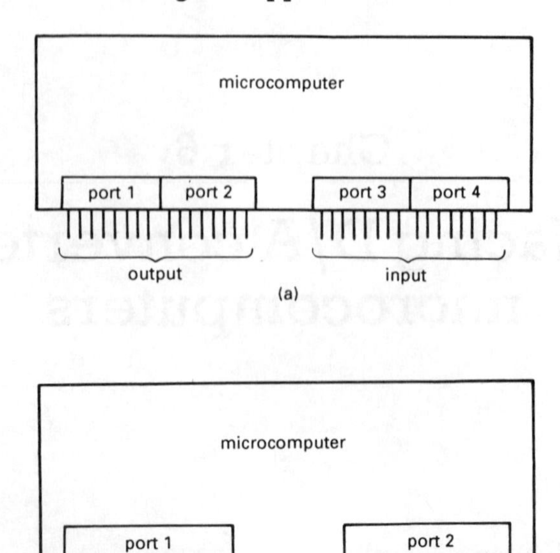

Fig. 6.1 A microcomputer with I/O ports: (a) 8-bit ports; (b) 16-bit ports.

6.2 Interfacing D/A converters to microcomputers

In situations where an analogue signal needs to be supplied from a microcomputer to an analogue device, it must undergo a D/A conversion. This can be achieved by the use of a D/A converter. The analogue output of such a device is directly related to the digital value supplied to its input. Consequently, interfacing such a device to a microcomputer provides the latter with the ability to control external analogue devices. This can be achieved if the converter is correctly interfaced to the microcomputer. The method with which the interface is best implemented depends on many factors, such as the number of channels of the D/A converter, its resolution, and the number of digital-output lines available in the microcomputer.

When a D/A converter is to be operated under the control of a microcomputer, software needs to be written to perform such an operation. Consider a typical application where the microcomputer of Fig. 6.1(a) does nothing else except repeatedly execute a task. It calculates the value of an n-bit variable 'x' and supplies the result to the digital inputs of an n-bit D/A converter. The value is supplied via an output port immediately after completing the calculation. In such an application, the maximum rate of updating the output signal is inversely proportional to the time needed to perform the calculation.

Example 6.1: **Describe the procedure of operating a single-channel 8-bit D/A converter from each of the microcomputers of Fig. 6.1.**

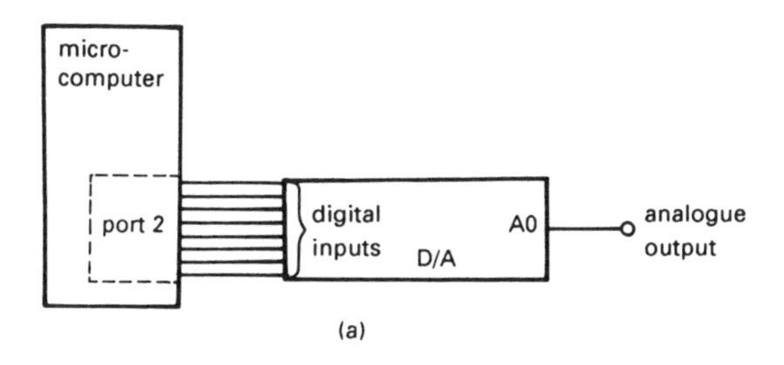

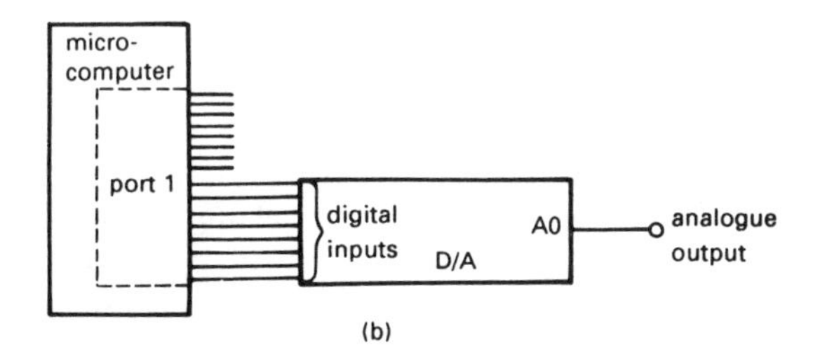

Fig. 6.2 Interfacing an 8-bit D/A converter to a microcomputer: (a) using an 8-bit port; (b) using a 16-bit port.

The connection of a single-channel 8-bit D/A converter to the microcomputer of Fig. 6.1(a) is shown in Fig. 6.2(a), where the eight digital inputs of the converter are supplied from port 2. Since port 2 has the same number of lines as those needed to supply the digital values to the converter, then the procedure is very simple and can be represented by:

1 CALCULATE VALUE
2 SEND VALUE TO PORT 2
3 CONTINUE

The address of port 2 depends on the type of microcomputer used as well as on the interfacing circuit, as explained in Chapter 2.

In a similar way, the interface of the 8-bit D/A converter to the microcomputer of Fig. 6.1(b) can be implemented as shown in Fig. 6.2(b). In this case, the microcomputer supplies information through a 16-bit output port where it writes to the 16 lines simultaneously. Lines 0 to 7 are used for driving the converter and the rest can be utilized for other purposes. The program should be arranged such that the state of lines 8 to 15 should not be altered when a value is sent to the converter. This can be accomplished by the following procedure, where an '&' character represents a software AND, a '|' character represents a software OR, and 'w' is a 2-byte word containing the value last supplied to the output port.

1 CALCULATE THE 8-BIT VALUE
2 w = w & 1111 1111 0000 0000
3 w = w | VALUE
4 SEND w TO OUTPUT PORT
5 CONTINUE

In this procedure, step 2 clears the lowest 8 bits of 'w' and step 3 writes the new value into these bits without altering the upper 8 bits.

The rate at which a microcomputer can update an analogue output efficiently depends on several factors. If, for instance, the microcomputer is employed to perform a simple task, such as generating musical tunes, then it can afford to be devoted to updating an analogue output at the maximum possible rate. Similarly, if the microcomputer is utilized as an intelligent controller for a model train, then it can spend a large part of its processing time waiting for an event to occur before supplying a new value to the D/A converter. This is not the case in the majority of applications, where certain time requirements need to be satisfied and the microcomputer needs to attend to other tasks beside updating the output. The choice of the best approach of updating an analogue output can be achieved by comparing solutions in a similar manner to that of Section 3.5.

6.3 Testing D/A converters

Before using a data converter, it needs to be tested to ensure that it operates correctly and achieves the required performance. Clearly, the method of testing and the instrumentation needed depend on the functions to be tested and the required accuracy and speed. Therefore, the objectives of the test need to be set first. An essential first step is to examine the hardware connections to ensure that no damage to the microcomputer will result from connecting the D/A converter to its port. Let us consider performing a useful test. The object is to verify the relation between the input and output of the converter. In addition to measuring the offset and accuracy of the converter, such a test may point out a hardware fault which was not detected during the inspection of the hardware.

6.3.1 Testing without a microcomputer

A circuit such as that of Fig. 6.3 can be employed for this test where SPST (single-pole single-throw) switches are utilized to supply the inputs of an 8-bit D/A converter with digital values. The output signal can be measured by an accurate digital voltmeter. If the D/A converter is unipolar and its output can ideally be varied between 0 and 5 V, then the equivalent digital representation varies from 00000000 to 11111111 binary (equivalent to the hexademical value 00H to FFH). The test can start with all switches in the off position (contacts open). This results in supplying the converter with a digital value of 00H; the generated analogue output can be recorded and the test continues. If the object of the test is to obtain the transfer function of the converter, then the test can take the form of closing the switches in a binary sequence until the converter is supplied with the digital value FFH. A graph can be drawn representing the relation between the input and output of the converter. The obtained transfer function can be compared with the ideal one (see Section 5.7).

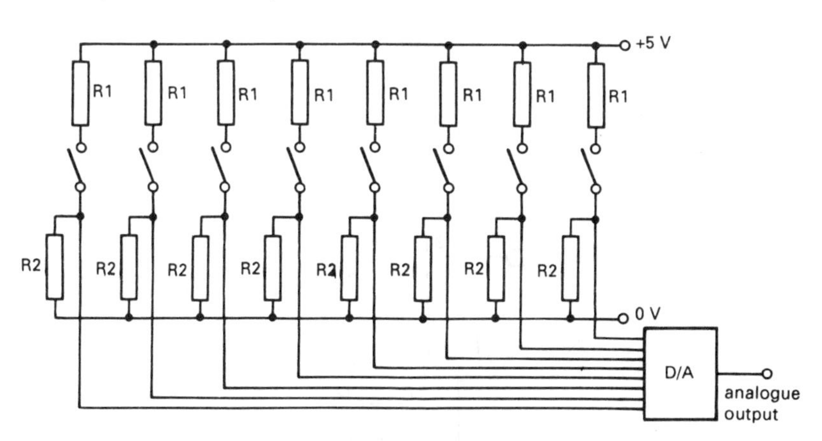

Fig. 6.3 Using SPST switches to supply a D/A converter with digital values.

Let us consider a typical fault which can be found by such a test. Consider a situation where the relation between the input and output is found to be true only for even values of the input signal. Further tests reveal that the output does not alter when the least significant bit (LSB) of the input is varied. Testing the LSB pin of the converter indicated that it is always at a logic 0 state, whether its corresponding switch is opened or closed. What do you consider the most likely cause of this fault? As you may have deduced correctly, a likely cause is incorrect wiring, or a short-circuit between the line carrying the LSB signal and another at a logic 0 state.

Let us now consider a situation where a microcomputer is employed to control a system by supplying an analogue voltage. Assume that the microcomputer-generated analogue voltage does not cover the range given in the design specification of the system. This implies that the output needs to be amplified. A non-inverting amplifier, such as that of Fig. 6.4(a), can be utilized to achieve the amplification and to serve as an output buffer. The gain of such an amplifier is equal to (1 + R2/R3) and can be set by adjusting R2. The amplifier can also be used to remove any offset and to calibrate the output by the inclusion of an offset-correction potentiometer (R0 in Fig. 6.4(b)).

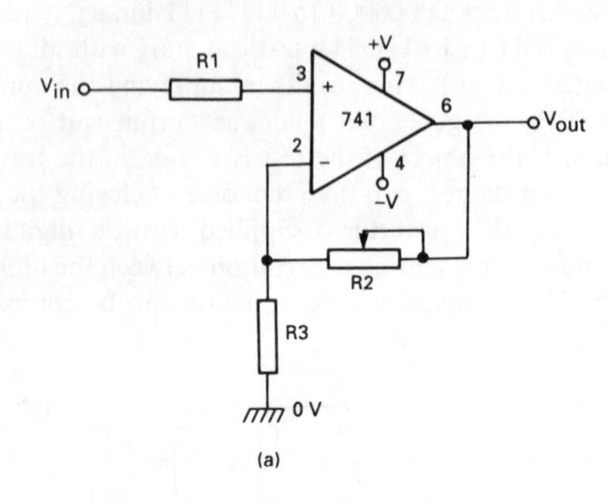

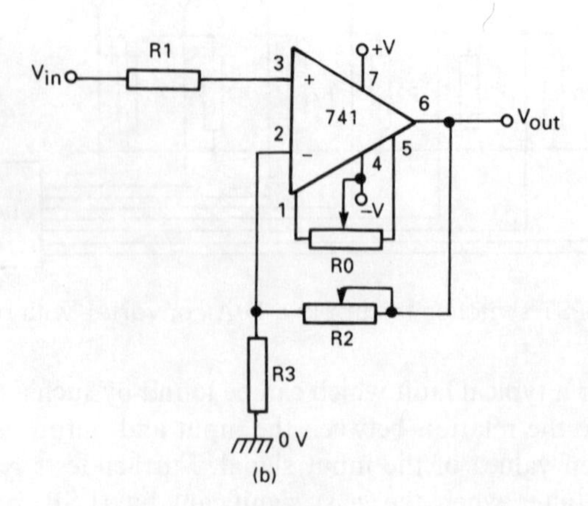

Fig. 6.4 Non-inverting operational amplifier.

If an amplifier is employed, then the above-described test can be carried out by using the combination of the D/A converter and the amplifier. The offset needs to be removed and the converter calibrated before starting the test. The offset can be observed when all the switches are open. If under this condition the output of the amplifier is not zero, then that offset can be removed by adjusting R0. Following this, all the switches should be closed and the output measured. If it does not correspond to the required maximum magnitude, then R2 needs to be adjusted to obtain the required value. These two steps may need to be repeated several times to obtain the correct settings. The amplifier can be a 741 IC and the value of the resistors depends on the required gain. Note that for a non-inverting amplifier, the offset can be reduced by selecting the same value of R1 as that of the parallel combination of R2 and R3.

Although the test is carried out for a unipolar D/A converter, Fig. 6.4 shows that the 741 IC is connected to −V as well as to +V power supply. The negative voltage is required to ensure the elimination of the offset, thus enabling the D/A converter to produce 0 V when all the switches are open. If the terminals of the microcomputer do not include a negative supply, then the designer can either introduce such a supply to the microcomputer (Section 16.6), or use an amplifier such as the 3140E IC instead of the 741. This IC has the same pinout as the 741. It can, however, be operated from a single or a dual supply. Consequently, pin 4 can be connected to the 0-V terminal and pin 7 to the +V volts terminal. The disadvantage of not using the negative supply is that it may not be possible to remove the offset completely.

6.3.2. Testing with a microcomputer

This section presents two examples illustrating how the use of a microcomputer can ease the implementation of the above-described test.

Example 6.2: **Describe the procedure of testing the relationship between the input and output of an 8-bit D/A converter controlled by the microcomputer of Fig. 6.1(a). Allow an operator to select any of the digital values through switches connected to the input port of the microcomputer.**

The D/A converter can be interfaced to the microcomputer through output port 2 as shown in Fig. 6.2(a). Rather than using the eight switches employed in the previous section, the digital values can be supplied by employing three PTM (push-to-make) switches connected in a similar manner to that of Fig. 4.1(b) and interfaced to port 3.

The test can start from a default digital value, e.g. 0. This value is incremented each time the first switch is pressed, and is decremented after each press of the second switch. The up-to-date value can be displayed by the microcomputer to permit an operator to read that number and alter its value. When the operator is satisfied with the displayed value, he or she can press the third switch to inform the microcomputer of the need to supply that value to the converter. The steps required to implement this task can be represented by the following procedure. It is assumed that switches 1, 2 and 3 are connected to pins 0, 1 and 2 of part 3 respectively, and character '&' represents a software AND:

```
1  LET VALUE = 00H
2  DISLAY VALUE
3  READ PORT 3 AND STORE READING IN y
4  IF (y & 0000 0001) = 0,
       THEN INCREMENT VALUE,
            DISPLAY VALUE,
            AND GO TO STEP 8
5  IF (y & 0000 0010) = 0,
       THEN DECREMENT VALUE,
            DISPLAY VALUE,
            AND GO TO STEP 8
6  IF (y & 0000 0100) = 0,
       THEN SUPPLY VALUE TO PORT 2 AND GO TO STEP 8
7  GO TO STEP 3
8  WAIT FOR 50 MILLISECONDS
9  GO TO STEP 3
```

In this procedure, the microcomputer is employed to continuously monitor the switches. During the process, it spends most of its time executing steps 3 to 7, repeatedly waiting for a closure of a switch. It executes step 8 only after the detection of such a closure. The delay introduced in step 8 is required to ensure that only a single increment or decrement of the value results from a momentary press of a PTM switch. It also replaces the need for the debouncing capacitor.

In the above procedure, pressing switches 1 and 3 alternately causes the value to be incremented and supplied to the output. Since the count consists of 8 bits, it can be incremented until it reaches the hexadecimal value FFH. By allowing the count to roll over, the next increment makes the count equal to 00H. Therefore, the continuous repetition of pressing switches 1 and 3 alternately results in producing a low-frequency sawtooth waveform at the output of the D/A converter.

Example 6.3: **Assume that a keyboard is available for the microcomputer of Fig. 6.1(a). How can the procedure of Example 6.2 be altered to allow the user to enter digital values from the keyboard?**

The solution is similar to that of the previous example. The inputs of the D/A converter are supplied from port 2, whereas the use of port 3 and the PTM switches is replaced by the keyboard. A program needs to be written to enable the microcomputer to read the user-supplied values and to transfer each value to the converter. The values are supplied by pressing the numerical keys of the keyboard and are followed by pressing the *enter* key. Since an 8-bit converter is employed, each digital value should be in the range 0 to 255 decimal (i.e. 00000000 to 11111111 binary). The procedure can take the form of continuously testing the keyboard and waiting for *three* numeric keys to be pressed before the *enter* key. This can be implemented by the following steps where the value is represented by the variable 'v':

```
 1  LET y = 0, v = 0 AND DISPLAY 0
 2  WAIT FOR A KEY TO BE PRESSED
 3  y = y + 1
 4  IF IT IS A NUMERIC KEY,
          THEN DISPLAY v = (10 * v) + (key value)
               AND GO TO STEP 7
 5  IF IT IS THE '+' KEY
          THEN GO TO STEP 13
 6  GO TO STEP 1
 7  WAIT FOR 100 MILLISECONDS
 8  IF y < 3,
          THEN GO TO STEP 2
 9  IF v > 255
         THEN INDICATE AN ERROR AND GO TO STEP 1
10  WAIT FOR A KEY TO BE PRESSED
11  IF IT IS NOT THE 'enter' KEY
          THEN GO TO STEP 1
12  SEND v TO CONVERTER
13  END PROCEDURE
```

With this procedure, the microcomputer repeatedly executes step 2 until a key is pressed. The (+) key is used to allow the user to exit from the procedure without having to reset the microcomputer. If the wrong key is pressed, then step 6 restarts the procedure. The value will be supplied to the D/A converter (as an 8-bit binary number) only if it is equal to or less than 255 decimal.

Suppose that you want to carry the above test. Will this procedure satisfy your requirements? What alterations will you introduce?

6.4 Examples of using single-channel D/A converters

Example 6.4: **What is the procedure required for producing a square wave at the output of a microcomputer-controlled 8-bit D/A converter? Compare**

the solution to those of Section 3.5 where the waveform is generated from a digital-output line.

In Section 3.5 a single digital-output line is utilized to generate the square wave. This results in producing a waveform with a fixed amplitude. Rather than employing a digital-output line, a square wave can be generated as a result of supplying the inputs of a D/A converter with two consecutive digital values, such as 00H and FFH. The amplitude of the waveform depends on the difference between the two values. Its frequency depends on the rate of updating the output. Offsetting the waveform can be achieved by software through altering the two digital values. The output of the D/A converter, however, requires a finite time to settle (e.g. 1 or 2 μs). This may become noticeable when producing a very high frequency waveform.

Similar to the examples given in Section 3.5, generating the waveform by a D/A converter can be performed in several ways, e.g. the use of the program-controlled or the interrupt-controlled method. It can therefore follow any of the procedures presented in the examples of Section 3.5, except that here the output alternates between two selectable values. These values can, for instance, be placed in two consecutive memory locations and the program arranged to examine the state of a bit (c) to identify which of the two values needs sending to the converter. It will complement the state of that bit after sending each value, as follows:

```
1  IF c = 0
        THEN TRANSFER VALUE FROM LOCATION y TO OUTPUT
        PORT
   ELSE
        TRANSFER VALUE FROM LOCATION y + 1 TO OUTPUT
        PORT
2  COMPLEMENT c
```

As an exercise, write the procedure for generating a square wave by including the above two steps in each of the procedures of Examples 3.1 and 3.3.

The above examples employ a unipolar D/A converter. Let us now consider the difference between using unipolar and bipolar D/A converters. If both have an output range of 5 V, then the output of the unipolar is between 0 and 5 V, while the output of the bipolar is between −2.5 to +2.5 V. The 0 V of the bipolar type is therefore in the middle of the range. For example, one can use an 8-bit bipolar converter which produces +2.5 V analogue output when supplied with a digital value of FFH. The magnitude of the output will change to −2.5 V when supplied with 00H. In such a converter, an output of 0 V can be produced when the D/A is supplied with 80H. Although interfacing the converter to the microcomputer is the same for both unipolar and bipolar converters, the digital value required to produce a certain analogue amplitude is different.

Example 6.5: **Consider an application where a 5-V 8-bit unipolar D/A converter is connected to a microcomputer as shown in Fig. 6.2(b). It is required to supply a bipolar output voltage in the range −2.5 V to +2.5 V. How can the unipolar converter be modified to satisfy the requirements?**
The analogue output of the unipolar D/A converter is in the range 0 to 5 V and therefore needs to be offset by half of full-scale, i.e. 2.5 V. This can be achieved by using the configuration of Fig. 6.5(a), where R1 = R2 = R4 and R3 equals the parallel combination of R1, R2 and R4. Then −2.5 V is obtained by adjusting R_r (the −2.5 V is added to the output of the D/A to produce the required range). Note that the configuration inverts the sign of the resulting sum. One way of correcting the sign is to use a unity-gain inverting amplifier (Fig. 6.5(b)) with R1 = R2 and R3 equal to the parallel combination of R1 and R2 following the summing amplifier. Alternatively, a software inversion can be used before supplying the digital value to the D/A converter.

An alternative way of offsetting the signal is to employ a differential amplifier instead of the summing amplifier. In this case, a value of +2.5 V is subtracted from the input signal as shown in Fig. 6.5(c), where R1 = R2 = R3 = R4. Note that the use of precision resistors is recommended for such an amplifier to achieve the best common mode rejection.

6.5 Relation between resolution of a D/A converter and type of interface

This section considers the effect of the resolution of a D/A converter on its interface to each of the microcomputers of Fig. 6.1. As explained in Example 6.1, interfacing an 8-bit converter to the microcomputer of Fig. 6.1(a) is a straightforward task. Similarly, a 16-bit converter can be directly interfaced to the microcomputer of Fig. 6.1(b). Interfacing converters of other resolutions to each of these microcomputers is different, as shown below.

6.5.1 16-bit interfacing

Let us first consider interfacing a D/A converter, whose resolution is less than 16 bits, to the microcomputer of Fig. 6.1(b). In this case, the converter can be directly connected to the output port. It has, however, been shown in Example 6.1 that the microcomputer writes to the 16 output lines simultaneously, and simple software techniques can be introduced to permit

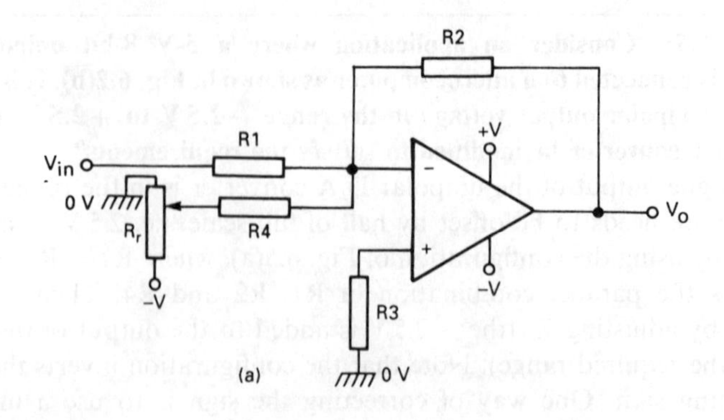

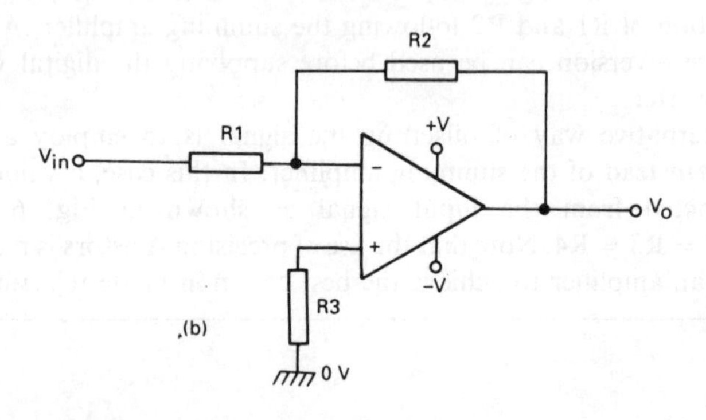

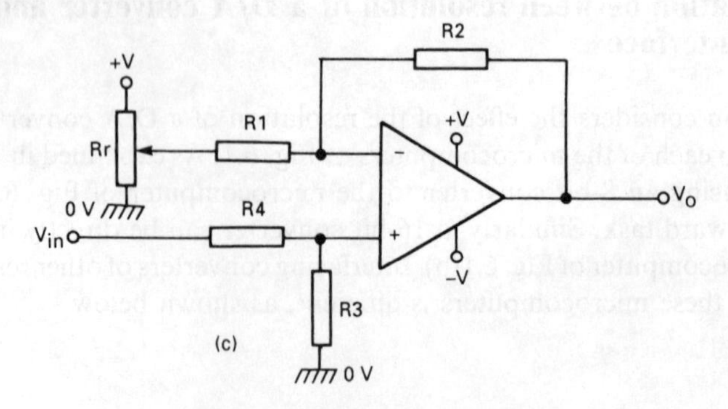

Fig. 6.5 Analogue circuits: (a) summing amplifier; (b) inverting amplifier; (c) differential amplifier.

using the rest of the lines for other purposes. Clearly, if a 16-bit D/A converter is connected to the 16-bit port, then the microcomputer writes the 16-bit value directly to the port without the need for executing the additional software instructions.

6.5.2 8-bit interfacing

In many applications, 10-bit, 12-bit or 16-bit D/A converters need to be interfaced to microcomputers such as that of Fig. 6.1(a). This is more difficult than their interface to the microcomputer of Fig. 6.1(b), since the digital-input lines of the converter exceed the number of lines available in a single output port of the microcomputer. There exist several ways of fulfilling the requirements, each with its advantages and limitations. Consequently, alternative methods are given in the rest of this section to demonstrate a few solutions.

Let us consider the interface of a 10-bit D/A converter to the microcomputer of Fig. 6.1(a). Inside the microcomputer, the 10-bit digital number is normally stored in two successive bytes in memory. One byte contains the eight lower bits, while the other byte contains the upper two bits. The following example shows how the 10-bit value can be supplied to the inputs of the converter via two output ports of the microcomputer. They are allocated as 8 bits from port 2 and 2 bits from port 1, as shown in Fig. 6.6(a).

Example 6.6: **Describe alternative methods of operating a 10-bit D/A converter from the two 8-bit output ports of the microcomputer of Fig. 6.1(a).**

Solution 1

Since the microcomputer of Fig. 6.1(a) can only supply values to one 8-bit port at a time, it supplies the upper two bits first (through port 1) and then transfers the lower eight bits from memory to port 2. At the interval between supplying the upper and the lower bits, the output of the D/A converter is not correct. Although this may only appear for a short time, it may not be acceptable in many applications.

If, for example, at instant t1 a digital value of 0011111111 (i.e. 0FFH) is supplied to the D/A converter, a corresponding analogue value is produced at its output. Therefore, at instant t1 the upper two bits are supplied with 00 while the lower eight bits are supplied with 11111111. If at instant t2 the value is to be incremented by 1, then a value of 0100000000 (i.e. 100H) is to be supplied to the D/A converter. Since the configuration of Fig. 6.6(a) is used, the value can only be supplied in two stages. Therefore, at instant t2 the upper two bits are supplied as 01. At that instant, however, the lower

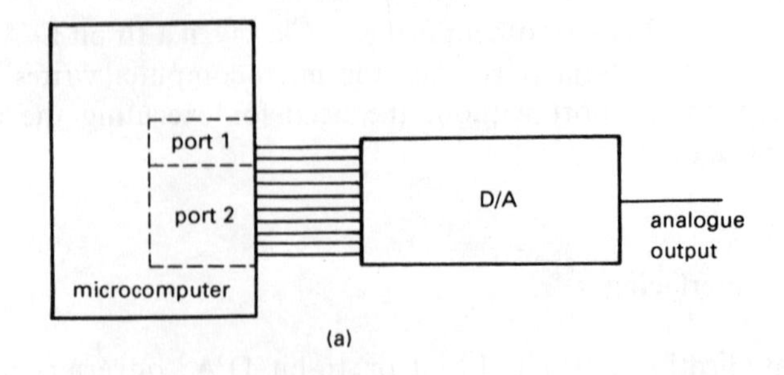

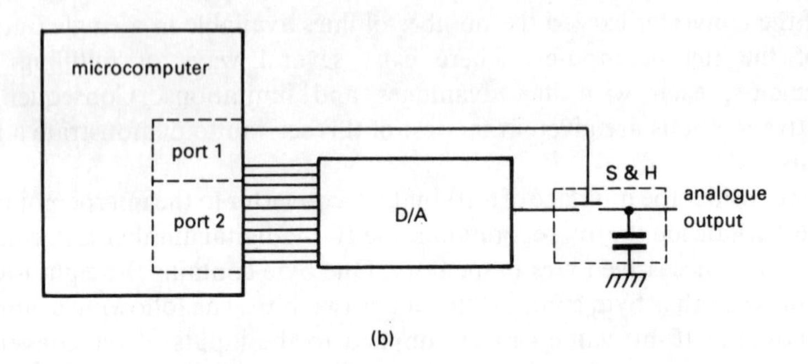

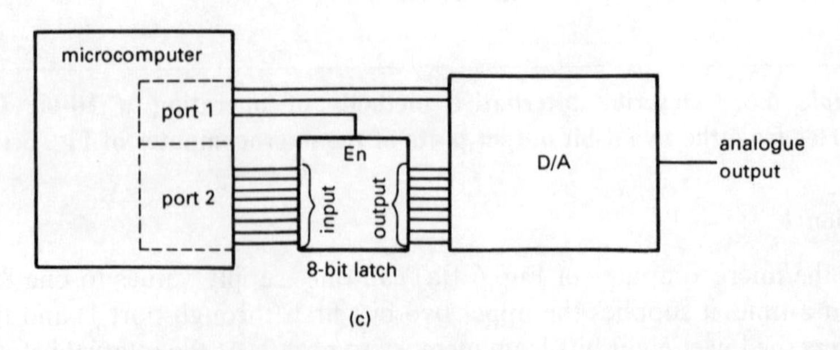

Fig. 6.6 Interfacing a 10-bit D/A converter to two 8-bit ports: (a) no additional hardware; (b) using an S&H; (c) using an additional latch.

eight bits are still supplied with the value 11111111. After that, at instant t3, the lower eight bits are supplied with the value 00000000 and result in providing the D/A converter with the required value 0100000000.

One should note that at the interval t2 to t3, the D/A converter is supplied with the value 0111111111 (i.e. 01FFH), which is almost twice the required value. Interval t2 to t3 is, however, short and a glitch may appear at the output. The glitch can be reduced by the use of a filter at the output of the

D/A converter. This is a simple solution which may not satisfy the requirements of many systems.

Solution 2

Let us consider a solution based on using a S&H at the output of the D/A converter as shown in Fig. 6.6(b). An additional output line is employed from port 1 of the microcomputer to control the operation of the S&H (see Section 5.3). At start, the switch of the S&H is open and the digital value is supplied to the converter in two stages, as described in Solution 1 above. After the D/A produces the required analogue signal, the microcomputer commands the switch of the S&H to close, causing the associated capacitor to charge to the correct value. The switch will remain closed until before supplying the next value. The analogue output is therefore equal to the voltage stored in the capacitor. If the output of the S&H is not buffered, then an external buffer is required to prevent discharging the capacitor by the analogue circuit.

Solution 3

Instead of adopting the above solutions, let us introduce a third solution to perform the correct transfer of the 10-bit value to the inputs of the converter. This approach makes use of an external latch to store the lower eight bits of the digital value. The interface of the latch to the microcomputer is similar to that of an output port (Section 2.4) except for two differences. The first is that the inputs of the latch are supplied from an output port of the microcomputer rather than from the data bus. The second difference is that the enable line is controlled through a digital-output line rather than from a device-selection signal. The interface is shown in Fig. 6.6(c). In this case, the software is arranged to begin the process by delivering the lower eight bits from port 2 when the external latch is not enabled. This is followed by supplying, through port 1, the upper two bits and the enable signal of the external latch. This results in delivering the 10 bits to the inputs of the D/A converter simultaneously.

Discussion

The above solutions employ different means for satisfying the requirement, and the last two involve the introduction of additional hardware. Which of these approaches do you prefer to use? You may prefer to adopt a different approach, such as utilizing a serial-to-parallel conversion IC which will not update the output until it receives a complete set of bits corresponding to all the parallel output lines.

6.6 Multichannel D/A converters

6.6.1 Methods of multichannel conversion

Rather than using a multichannel D/A conversion IC (Section 8.4.1), a multichannel D/A converter can be constructed by employing either of two methods. The first is based on using several individual D/A converters, each representing one channel. The second method employs a single-channel D/A converter, S&Hs and a digitally controlled channel selector. These methods are briefly explained in Section 5.6.1. The first method is widely used and is feasible because the price of D/A converters is relatively low, especially those of low resolution. As the resolution of the D/A converter increases, its cost rises, making the second method more attractive in many applications. The second method is also of interest when a multichannel analogue output is required and a single-channel D/A converter is already interfaced to the microcomputer. In this case, a digitally controlled channel selector and capacitors can be connected to the output of the D/A converter to form a multichannel converter, as explained below.

6.6.2 Microcomputer-controlled multichannel D/A conversions

When a microcomputer controls a multichannel D/A converter, it supplies each of the channels with a corresponding digital value. These values may have been obtained by other parts of the program and are stored in sequential memory locations. If all the channels are to be updated in sequence, then use can be made of an incremental software pointer. At the start, such a pointer points to the memory location corresponding to the first channel and is incremented each time a value is supplied to a channel. After updating all the channels, the pointer resets to the start location and the process is repeated.

If the multichannel converter is designed to operate in any of several applications, some of which do not require the use of all the channels, then execution time can be saved if the program is modified to bypass the redundant channels. This, however, requires a software change which may not be economical to implement. Alternatively, onboard links or switches can be introduced to indicate to the microcomputer which of the channels is to be bypassed. For example, the use of an *N*-channel converter can be accompanied by connecting *N* links to *N* digital-input lines of the microcomputer (a single link per line). These lines can be provided with pull-up resistors (e.g. 47 kΩ) to the +5-V supply, so that each line is at a logic 1 state when a link is not placed. The state of a line can be changed to a logic 0 by placing a link to connect that line to the 0-V rail (instead of using links, DIL switches can be used in the same way). The microcomputer examines the input lines and takes a decision according to their logic state. If a line is at a logic state 1, then its corresponding channel will be converted, otherwise it will be bypassed.

6.7 Employing multichannel D/A converters

The interface of several single-channel D/A converters to the microcomputer is similar to the interface of one single-channel converter described in the previous examples. In such a case, each converter is supplied from an output port of the microcomputer, or it can be connected directly to the data bus if it possesses a 3-state latch (Chapter 8). If, on the other hand, a multichannel D/A converter is constructed from the time-sharing of a single D/A converter, then the interface is slighty different, as shown in the following examples.

Example 6.7: **Describe the approach of constructing an 8-bit 8-channel D/A converter and operating it under the control of the microcomputer of Fig. 6.1(a) using the time-sharing approach.**
Figure 6.7 gives the required configuration; the eight channels are numbered from 0 to 7. Port 2 is employed to supply the digital value, and port 1 is used to source the channel number as well as the enable signal to the 8-channel analogue demultiplexer. As explained in Section 5.6.1, the analogue value of each channel is stored in the corresponding capacitor. The selected channel is the one whose number is equal to the binary combination of the logic state of the three selection lines. Consequently, the analogue output of the single-channel D/A converter can be diverted to any of the eight channels by controlling the selection lines.

Now we need to decide on how to arrange the software of the microcomputer to fulfil the requirements. Keep in mind that the microcomputer can update one port at a time. The configuration imposes the requirement of supplying the analogue values to all the channels regularly even when no change has

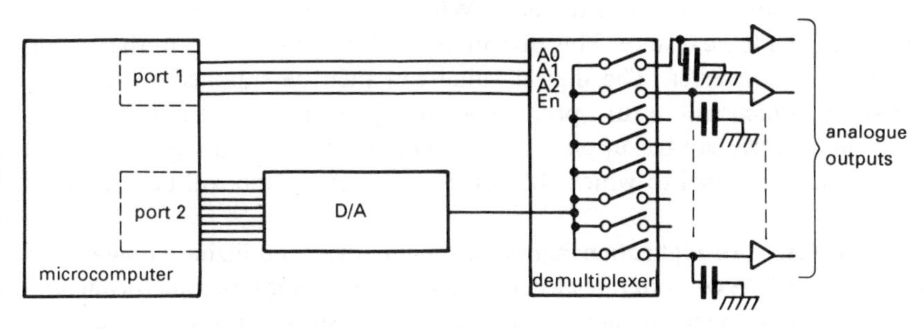

Fig. 6.7 An 8-channel D/A converter.

taken place in the digital values. This refreshes the stored charge on the capacitors to compensate for any leakage. The microcomputer is therefore regularly switching from one channel to another.

In order to consider the importance of the enable line of the demultiplexer, let us consider two solutions.

Solution 1 Enable line not used

This solution considers the procedure of switching between the channels without using the enable feature of the dumultiplexer. The enable pin is therefore set to a logic 1 state (always activated).

Consider a situation where following the selection of channel x, channel x + 1 is to be selected and a new value sent to it. Since the microcomputer of Fig. 6.1(a) is utilized, the procedure is in two parts: the first part supplies the value and the second part supplies the channel address. Which of these two parts is to take place first?

If the value is supplied before selecting channel x + 1, then the new value will alter the charge on the capacitor of channel x (because the output of the D/A converter is connected to channel x until the microcomputer supplies the address of channel x + 1). If, on the other hand, the address is supplied first, then the channel selection will change, but the D/A converter will still be supplying the value corresponding to channel x. This will start to alter the value of the charge of the capacitor corresponding to channel x + 1 until the microcomputer supplies the correct value. The second approach is therefore better than the first, because the alteration of the charge will be quickly corrected by the new value. The combination of the contact resistance of the demultiplexer and the capacitor acts as a filter which reduces the rate of change of the charge and hence minimizes the undesirable variation.

Solution 2 Using the enable line

The undesirable alteration of the charge of the capacitor can be avoided by using the enable line. When that line is set to a logic 0, no switch will be closed. Consequently, an intermediate step can be introduced during the change from one channel to the next. When a change is required, the enable line is set to a logic 0 state. This disconnects all the channels from the output of the D/A converter. The digital value can then be supplied to the D/A converter, which will in turn produce a corresponding analogue value. After that, the address can be supplied and the enable line set to a logic 1 to select the required channel to which the correct analogue value will be supplied.

Example 6.8: **An application imposes the requirement of utilizing a time-shared multichannel D/A converter. Under certain operating conditions, a microcomputer is required to start supplying an analogue value to two channels simultaneously. How can this be accomplished?**

The demultiplexer of Example 6.7 can only select one channel at a time, and will therefore not be employed for this application. On the other hand, it is explained in Chapter 5 that analogue switches are available in IC form. Each individual switch can be enabled and disabled through its control line independently of the other switches. This feature makes analogue switches attractive to use in the application under consideration. All the switches can therefore be opened during the transition from one channel to the next. During that time, the digital value can be supplied to the D/A converter before enabling the control lines of the required analogue switches. This approach differs from that of Example 6.7 in that it uses eight instead of four digital-output lines from the microcomputer for the selection of the channels (one per channel).

Example 6.9: **Consider an application where the microcomputer of Fig. 6.1(b) needs to supply five digital and 16 analogue signals to a device. The analogue input of that device includes a 16-channel analogue demultiplexer controlled by four selection lines, as shown in Fig. 6.8. How can the requirements be fulfilled by employing an 8-bit D/A converter and without introducing another output port?**

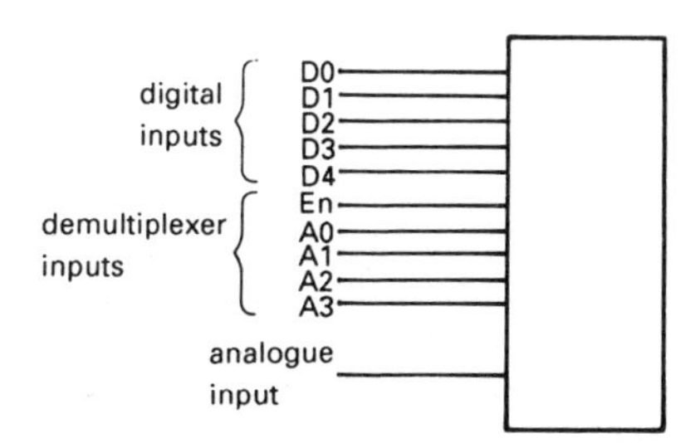

Fig. 6.8 A device with five digital inputs and 16 internal analogue channels.

- *Analogue lines*
  The 16 analogue signals can be provided from one 8-bit single-channel D/A converter to the single analogue-input line of the demultiplexer. The D/A converter can be driven from lines 0 to 7 of the output port of the microcomputer, in a similar manner to that of Fig. 6.2(b). The microcomputer needs to provide four channel-selection lines to divert the analogue signals to their corresponding channels.
- *Digital lines*
  In addition to the four selection lines, another digital-output line is needed as an enable to the demultiplexer. Furthermore, five lines are required to deliver the five digital signals. This brings the total to 18 output lines (8 + 4 + 1 + 5).

- *Using available lines*
  The microcomputer in Fig. 6.1(b) has only 16-digital-output lines. Since eight lines are reserved for supplying the D/A converter with digital values, the requirement must be satisfied by the remaining eight lines. We start by using five of these lines to provide the five digital signals, and the sixth line delivers the enable signal to the demultiplexer. The next step is considering how to use the remaining two lines to supply the 16 channel addresses.
- *Introducing additional hardware*
  The provision of the additional lines can be implemented by employing additional hardware in the form of a binary counter. This counter is employed to supply the channel number (the address) to the demultiplexer as shown in Fig. 6.9. The channels are therefore selected in sequence. The interface between the microcomputer and the counter is through the remaining two lines. The first line provides the microcomputer with the ability to reset the counter, thus allowing the conversion sequence to start from channel 0. The second line provides the counter with clock pulses. These pulses are counted and the resulting binary count is given at the output of the counter. The binary value of that output represents the address of the channel to be selected. An example of a suitable counter is the 74LS93 IC (one of the counters of Fig. 7.1). The 74LS93 is a 4-bit binary counter whose count is incremented sequentially from 0000 to 1111 and then overflows to 0000.

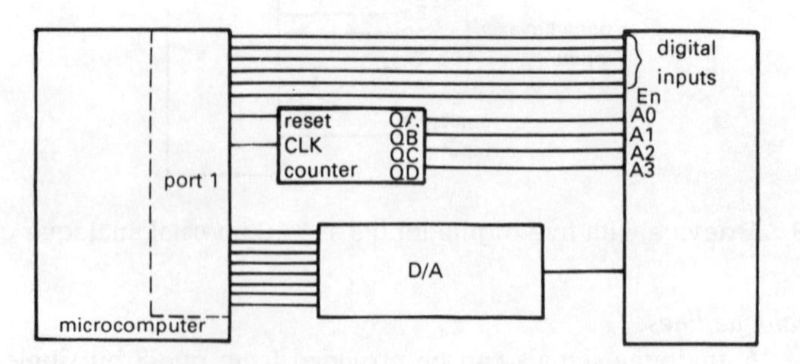

Fig. 6.9 Channel selection by a binary counter.

- *Microcomputer-controlled operation*
  At the beginning, the microcomputer resets the counter to select channel 0. Each time an analogue value needs to be supplied to a different channel, the microcomputer takes the following steps. It disables the demultiplexer through the enable line before supplying a pulse (a logic 1 followed by a logic 0) to the clock input of the counter. This pulse increments the count in a binary fashion, and delivers an address to select the next channel. The microcomputer then supplies the digital value to the inputs

of the D/A converter before enabling the demultiplexer. Clearly, the interval between the selection of one channel and the next is controlled by the microcomputer, as it supplies the counter with the pulses.

- *Providing flexibility*
With this approach, the analogue-output channels are connected to the output of the single-channel D/A converter sequentially. This constrains all channels to sample at the same rate. If such a configuration is employed in an application where only some of the channels are needed, then consecutive pulses can be delivered from the microcomputer to the clock input of the counter to bypass the redundant channels.
- *Fault detection and rectification*
When the configuration of Fig. 6.9 is utilized, it is advisable to check the output of the counter by the microcomputer to ensure correct selection of the channels. This is because noise effects may cause the counter to be incremented. If this happens, it will result in supplying values to the wrong channels.

There are several methods of performing the check. One approach is to feed the four channel-selection lines to four input lines of the microcomputer to permit the latter to examine their state. This approach, however, uses four lines which may be needed for other tasks. If this is the case, another approach can be adopted.

This is based on connecting a single digital-input line of the microcomputer to the line carrying the least significant bit of the selection address (line QA of the counter of Fig. 7.1). The test should reveal that the line is at a logic 0 state after reset and should only change state when a pulse is supplied from the microcomputer to the counter. It should be at a logic 0 state when an even channel is addressed, and at a logic 1 state when addressing an odd channel number.

A decision is required on what action to take if a fault is detected. One solution is to allow the microcomputer to reset the counter and supply consecutive pulses to select the required channel.

Chapter 7

Interfacing A/D converters to microcomputers

7.1 Introduction

The previous chapter demonstrated 8-bit and 16-bit interfacing of D/A converters to the two microcomputers presented in Fig. 6.1. Each of these microcomputers provides 16 digital inputs and 16 digital outputs. The first uses 8-bit interfacing and the second uses 16-bit interfacing, where each microcomputer reads from, or writes to, one port at a time. A/D converters can be interfaced to such microcomputers to provide them with the ability to measure analogue signals, monitor analogue events, or operate as part of a control process containing analogue signals.

This chapter utilizes the microcomputers of Fig. 6.1 to demonstrate interfacing A/D converters to microcomputers through I/O ports. Examples of interfacing A/D converters directly to the data bus are presented in the next chapter. This chapter is intended to familiarize the reader with the interface of A/D converters to microcomputers. Consequently, it presents devices such as multiplexers, S&Hs and data converters in a block diagram form. Examples of commercial devices are given in Chapters 5 and 8.

It is explained in Chapter 3 that some microcomputers include one or more data converters (it may be a built-in or a previously interfaced converter). Although such converters can be very suitable for many applications, they may not fulfil the requirements of many other applications. If, for example, a microcomputer possessing a single-channel A/D converter is employed to control the operation of a robot arm, then an analogue feedback signal may need to be received from each of the motors within the arm. For such an application, a multichannel converter is required. The examples presented in this chapter demonstrate how a single-channel converter can be modified to perform multichannel conversions.

Furthermore, an in-computer converter may not be fast enough to satisfy the requirements of many applications. For instance, an A/D converter designed to detect movements of a joystick does not need to be fast. However, it may not be suitable for obtaining detailed information on a transient signal in an industrial application. For such a case, this chapter provides some solutions. For instance, it shows how an additional converter can be designed and/or interfaced to a microcomputer. An alternative solution is presented in Example 7.11. It shows how a limited number of samples can be obtained at high speed from a slow, single-channel A/D converter. Example 7.10 shows how simultaneous samples can be obtained from a small number of channels by interfacing additional circuitry to a single-channel converter. Further examples are given in Chapters 8 and 16. For instance, Section 16.5 explains how an input port can be multiplexed to accept digital values from two A/D converters.

Once the interface algorithm has been developed, a program can be written to maintain the correct order of timing events to achieve communication between the memory of the microcomputer and the analogue world. Such a program can be configured as a subroutine. In the case when a value is sent to the outside world, such a program reads the value from one or more specified bytes in memory and transfers it to the input terminals of a D/A converter. Similarly, when reading an analogue value, the program transfers that value from the output of an A/D converter to one or more specified bytes in memory. A user-written program requiring communication with the analogue world need not be involved with producing timing signals required for the interface. Instead, it can call that subroutine and only deal with the associated memory locations.

7.2 Interfacing A/D converters to microcomputers

When an analogue signal, such as the output of a transducer, needs to be processed by a microcomputer, it requires conversion to the digital form. An A/D converter performs such a conversion and provides the microcomputer with a digital value equivalent to the amplitude of the analogue signal at the time of conversion. The rate at which the A/D converter samples the continuous analogue signal affects the information contents of the sampled signal. This can be observed if the digital signal is reconverted to an analogue form using a D/A converter (Section 8.6). The maximum sampling rate of microcomputer-controlled conversions is limited by the execution time required to complete various tasks of the microcomputer efficiently.

7.2.1 Typical applications

The operation of numerous systems is based on obtaining a single result regularly through an A/D converter. This result is used as an input to a software task which is executed immediately after the end of the A/D conversion process. Some applications impose the requirement of a higher signal-to-noise ratio. Such a requirement can be satisfied by averaging several samples of the same signal. This, however, consumes additional processing time. In some applications, the additional time constraint is removed by utilizing a hardware filter to achieve the averaging. Examples of interfacing and applying A/D converters are given in this chapter as well as in Chapters 8 and 16.

In many systems, several readings are required from the same or different A/D conversion channels before commencing the execution of a task. Such readings may be required after the occurrence of a certain event. If that event is, for example, a major fault in an industrial system, then after detecting the fault the microcomputer can start executing a program which suspends the execution of predetermined tasks and stops the system safely. At the same time, the microcomputer can concentrate on obtaining as many readings as possible. These readings can later be used to analyse the cause of the fault. This is important in detecting the faulty part and may be useful in modifying the procedure of operation so that such a fault can be prevented in future. The detection of the cause of a fault can become easier if a history record is kept in memory on how the value of some, or all, parameters is changing during operation. The procedure of collecting a large number of results and the limitation imposed by the storage capability of the microcomputer are described in Sections 17.1 and 17.2 respectively.

7.2.2 Operating A/D converters

Most A/D converters are designed to accept a command from a controlling device, such as a microcomputer, before commencing a digitization process. This command is called the start of conversion (SC). It may be in the form of a pulse, a logic level, or a software command. The provision of such a command is very useful, particularly when digitizing waveforms. It allows a controlling device to read the magnitude of an analogue signal at a particular instant of time. It also makes it possible for that device to reconstruct the signal.

Consider two methods of operating an A/D converter. The first is operation under the control of a microcomputer. In this method, the microcomputer starts an A/D converter by supplying the SC command. The second method is often referred to as the free-running operation. It is based on operating the converter without the intervention of the microcomputer. In this method, the SC command can be supplied to the A/D converter from external, or onboard, circuitry. Alternatively, the converter can be connected so that it does not need an SC command, i.e. one conversion starts immediately after the completion of another. Use can also be made of a converter that does not need an SC signal, e.g. a flash A/D converter (its operation is based on the simultaneous-conversion method described in Section 5.5.1).

Many A/D converters produce an end-of-conversion (EOC) signal to indicate the completion of a conversion. A microcomputer can therefore examine that signal to detect when to-read the digital result.

7.3 Alternative solutions to an interfacing problem

Several methods of A/D conversion are presented in Chapter 5. Let us consider an application where a microcomputer needs to be provided with the ability to read an analogue signal. This section presents several solutions in order to demonstrate how the analogue interface requirements can be met by the use of any of several approaches. Each approach has advantages and limitations and is best suited for some, but not all, applications.

7.3.1 Designing a computer-controlled ramp A/D converter

Let us start by considering the use of an 8-bit ramp A/D converter such as that described in Section 5.5.2 and shown in Fig. 5.9. The operation of that converter is based on supplying the digital inputs of an 8-bit D/A converter from an 8-bit binary counter. This type of counter can be constructed from cascading two 4-bit binary counters, such as the 74LS93 ICs shown in Fig. 7.1. When the reset lines of that counter are driven to a logic 1 state, all the output lines reset to a logic 0 state. Driving the reset lines back to a logic 0 state allows the counter to start counting from 0. The reset signal is therefore an active-high pulse which serves as the SC (start of conversion) command of this type of A/D converter. The conversion takes place in the manner described in Section 5.5.2 until the digital result is obtained. This is indicated by a change in the state of the signal produced by the comparator from a high logic level to low. This stops the supply of pulses to the counter and can therefore be considered as the EOC signal.

Example 7.1: **How can the 8-bit ramp A/D converter described above operate under the control of the microcomputer of Fig. 6.1(b)?**

The interface of the A/D converter to the microcomputer of Fig. 6.1(b) is shown in Fig. 7.2, where the conversion result is supplied to the microcomputer through lines 0 to 7 of port 2. The conversion is initiated when line 0 of port

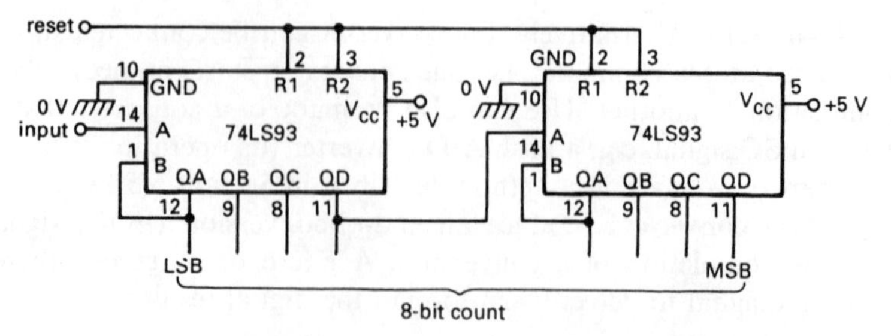

Fig. 7.1 An 8-bit binary counter.

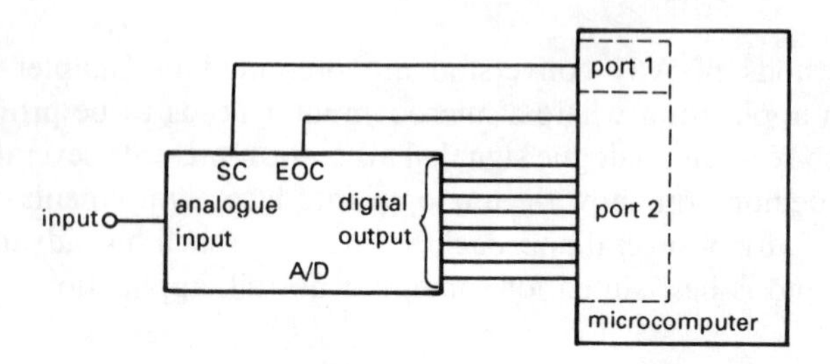

Fig. 7.2 Interfacing an 8-bit A/D converter to a microcomputer.

1 supplies a pulse to the SC input of the A/D converter (the reset lines in the case of the above-described converter). The microcomputer can be programmed to detect the completion of conversion by monitoring the logic state of the EOC signal supplied to line 8 of port 2. After detecting a change of state, the microcomputer acquires the conversion result by reading the output of the counter. Following this, the microcomputer can issue another SC command to start the next conversion.

The procedure can therefore be presented by the following steps, where an '&' character represents a software AND, a '|' represents a software OR, and each of 'w' and 'y' is a 2-byte word; 'w' contains the value last supplied to the 16-bit output port and 'y' is used to store the result of reading the 16-bit input port.

```
1  w = w|0000 0000 0000 0001
2  OUTPUT w FROM PORT 1
3  w = w & 1111 1111 1111 1110
4  OUTPUT w FROM PORT 1
5  READ PORT 2 AND STORE RESULT IN y
6  IF (y & 0000 0001 0000 0000) > 0
       THEN GO TO STEP 5
7  DIGITAL RESULT = (y & 0000 0000 1111 1111)
```

In this procedure, steps 1 to 4 generate the SC command without altering the state of other lines within output port 1. A very short delay may be required after step 4 and before reading the input port at step 5. During the conversion time, the microcomputer will be executing steps 5 and 6 repeatedly until bit 8 (the EOC signal) changes state, indicating the availability of the result.

As an exercise, design the interface of the above-described converter if the microcomputer is that of Fig. 6.1(a).

7.3.2 Using software to reduce hardware

The A/D converter presented above consists of a small number of hardware components. In many situations, such components can be replaced by software. The replacement, however, should only take place in situations where it is economical to do so, as shown in the following example.

Example 7.2: **How can software be used to minimize the hardware components in a ramp A/D converter (such as the one described in Example 7.1)?**
The microcomputer can perform the function of the digital parts of the converter by software. The hardware counter can be replaced by a software counter only if the microcomputer can afford to spend the time to do so. With this approach, a program needs to be written to supply the D/A converter with sequential digital values to generate the ramp. The program is in the form of a loop which starts by supplying a binary value of 00000000 to the D/A converter. It repeatedly increments that value and tests the output of the comparator until a change of state is detected. The circuit is shown in Fig. 7.3, where the output of the comparator is supplied to line 8 of port 2 of the microcomputer of Fig. 6.1(b).

The program can be represented by the following procedure, where 'w' contains the value last supplied to the output port, and the count is contained in the lower eight bits of 'COUNT'

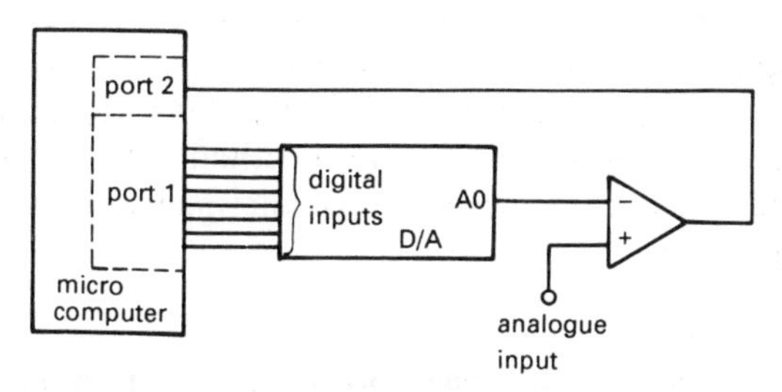

Fig. 7.3 Microcomputer-controlled A/D conversion.

```
1  LET COUNT = 0000 0000
2  w = w & 1111 1111 0000 0000
3  w = w| COUNT
4  OUTPUT w THROUGH PORT 1
5  READ PORT 2 AND STORE VALUE IN y
6  IF (y & 0000 0001 0000 0000) > 0
       THEN INCREMENT COUNT AND GO TO STEP 2
7  DIGITAL RESULT = COUNT
```

The procedure is similar to that of Example 7.1. Step 2 clears the previous value and step 3 stores the new value in its place. A very short delay between steps 4 and 5 may be required to allow the output of the comparator to settle. During conversion, the microcomputer executes steps 2 to 6 repeatedly.

With this approach, the hardware is reduced, eliminating the need for a counter, an AND gate and a clock. This leads to a low-cost A/D converter. However, the microcomputer does not have time to execute other tasks during the conversion. This may be acceptable in some applications, such as when the conversion is not repeated at a high rate and the microcomputer can afford to wait for the conversion to complete. For example, if the microcomputer is employed to control the temperature of a room, then testing the temperature does not need to be repeated at a high rate. This is because the temperature is a slowly changing variable and it is not necessary for the microcomputer to react immediately when the required temperature is reached. The microcomputer can therefore execute other tasks before repeating the A/D conversion.

7.3.3 Employing a different conversion method

When compared to other conversion methods, the conversion time of the ramp method is relatively long and it depends on the resolution of the converter as well as on the amplitude of the analogue signal. It takes up to 256 steps, for instance, to perform an 8-bit conversion. Although achieving the count by a microcomputer reduces the cost of hardware, it can be considered as a poor use of the capabilities of the microcomputer. This is because this approach requires the involvement of the microcomputer for the time of conversion, which is normally long. One way of saving valuable processing time is to adopt a conversion method with a lower digitization time. It is possible to achieve this by simply altering the software of the above example, as shown below.

Example 7.3: **By utilizing the same hardware as that of the previous example, how can the A/D conversion approach be altered to reduce the conversion time?**

In the previous example, the ramp is generated by incrementing the count by 1 each time. Rather than using small steps, the successive-approximation approach can be adopted. With this approach, binary-weighted steps are introduced as described in Section 5.5.2. The procedure starts by supplying a value equal to half the range to the D/A converter (this is achieved by setting the most significant bit to a logic 1 and keeping the rest of the bits at logic 0). The resulting analogue voltage is compared with the analogue-input signal by the analogue comparator. If the output of the comparator changes state, then the microcomputer will know that the result is located at the lower half of the range; otherwise the result is within the upper half. In this way, half of the range is eliminated in one step, and the procedure can therefore continue by considering only the remaining half and repeating the process to eliminate half of that.

The successive-approximation approach performs an 8-bit conversion in eight steps instead of the 256 steps required by the ramp method. Therefore, the successive-approximation approach leads to a considerable reduction in the conversion time, making this method of conversion very attractive to use in numerous applications. However, high accuracy requirements create the need for using a S&H device.

The conversion time can be minimized if the selected conversion approach is of the open-loop type, such as the simultaneous-conversion approach (Section 5.5.1). With this approach, the result is immediately available for the microcomputer to read.

7.3.4 Reducing the involvement of the microcomputer

As with the ramp method, the involvement of the microcomputer with every step of the successive-approximation conversion can be replaced by the introduction of additional hardware, a successive-approximation register. Alternatively, an A/D converter IC can be utilized, as shown in the next chapter. If the hardware rather than the software approach is adopted, then the hardware should be capable of informing the microcomputer of the availability of valid data. This can be implemented in several ways.

One method of fulfilling the requirement is to allow the EOC signal to interrupt the microcomputer. This approach permits the microcomputer to read the digital value as soon as it is obtained without introducing unnecessary delays. The interrupt activates an interrupt-service routine, which can be very short, as all it needs to do is to read the digital result and store it in a specified memory location. The interrupt routine can also set a software flag to indicate to other parts of the program the availability of a new value. The procedure consists of two parts, the initialization and the interrupt-service routine, in the same way as that of Example 3.3. The interrupt-controlled approach, however, can only take place if there is access to a hardware interrupt line of the microcomputer to which the EOC line can be connected. The use of correctly organized interrupts is attractive, especially in real-time control applications (Chapter 13). Section 1.9.1, however, explains that the use of interrupts may not be attractive when the rate of their occurrence is very high.

Instead of employing the interrupt-control approach, a change in the state of the EOC signal can be detected by the use of the program-controlled input method. One way of applying this method is to program the microcomputer to continuously interrogate the EOC line, seeking the detection of a change of state. This, however, leaves the microcomputer waiting for the conversion to complete. If the conversion time is long, then enhanced performance can be realized if the microcomputer is permitted to execute useful tasks while the conversion is taking place. The following example describes how this can be implemented.

Example 7.4: **How can a microcomputer adopt the program-controlled approach to detect the EOC signal generated by an A/D converter and be able to execute useful tasks during the conversion time?**

Solution 1

One way of fulfilling the requirement is to adopt an approach based on estimating the time needed for the converter to perform a conversion. The microcomputer can therefore execute a suitable routine after supplying the SC command to the converter. When this routine is complete, the microcomputer can examine the EOC line before reading the digital value.

The success of the implementation of this approach depends on the correct estimation of the conversion time. In a ramp A/D converter, the conversion time depends on the magnitude of the analogue-input signal, which is unknown prior to the conversion. In other conversion methods, such as when using a hardware-based successive-approximation conversion (Chapter 8), the conversion time can be estimated by the designer. If it takes a relatively long time, the designer can arrange the software to make the microcomputer execute a suitable routine during the conversion period.

Solution 2

It is possible to adopt an alternative approach which is independent of whether or not the conversion time can be estimated. It is based on storing the digitized result after the end of conversion without the intervention of the microcomputer. If the converter has no storage ability, then an external latch can be included. The operation of the converter can be initiated by an

SC command from the microcomputer. When the conversion is accomplished, the EOC signal instructs a latch to store the results without the intervention of the microcomputer. The hardware connection of the latch is very similar to that of Fig. 2.15, except that the EOC signal replaces the device-selection signal and the latched value will be read by an input port. (Clearly, an inverting gate may be required if the logic state of the EOC signal is opposite to that required to enable the latch.) The up-to-date conversion result is therefore held in the latch and will be updated after completing the next conversion (which will only start by a command from the microcomputer).

With such an approach, the microcomputer does not need to read the digital value as soon as it is produced. Instead, it can command the start of conversion and then execute a suitable routine. The time required to process that routine can be longer than the maximum specified conversion time. If the interval between the microcomputer applying the SC command and being ready to read the result is relatively long, there will be no need to examine the EOC line. This reduces the processing time and frees the associated input line of the microcomputer to be used for another purpose.

7.3.5 Adopting the free-running approach

The above examples consider operating A/D converters under the control of a microcomputer. The following example describes the free-running approach in which an A/D converter digitizes a value without receiving commands from a microcomputer. Such an approach removes from the microcomputer the need for generating the timing signals required for the operation of the converter. This, however, implies that the microcomputer will no longer be able to specify the instant of starting a conversion.

Example 7.5: **Describe a way of generating a SC signal for an A/D converter without the intervention of a microcomputer. The conversion process should produce a digital value which can be read by the microcomputer.**

In such a situation, the SC signal can either be supplied from a source external to the microcomputer, or derived from the clock of the microcomputer. The external source can be an external device, an onboard clock, or an onboard processor. For example, an onboard clock can be produced by using the square-wave generator of Fig. 3.1. The duration of the SC signal is defined by the mark-to-space ratio of the generated clock, which can be set by choosing the value of the components, as explained in Example 3.2.

On the other hand, the derivation of the clock from that of the microcomputer requires an access to such a signal, which may not be available for the designer of the interface. If it is available, then it is likely to possess a frequency higher than the required conversion rate. The required frequency can be produced by utilizing a divide-by-N counter, such as the 74LS90 (divide-by-10 counter). The operation of this counter is similar to that of the 74LS93 (4-bit binary counter) employed in Section 7.3.1 to produce the ramp A/D converter. Suppose that an A/D converter needs to digitize an analogue signal once every 100 μs. An SC signal is therefore required at a rate of 10 kHz. If that signal is to be derived from a microcomputer clock possessing a frequency of 1 MHz, a division by 100 is required. This can be implemented by cascading two divide-by-10 counters, as shown in Fig. 7.4.

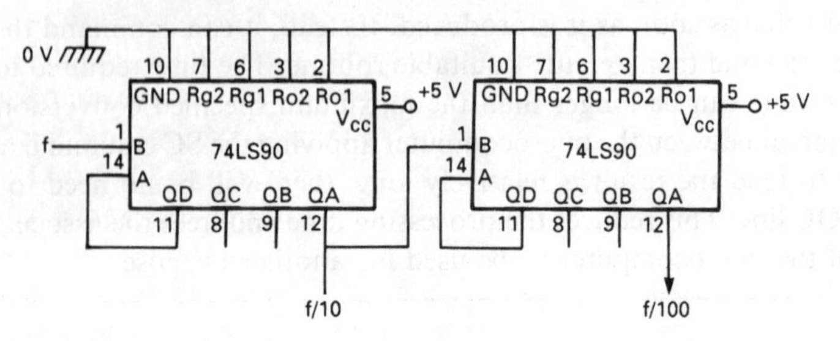

Fig. 7.4 Reducing the frequency of the clock.

Since the digitization of a free-running A/D converter is performed independently of the microcomputer, the latter has no indication of the instant of starting a conversion and hence it cannot predict the EOC instant. Therefore, the microcomputer can detect the availability of the digital result by any of the methods described in the previous example. It can, for instance, adopt the method of using the EOC signal as a hardware interrupt, or the method of testing the EOC line until it changes state. Alternatively, the microcomputer can read the most up-to-date result by adopting the method of storing the conversion result in a latch enabled by the EOC signal.

7.3.6 Discussion

The previous sections give methods of digitizing analogue signals under the control of a microcomputer without using commercial A/D conversion ICs. Utilizing a counter to implement the ramp method provides a simple solution. In practice, however, a commercial D/A conversion IC with an on-chip counter can be used to perform the operation as shown in Section 8.3.1. There are, however, some situations where the use of a separate counter is attractive. For instance, the microcomputer may be employed in an application where a counter is already employed in an existing circuit. Assume that it has a 12-bit resolution. It may therefore be possible to meet the requirement of providing an 8-bit A/D conversion channel by sharing that counter through a simple additional circuit. The lowest eight bits of the counter can, for example, be supplied to the inputs of an 8-bit D/A converter. The microcomputer can then be programmed to perform the A/D conversion at intervals where the counter is not used by the original system. This can be attractive in applications having no requirements for fast sampling, nor having to perform the conversion at a particular instant of time. This can be the case when the microcomputer is required to measure the temperature of a room and decide on whether to turn a heater on or off.

Although the use of the counter can be replaced by software, the monopolization of the microcomputer during the conversion period is not attractive in situations where the conversion time is long. Example 7.3 shows how to reduce the conversion time by employing the successive-approximation approach. A software-based successive approximation not only gives the microcomputer the ability to perform A/D conversions, but it also makes it possible to perform D/A conversions by utilizing the same hardware at different times (see Section 8.6.2).

Section 7.3.4 describes how the microcomputer can be provided with the ability to perform tasks while waiting for a conversion to complete. The requirement of controlling the operation of an A/D converter can be removed from the microcomputer by adapting the free-running approach as described in Section 7.3.5, where it is described how this approach removes from the microcomputer the ability to specify the instant of starting a conversion (unless additional hardware is included).

Other methods of conversion can be applied in a similar manner. For example, the implementation of the tracking method is very similar to the implementation of the ramp method, except an up/down counter needs to replace the binary counter. The use of such a counter is presented in Example 12.3.

7.4 Testing A/D converters

The construction and operation of A/D converters are considered in the above examples. Let us now consider a method of testing such converters.

Example 7.6: **How can the relation between the input and output of an 8-bit A/D converter be verified without the use of a microcomputer?**
The test can be performed by supplying the input of the converter with an analogue signal of an adjustable magnitude from a stable power supply. After a conversion is accomplished, the converter produces a digital output which can be examined to verify the relation between the input and output. Without employing a microcomputer, external hardware is required to provide the

converter with timing signals (as in Example 7.5). The test can be performed by using discrete LEDs to display the conversion result. Figure 7.5 shows a configuration for testing an 8-bit A/D converter. It utilizes eight LEDs, each of which is driven from a digital-output line of the A/D converter. Inverting gates (two 74LS06 hex-inverting ICs) are employed to provide sufficient drive to the LEDs through 470-Ω resistors. The two ICs contain 12 inverting gates. Of these, only eight are used, leaving four gates available for another application. A LED will be ON when the A/D converter sets the corresponding bit to a logic 1 state and will be OFF when the converter sets that bit to a logic 0 state.

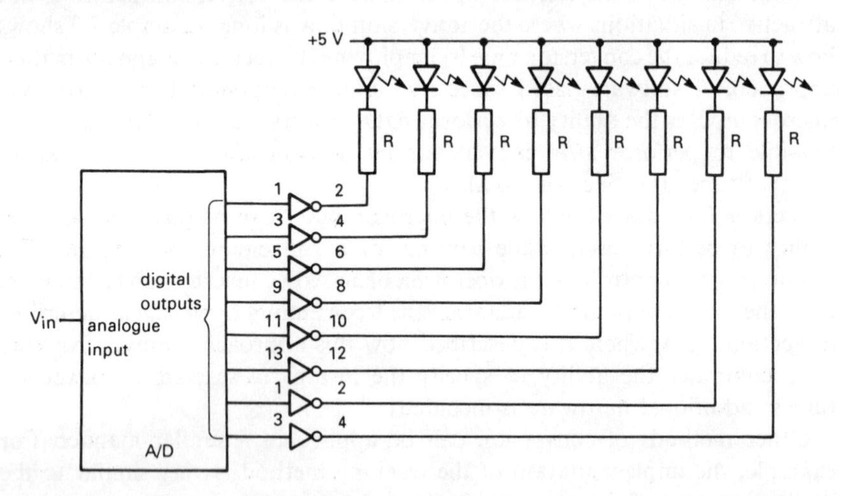

Fig. 7.5 Testing an 8-bit converter.

Testing the relation between the input and output of the A/D converter is based on varying the analogue voltage at the input of the converter and reading the digital results. During the test, the analogue-input voltage can be measured by employing a digital voltmeter (if such a reading is not available from the adjustable power supply). The resulting digital output is read from the LEDs.

The test should show that for a given input value, the output corresponds to that expected from the converter. If required, the actual transfer function of the converter can be obtained from the test and compared to the ideal one. Such a comparison reveals the converter errors described in Section 5.7. The test should show that all possible bit combinations can be produced at the output by varying the input voltage. In some situations, however, missing codes are observed. This could be due to a hardware fault that was not detected when the hardware connectors were inspected, or it may be due to the converter itself (see Section 5.7). For instance, the test may reveal that the conversion results only include even numbers. What is the likely course of such an error? In this case, the even values obtained imply that the least significant bit is always low (logic 0). This could be due to a wiring fault, or the LED may not be functioning. If neither of these is the source of the fault, then the A/D converter is likely to be faulty. The method of correcting the fault obviously depends on the type of converter used and on whether it is in an IC form, or it is constructed by the designer as described above.

To perform the above test by the use of a microcomputer, the latter can supply the timing signals and display the results. A suitable program can be written to allow the microcomputer to wait for an operator to press a key before it initiates a conversion and reads the result. The procedure is similar to that in Example 6.3. As an exercise, write such a procedure.

7.5 Examples of using single-channel A/D converters

We have already considered the difference between unipolar and bipolar D/A converters in the previous chapter. The theoretical relation between the analogue and the digital values is obviously the same for A/D and D/A converters. Let us now consider an example of producing a bipolar A/D converter.

***Example** 7.7:* **What are the alterations needed to allow a unipolar A/D converter to read bipolar signals?**
Similar to the case of utilizing a unipolar D/A converter to generate bipolar signals (Example 6.5), a bipolar A/D converter can be produced by offsetting the analogue input to a unipolar A/D converter by half of full-scale. This can be achieved by adding a voltage level of +2.5 V to the analogue-input voltage. A configuration such as that in Fig. 6.5(a) can be used, except that the −V reference is replaced by +V. The inversion of the sign introduced by the summing amplifier can be corrected by employing an inverting amplifier, such as that of Fig. 6.5(b). The introduced offset makes the signal look to the converter as if it is unipolar. This offset must be taken into account when processing the digital results in the microcomputer.

Consider, for example, a unipolar 8-bit A/D converter which produces 00H when supplied with an analogue value of 0 V and delivers a digital value of 80H when supplied with an analogue value of +2.5 V. This implies that offsetting the input signal by +2.5 V will result in representing a 0-V value by 80H. The offset can be removed from the digital result by software where a value of 80H is subtracted from each digitization result. If, for instance, the microcomputer receives a value of FFH from the converter, then it will subtract 80H and use the result, which is 7FH. This is equivalent to a +2.5-V value of the analogue-input signal.

The examples presented so far do not consider the relation between the range covered by the analogue signal and that covered by the A/D converter. Signals received from transducers may not cover the same range as the A/D converter. For example, an A/D converter possessing an input range of 0 to +5 V may be supplied with an analogue signal from a transducer covering the range 0 to +1.25 V. In such a situation, the signal only covers 25% of the range of the converter. If such a converter has a resolution of 8 bits, the signal will only be represented by 6 bits. Better representation can therefore be achieved when that signal is amplified by a factor of 4. An example of an amplifier which can fulfil the requirement is the non-inverting amplifier shown in Fig. 6.4. Amplification can therefore be used to help in reducing the quantization uncertainty and lead to digitizing the signal more accurately.

Example 7.8: **Consider a microcomputer-controlled A/D converter employed for operation in any of several modes. In each mode, the signal covers a fraction of the range. It is therefore desirable to amplify the signal in all modes. How can the signal be amplified, assuming that the optimum magnitude of amplification varies from one mode to another?**
The requirement can be fulfilled by the use of an approach based on employing a variable-gain amplifier such as that shown in Fig. 7.6. The amplifier can operate under the control of a microcomputer. The figure shows that the amplifier includes four switches. When these switches are open, the gain of the amplifier is given by 1 + (R2/R3). Instead of using a potentiometer to vary the gain manually, the four switches can be controlled by the microcomputer to vary the gain in stages. The value of the gain depends on

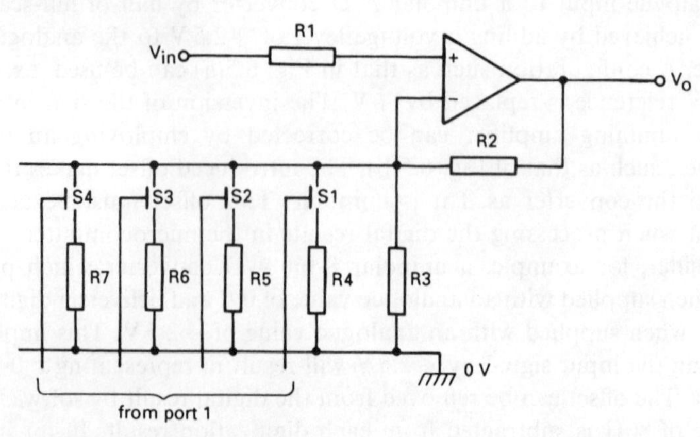

Fig. 7.6 A variable-gain amplifier.

which of the switches is closed, if any. The change of gain can either be implemented by closing one switch at a time, or by closing the switches in a binary fashion. The latter approach provides 16 gain stages.

The method of varying the gain of such an amplifier is based on the fact that the magnitude of the resistance of a given path can be changed by placing another resistance in parallel with it. For the purpose of illustration, consider Fig. 7.6, in which R3 > R4 > R5 > R6 > R7. Assume that, at the beginning, switches S1, S2, S3 and S4 are open. In order to obtain 16 amplification stages, the switches will be closed in a binary sequence. The magnitude of the parallel combination of the resistors will decrease as the binary number increases. The decrease in the value of the combined resistors results in increasing the gain of the amplifier.

Depending on the application, the gain can either be software-selectable (using analogue switches as in Fig. 7.6) or hardware-selectable (by employing onboard links or switches). The designer needs to select a suitable number of switches and to define the range of allowable gain variations. Noise effects may impose an upper limit on the allowable gain and the value of the resistors can be chosen accordingly. After constructing the amplifier, a user can easily select any of the allowable gains. The selection can be based on the knowledge of the range covered by the signal.

In the configuration of Fig. 7.6, four analogue switches are employed. This corresponds to the use of a single IC (see Section 5.2) and leads to 16 amplification stages to assist in providing an input signal with a range closely matching that of the A/D converter.

7.6 Relation between type of interface and resolution of an A/D converter

When connecting an A/D converter to an input port of a microcomputer, consideration should be given to the number of digital-output lines of the converter and the lines available in the port.

7.6.1 16-bit interfacing

Let us consider interfacing A/D converters to the microcomputer of Fig. 6.1(b). If the resolution of the converter is less than 16 bits, the input port can be used to monitor the EOC signal (if needed) as well as to read the result of conversion and other digital inputs (if any).

If, on the other hand, a 16-bit converter is employed, then the digitized value is supplied by the same number of lines as those available in the input port of the microcomputer. These lines can therefore be directly connected to the input port. A decision is needed on how the microcomputer can recognize the end of conversion. Let us consider a few solutions:

1 Using the interrupt-controlled approach where the EOC line is employed to generate a hardware interrupt.
2 Multiplexing line 0 of the input port to read bit 0 of the digital result as well as the EOC line under the control of the microcomputer. Figure 7.7 shows two ways of implementing the multiplexing:

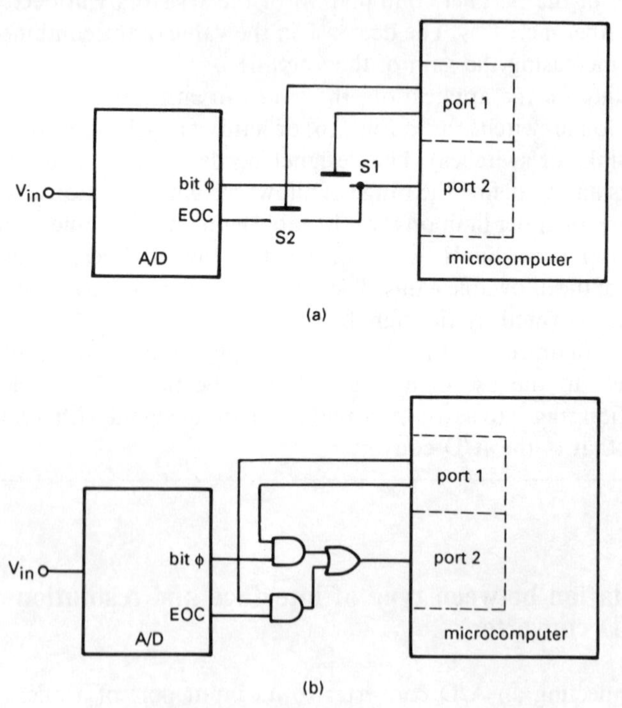

Fig. 7.7 Multiplexing an input line: (a) using analogue switches; (b) using logic gates.

- Analogue switches can be used as shown in Fig. 7.7(a). During conversion, the microcomputer opens switch S1 and closes switch S2 to examine the state of the EOC line. When the end of conversion is detected, the microcomputer opens S2 and closes S1 before reading the digital result. The logic state required for opening or closing an analogue switch depends on the type of switch selected (see Section 5.2).
- Instead of using analogue switches, logic gates can be utilized as shown in Fig. 7.7(b). The operation takes place in a manner similar to that of using the analogue switches. The microcomputer selects

one of the two lines by delivering a logic 1 signal to the input of its corresponding AND gate while supplying the other AND gate with logic 0.

3 If possible, a new input port can be introduced, as described in Chapters 2 and 3.

4 A simple, but not a very attractive, solution is to neglect the lowest bit of the A/D converter, i.e. to use it as a 15-bit converter. In this case, the digital-output lines of the converter can be connected to lines 0 to 14 of the input port. The EOC line can therefore be connected to line 15 of the same port.

7.6.2 8-bit interfacing

The previous examples show that interfacing an 8-bit A/D converter to an 8-bit port is easy to implement. If the converter provides the digital result on more lines than those available in an input port, then the interface is slightly more difficult. Here, the interface of a 10-bit A/D converter to the microcomputer of Fig. 6.1(a) is taken as an example. In this case, reading the digital result involves using both input ports of the considered microcomputer, as shown in the following example.

Example 7.9: **How can the microcomputer of Fig. 6.1(a) be employed to read the conversion result of a 10-bit A/D converter when the interface is as shown in Fig. 7.8?**

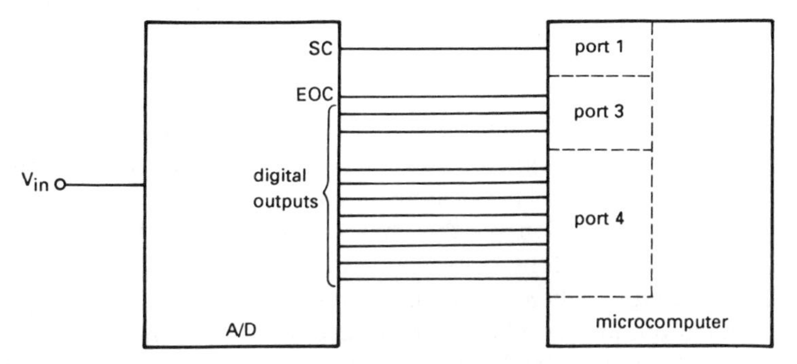

Fig. 7.8 Interfacing a 10-bit A/D converter to two 8-bit ports.

It can be seen from Fig. 7.8 that the lower eight bits are supplied to port 4, whereas the two upper bits are supplied to port 3. Both of these ports are non-latching and are read individually by the microcomputer. Therefore, the digital value must be available at these ports for a time long enough for the microcomputer to read it.

Figure 7.8 assumes that the converter will latch the results. If this was not the case, that result can be stored in an external latch (as in Solution 2 of Example 7.4). In addition to activating the latch, the EOC line can be connected to an input line for the microcomputer to provide information on the state of conversion.

When the microcomputer needs to read the digitized value, it transfers that value to two consecutive memory bytes. Once the transfer is accomplished, the microcomputer can issue a command to start the next conversion. Note that the process of transferring the value from the converter to memory is in two steps. In the first step, the microcomputer transfers the lower eight bits from port 4 to a RAM location (a byte). This is followed by the second step, which transfers the upper two bits from port 3 to the next location. This configuration only uses three lines from port 3. When reading the digital value, or when examining the EOC signal, the microcomputer reads port 3 as a byte. The bits of interest need to be isolated from the other bits before testing or storing. This can be achieved by the use of a software AND as shown earlier.

7.7 Multichannel A/D converters

This section considers two approaches to designing a multichannel A/D converter. The first is the time-sharing approach and the second employs a single converter per channel. Whatever approach is adopted, the channel-selection process can either be sequential or random, according to the requirements.

Care should be taken when designing an interface for a multichannel converter, as a poorly designed system may reduce the speed of response. For instance, the sequence of performing the conversions may not be correctly arranged, or the conversion process may require the involvement of the microcomputer during a long conversion period.

Consider a situation where a microcomputer reads three analogue signals, solves an equation, and decides on the direction and distance of moving a robotic mechanism accordingly. Assume that the signals are obtained from a multichannel converter which converts the channels in sequence, one channel per 100 μs. If the signals are supplied to channels 1, 4 and 8, the calculation will take place after completing the conversion of channel 8. By this time, the reading obtained from the first signal (the one supplied to channel 1) is 800 μs old. A better response can be obtained if the signals are supplied to channels 6, 7 and 8. In this way, signal 1 is obtained 300 μs prior to the start of the calculation, thereby reducing the sampling delay and improving the response.

It is therefore important to correctly arrange the conversion procedure and to select devices that satisfy the timing and financial requirements. Keep in mind that without using local intelligence in the interface circuitry, the process of digitizing multiple inputs is one of the tasks handled by the microcomputer. Thus, the maximum possible rate of digitization depends on the time needed to execute other functions and to respond to external events.

Certain multichannel applications create a need for providing a microcomputer with the ability to digitize one channel more frequently than others. These and other requirements are easy to fulfil when the multichannel A/D conversions are implemented under the control of a microcomputer, as will be explained. The time required to perform these operations may lead to a reduction in the system performance, thus creating the need for employing additional hardware. The add-on circuitry may perform the function of a time-consuming part of the conversion task, or it may control the multichannel conversion process. With the latter approach, the microcomputer only reads the digitized values without being involved with producing various signals needed for the conversions.

7.7.1 Using the time-sharing approach

One approach to designing a multichannel A/D converter is based on the time-sharing of a single-channel A/D converter between the channels (as briefly described in Section 5.6.2). A multiplexer is utilized to perform the task of selecting any of several input signals for digitization by the single-channel converter. Rather than using this approach, the designer may decide to employ a multichannel A/D converter IC as shown in Section 8.4.2.

Figure 7.9 shows an 8-channel A/D converter utilizing the time-sharing approach and operating under the control of the microcomputer of Fig. 6.1(b). Let us briefly consider the microcomputer-controlled operation. The microcomputer selects an input channel by supplying its corresponding address to the multiplexer through port 1. Following this, the microcomputer issues two commands through the same port. The first command instructs the S&H to capture the signal. This is followed by the second command, which instructs the A/D converter to start the conversion process. A waiting interval is required in between these two commands to allow the capacitor of the S&H to charge to the correct value (Section 5.6). The microcomputer can read the resulting digital value after receiving an indication through the EOC line in a similar manner to that of the previous examples.

In the time-sharing approach, the time required to sample and digitize each input consists of four parts. These are:

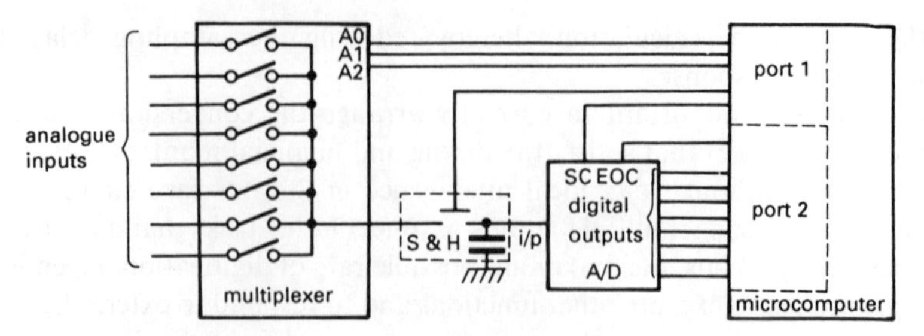

Fig. 7.9 A multichannel A/D converter.

- The period needed for the multiplexer to switch between channels.
- The delay required to allow the S&H capacitor to charge and hold the new value.
- The time needed to perform an A/D conversion.
- Reading the digital result by the microcomputer.

The software of the microcomputer can be arranged in such a way to minimize the idling time of the microcomputer during these periods. For example, the address of the following channel can be supplied to the multiplexer while a conversion is taking place. This will have no effect on the conversion, since the sample undergoing conversion is held by the S&H. Furthermore, useful tasks (such as processing and storing the digital results acquired by the last conversion) can be executed during the waiting intervals.

The advantage of the time-sharing approach is that it is cheaper than using several A/D converters. This is because the costly A/D converter is shared between the channels. The drawback of this approach is that only one channel can be digitized at any one time. This reduces the maximum attainable conversion rate. There are numerous applications where the analogue inputs do not need to be digitized at a very high rate. For such applications, the time-sharing approach will be very suitable. In many other applications, the analogue inputs need to be continuously read or simultaneous conversion of several channels is required. A decision is needed on whether it is possible to fulfil the requirements with the adoption of the time-sharing approach, or to use an A/D converter per channel.

There exist applications which create the need for a high sampling rate of only one of the channels. One method of satisfying the requirement of such applications is to employ a separate converter for that channel and to use the time-sharing approach for the rest of the channels. Alternatively, all the input signals can be supplied to a time-sharing converter, with that particular channel converted at a rate higher than the others. This is easy to arrange by software, as that channel can, for example, be converted after completing the conversion of each of the other channels.

7.7.2 Using one converter per channel

In some applications, the timing requirements impose the need for employing a single-channel A/D converter for each analogue input. The interface of each channel is therefore similar to that of a single A/D converter, the difference being the need for sufficient ports, and sufficient selection signals to achieve the interface. If these are not available, it may be possible to introduce them to the microcomputer as shown in Chapter 2. Alternatively, existing ports can be multiplexed to accommodate the required converters, as shown in Section 16.5.

When using a single converter per channel, you are recommended to choose an A/D converter capable of operating with minimum intervention from the microcomputer. This is because the latter may need to initiate several converters to digitize several analogue signals simultaneously and to be ready to read the digital results.

7.7.3 Selecting the optimum number of channels

When designing a multichannel converter, a decision is needed on the number of required channels. If the multichannel converter is designed for a specific application, the number of channels is determined by the application requirements. In some situations, however, it may prove more economical to use a larger number of channels in order to allow for future expansion.

If the multichannel requirements impose the need for using one converter per channel, the allowance for future expansion may take the form of leaving empty sockets on the interface board for the inclusion of additional converters should the need arise in future. On the other hand, in a time-shared multichannel A/D converter, the decision on the optimum number of lines depends on the type of application for which the converter is designed, and on the suggested number of channels. For example, the time-sharing approach normally employs four or eight channels, since it makes use of a 4-channel or an 8-channel multiplexer.

Consider an application where the number of the required channels is six. A 6-channel converter can be designed by the use of six analogue switches in a similar way to that of Example 6.8. Alternatively, the configuration of Fig. 7.9 can be used, where the additional two channels can either be ignored or prepared for future use. Consider a situation where these channels are to be ignored and the rest of the channels converted sequentially. One solution is to write the software so that it increments the channel-selection address from 0 to 5 and then resets to 0 and repeats the procedure. If, however, the software is not to be changed whenever the number of unused channels changes, an indication is required so that the microcomputer does not spend valuable execution time digitizing unused channels. The bypassing of unused channels can be accomplished by employing onboard links or switches. These tell the microcomputer whether or not to convert either of the additional two channels (the approach of using links or switches is explained in Section 6.6.2). This approach, however, reserves some of the digital-input lines of the microcomputer to achieve the testing.

In some applications, additional digital inputs are required to read external digital signals. Instead of including additional ports, these signals can be supplied to existing digital-input lines, replacing the onboard links. This implies that the unused analogue channels will be digitized although they are not needed. Rather than leaving them redundant, each channel can be employed to convert the logic level of one of the external digital signals. After conversion, the microcomputer examines the result and decides whether the converted value corresponds to a logic 1 or a log 0 state.

7.8 Examples of multichannel A/D conversion

It is explained in Section 7.5 that in a large number of applications analogue signals need to be amplified. The amplification needed for one channel of a multichannel converter may be different from that needed for other channels. One way of implementing the amplification is to place an amplifier at the input of each channel. Alternatively, a single amplifier with a digitally selectable gain (such as that described in Example 7.8) can be used when the time-sharing approach is adopted. It can be placed after the multiplexer to produce the required amplification under the control of the microcomputer, as shown in Fig. 7.10. In this figure lines G0, G1, G2 and G3 represent the control lines of the gain-selection switches (see Example 7.8). Let us now consider a different application.

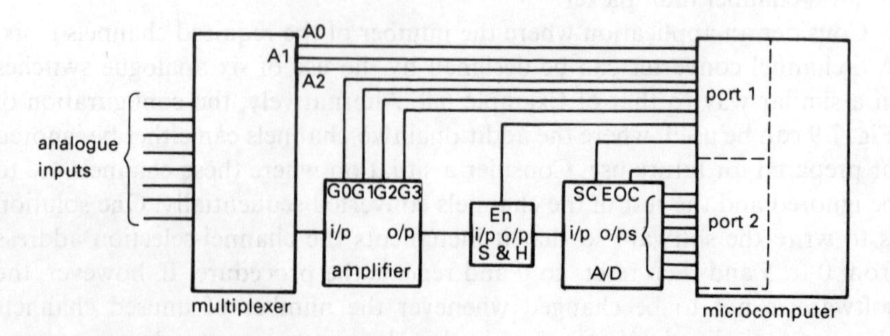

Fig. 7.10 AS multichannel A/D converter with a variable-gain amplifier.

Example 7.10: **Three analogue signals need to be read simultaneously after the occurrence of an event in an industrial application. What alterations are needed to allow the multichannel A/D converter of Fig. 7.9 to fulfil such a requirement?**

The configuration considered in Fig. 7.9 allows the selection and conversion of only one channel at any instant of time. Rather than using several A/D converters, a more economical approach can be adopted here. It is based on the use of three S&Hs at the input of the multichannel converter. Each S&H captures a single analogue signal and all the S&Hs need to be instructed to hold the signals at the same instant. After capturing the signals, the microcomputer can instruct the multichannel converter to digitize the channels sequentially.

This can be achieved by connecting all the enable lines of the S&Hs together to a single digital-output line of the microcomputer. Through this line, the microcomputer supplies the hold command to the three S&Hs. This, however, allows these three channels to operate only in the simultaneous-capture mode. More flexibility can be given to the microcomputer if the enable line of each S&H is operated from a unique output line of the microcomputer, as shown in Fig. 7.11. This obviously uses two additional output lines, but it gives the microcomputer the ability to sample the channels simultaneously or individually according to the application requirements.

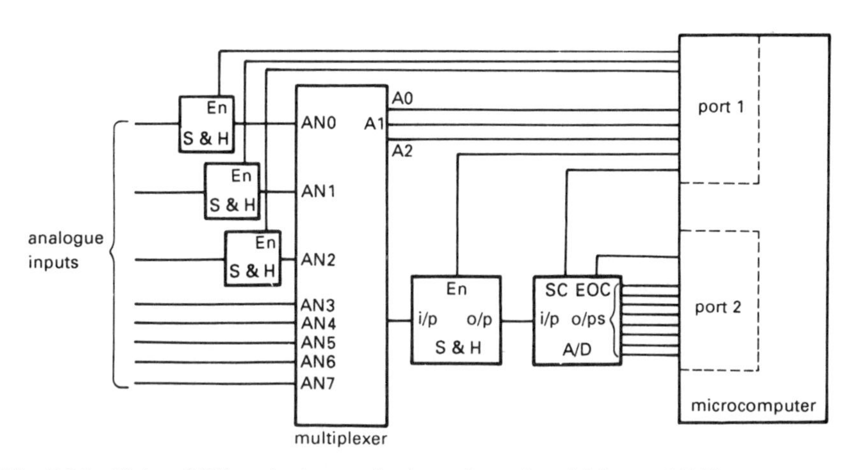

Fig. 7.11 Using S&Hs at the input of a time-shared multichannel A/D converter.

As can be seen from Fig. 7.11, the internal S&H of the multichannel converter is still in use. Although it is not needed for the digitization of channels 0 to 2, it is required for the correct conversion of the other five channels.

Example 7.11: **Consider an application where a microcomputer is to be used to monitor an analogue signal. After the occurrence of a certain event, three consecutive samples need to be obtained from that signal. These samples must be 10 µs apart. How can these samples be obtained assuming that an A/D converter with a 20-µs conversion time is employed?**

Assume that the microcomputer of Fig. 6.1(b) is utilized, and that the occurrence of the event is indicated by a change in the state of a digital signal. The signal can therefore be interfaced to an input line of the microcomputer. The latter monitors that line to detect the occurrence of the event. Following the detection, the microcomputer will not fulfil the requirement by sampling one channel at a time using the configuration of Fig. 7.9. Instead, the configuration of Fig. 7.11 can be employed. The input signal needs to be supplied to the three analogue-input channels, which include the additional S&Hs. The microcomputer issues three consecutive commands (through digital-output lines) to the three S&Hs to capture the values in sequence with a 10-µs interval in between. Once the three samples are captured, the microcomputer digitizes them sequentially.

This approach provides the microcomputer with the ability to obtain a limited number of samples at a high speed which is independent of the conversion time of the A/D converter.

Chapter 8

Using data-conversion ICs

8.1 Introduction

The increase in the use of data converters has led manufacturers to produce different types of A/D and D/A conversion ICs. Such converters differ in their resolution, conversion speed, price, etc. A designer can therefore choose the most suitable data converter for the particular need. Some converters are suitable for direct connection to the data bus of a microcomputer, while others need an interfacing device such as a 3-state input or output port.

Data-conversion methods and errors in practical data converters are presented in Chapter 5. Interfacing, operating and testing data converters are dealt with in Chapters 6 and 7, where many examples are given. This chapter considers commercial single-channel and multichannel conversion ICs. It describes their operation and interfacing to microcomputers. It presents application examples as well as making a reference to interfacing examples presented in Chapters 6 and 7. When using a data-conversion IC, it is advisable to keep digital and analogue grounds separate. They should be joined together at a single point. This, together with the use of power-supply decoupling capacitors and other recommendations given in Section 16.7, helps in reducing the effect of noise.

The last section of this chapter considers the effect of having analogue-input and analogue-output channels in a microcomputer-based system. It presents a few tests and proceeds to illustrate a technique for implementing low-cost analogue I/O channels. The technique is particularly suitable for systems having no requirement for a very high digitization rate.

8.2 Single-channel D/A conversion ICs

Chapter 6 deals with D/A converters and their interface to 8-bit and 16-bit microcomputers. This section gives examples of single-channel D/A conversion ICs. Multichannel conversion is the subject of Section 8.4.

8.2.1 8-bit D/A converters

Let us consider two 8-bit D/A converters. We start with a converter having no input latch and proceed to consider another possessing an on-chip 8-bit 3-state latch.

(1) A D/A converter with non-latching inputs

A widely used 8-bit D/A converter is the GEC Plessey Semiconductors™ ZN425E. It is a 16-pin IC with $\pm\frac{1}{2}$ LSB linearity error and a typical differential non-linearity error of $\pm\frac{1}{2}$ LSB. It has a typical settling time of 1 μs for a step change of one LSB. This increases to a typical duration of 1.5 μs when all the bits change simultaneously from 00000000 to 11111111, or vice versa.

The IC includes an on-chip 2.5-V reference voltage and an 8-bit binary counter. The use of the voltage reference and the counter are optional. The counter is included to allow employment of this IC as a ramp generator or as an A/D converter (Section 8.3.1). Figure 8.1 shows the connection needed to configure this IC as a D/A converter powered from a +5-V power supply. It converts an externally supplied 8-bit digital number to an equivalent analogue signal ranging from 0 to V_{ref}. Use is made of the internal voltage reference by joining pins 15 and 16 and employing a decoupling capacitor (C). The logic-select line (pin 2) is connected to the 0-V line to disable the counter.

An amplifier can be connected to the output of the converter to remove offset, provide buffering, permit calibration, and amplify the output signal to the required level. This is explained in Section 6.3.1, where a 741 IC is used. Figure 6.4(b) shows an example of such an amplifier where the +V

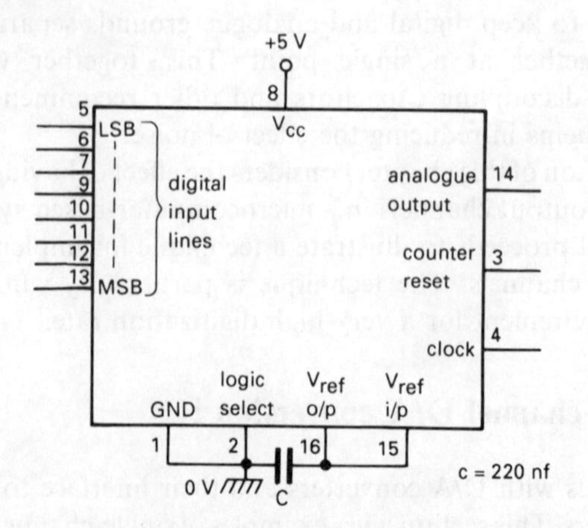

Fig. 8.1 The GEC Plessey Semiconductors ZN425E D/A converter.

and −V represent +5-V and −5-V power supplies. The amplification factor depends on the value of R2 and R3. Through R0 and R2, one can set the correct zero and full-scale values as described in Section 6.3.1. Since the input lines of the converter are non-latching, its interface to a microcomputer can be achieved through an output port. Figure 6.2 illustrates two types of interface, one through an 8-bit port and another through a 16-bit port.

(2) A D/A converter with latching inputs

The availability of a 3-state latch at the input of a D/A converter makes it suitable for direct connection to the data bus of a microcomputer. An example of such a converter is the GEC Plessey Semiconductors ZN428E shown in Fig. 8.2. It possesses an 8-bit 3-state input latch controlled through an active-low enable line (pin 4). The IC possesses 16 pins, provides a reference voltage of 2.5 V at pin 7, and has a typical settling time of 1.5 μs (when all input bits change from 0 to 1, or vice versa). It has a linearity error of $\pm\frac{1}{2}$ LSB. The configuration of Fig. 8.2 uses the internal reference by connecting pin 7 to the reference-input line (pin 6). The analogue-output voltage is supplied from pin 5, where it can be buffered and amplified as explained in Section 6.3.1. The digital and analogue grounds need to be joined together at a single point.

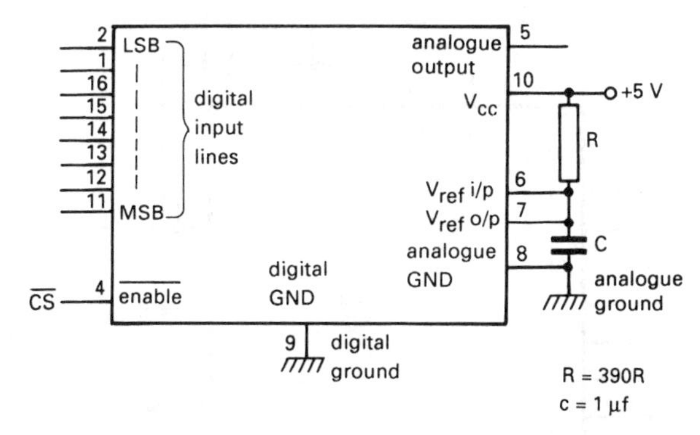

Fig. 8.2 The GEC Plessey Semiconductors ZN428E D/A converter.

When this converter is interfaced to a microcomputer, the enable line (pin 4) is utilized as the chip-select line. It allows a microcomputer to select the instant of transferring data to the converter. This line can be driven by a signal such as output CS2 of Fig. 2.3. The selection is performed when the enable line is set to a logic 0 state by the microcomputer. Supplying that line with a logic 1 signal results in latching the received digital value. The converter produces an equivalent analogue signal and continues to do so until another value is supplied or the power is switched off. Meanwhile, the microcomputer can use the data bus for other purposes.

8.2.2 12-bit D/A converters

The high-resolution requirements of a large number of applications can be satisfied by the use of low-cost 12-bit D/A converters. The Analog Devices™ DAC80 is one example. It is a monotonic converter which converts a complementary digital binary code to an analogue voltage. It has a maximum linearity error of $\pm\frac{1}{2}$ LSB and a maximum differential non-linearity of $\pm\frac{3}{4}$ LSB. It is packaged in a 24-pin IC which includes both an output amplifier and a 6.3-V internal voltage reference (with $\pm 1\%$ accuracy). The latter is supplied to pin 24 and is shown in Fig. 8.3 connected to the reference input line (pin 16). The device's data sheet recommends the use of a 1-µF capacitor to decouple the power supply which should be connected close to the converter. This is shown in Fig. 8.3, where the converter is operated from +15-V and −15-V power supplies and is configured for 0 to +5 V unipolar operation.

The range can be extended to cover 0 to +10 V if pin 20 is left open. Alternatively, a ±5-V bipolar operation can be achieved if pin 17 is connected

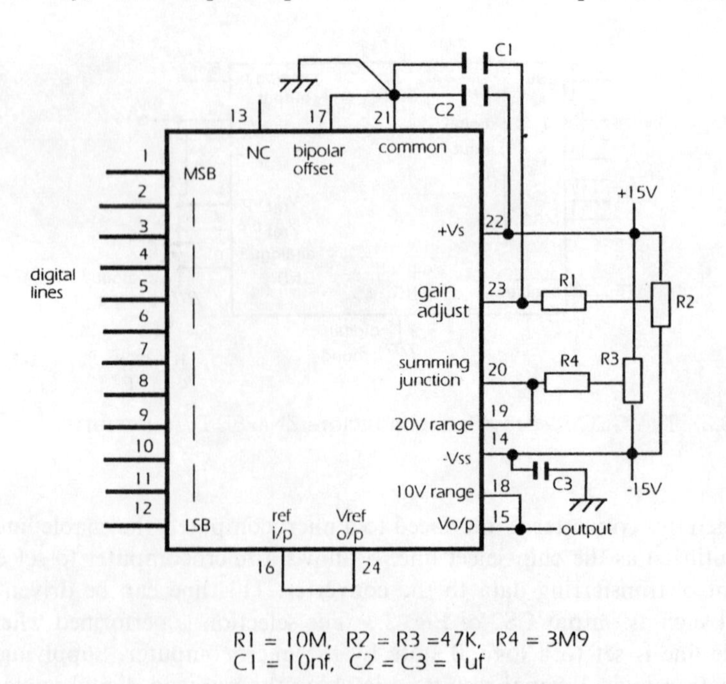

Fig. 8.3 The Analog Devices DAC80 D/A converter.

to pin 20 (instead of to pin 21) and pin 19 is left open. By connecting pin 19 to pin 15, the range of the bipolar operation will be extended to ± 10 V.

8.3 Single-channel A/D conversion ICs

8.3.1 8-bit A/D converters

Similar to Section 8.2.1, this section considers two converters: one possessing no input latches and the other including an on-chip 3-state 8-bit latch.

(1) An A/D with non-latching inputs

In Section 8.2.1 the ZN425E IC is used as an 8-bit D/A converter where the on-chip 8-bit binary counter is disabled. The configuration of Fig. 8.1 can be altered to operate the IC as a ramp A/D converter (this type of converter is described in Section 5.5.2). Configuring the ZN425E IC as an A/D converter can be accomplished by enabling the built-in binary counter. This is implemented by connecting the logic-select line (pin 2) of the IC to the +5-V supply line through a 1-kΩ resistor. This instructs the device to employ the eight outputs of the counter as the digital inputs to the D/A converter. When used for A/D conversion, the frequency supplied to the counter should not exceed 300 kHz. The resulting circuit is similar to that of Fig. 5.9 where, in this case, both the converter and the counter are within the same IC.

Examples of interfacing this type of converter to microcomputers are given in Section 7.3, where it is explained that the conversion rate depends on the frequency of the clock and is proportional to the amplitude of the input signal. This implies that the longest conversion time appears at full-scale. It corresponds to a count of 256. If, for example, a clock frequency of 128 kHz is used, then a count of 256 is obtained in 2 ms.

(2) An A/D with latching inputs

A popular 8-bit A/D converter IC is the GEC Plessey Semiconductors ZN427E. It is an 18-pin IC which performs A/D conversion by using the successive-approximation method (Section 5.5.2). It contains an on-chip voltage reference (2.56 V) whose use is optional, as the IC can also be operated from an external voltage reference in the range 1.5 to 3 V. The converter is provided with a 3-state output capability which makes it suitable for direct connection to the data bus of a microcomputer. Figure 8.4(a) shows the IC connected as a unipolar A/D converter. As it is a successive-approximation converter, its conversion time depends on the clock frequency, which can be derived from that of the microcomputer by employing a divide-by-N counter. The converter can achieve an A/D conversion in 10 μs when supplied

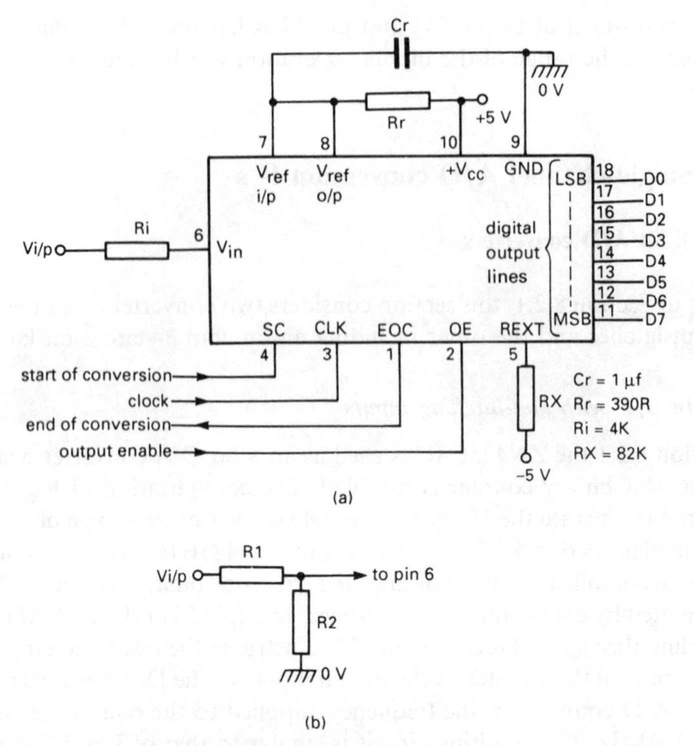

Fig. 8.4 The GEC Plessey Semiconductors ZN427E A/D converter: (a) unipolar connection; (b) including a voltage divider.

with a clock frequency of 900 kHz, and in 15 µs when the clock frequency is 600 kHz. It has no missing codes, a linearity error of $\pm\frac{1}{2}$ LSB, and a typical differential non-linearity of $\pm\frac{1}{2}$ LSB.

Each digitization process is initiated by an active-low SC pulse and progresses in the same way as that described in Section 5.5.2 and Section 7.3. The end of conversion is indicated by an active-high EOC signal which can be used to inform the microcomputer of the availability of the digital result (this line is at a logic 0 state during conversion). When the converter is interfaced to the data bus of a microcomputer, the microcomputer can read the conversion result by supplying the output-enable line (pin 2) with a logic 1 signal. This instructs the converter to transfer its digital result in the data bus.

The connection of Fig. 8.4(a) accepts an input voltage signal in the range 0 to V_{ref}. A series resistance R_i is utilized to provide matching with the internal resistance (4 kΩ). To permit operation with a wider input range, R_i can be replaced by two resistors connected in the form of a voltage divider as shown in Fig. 8.4(b). When utilizing these resistors, the converter will continue to receive a voltage in the range 0 to V_{ref}, but the input range is set according to the value of R1 and R2. The value of these resistors should also provide matching with the input resistance of the converter, i.e. their parallel combination should be close to 4 kΩ. For example, to increase the voltage range from +2.5 V to 5 V, each of R1 and R2 should have a value of 8 kΩ. Rather than using two fixed resistors, a potentiometer can replace R1. It can be adjusted to calibrate the converter and provide the required voltage range. The offset can be removed by using a 2-MΩ potentiometer between V_{ref} and V_{in}. Clearly, when employing a potentiometer, it is recommended to use it in series with a fixed resistor to prevent short-circuiting.

To connect the IC for a bipolar operation, a simple change is needed to the configuration of Fig. 8.4(a). The input needs to be offset by half of full-scale by placing an 8-kΩ resistor between the reference voltage and the input (i.e. pins 6 and 7) and changing the value of R_i to 8 kΩ (i.e. the parallel combination of R_i and the additional resistor should be equal to 4 kΩ to match the input resistance of the converter). If the input range is to be increased by the use of the voltage divider of Fig. 8.4(b), then the parallel combination of R1 and R2 should be equal to 8 kΩ.

In the case of a unipolar connection, changing the input value from 0 V to full-scale alters the digital results from 00H to FFH, with half of full-scale represented by 80H (assuming an ideal converter). On the other hand, a bipolar connection represents 0 V by 80H, positive full-scale by FFH, and negative full-scale by 00H.

8.3.2 10-bit A/D converters

After considering the interface of 10-bit A/D converters to 8-bit and 16-bit microcomputers in Section 7.6, this section is devoted to presenting commercial 10-bit A/D converters. Let us consider two 10-bit A/D converter ICs containing an on-chip 3-state output latch. They are the National Semiconductor ADC1005 and ADC1025. The former is a 20-pin IC designed for 8-bit interfacing. It possesses eight output lines through which digital results are supplied to an external device, such as a microcomputer. The ADC1025 IC is very similar to the ADC1005, except that it is a 24-pin IC which supplies its digital result through 10 output lines. It is therefore best suited for 16-bit interfacing. Both ICs operate from a single +5-V supply and accept an analogue-input voltage in the range 0 to 5 V.

Figure 8.5 shows the ADC1005 operating from a +5-V supply. It performs an A/D conversion in 50 μs when supplied with a clock frequency of 1.8 MHz. In Fig. 8.5, the clock input line is left open-circuit to allow supplying a clock signal from an external source, e.g. derived from the microcomputer's clock. The IC also possesses a clock generator which can be used by connecting a resistor (R) between pins 4 and 19, and a capacitor (C) from pin 4 to ground. The frequency of the generated clock is given by: $f_c = 1/(1.1RC)$. Rather than

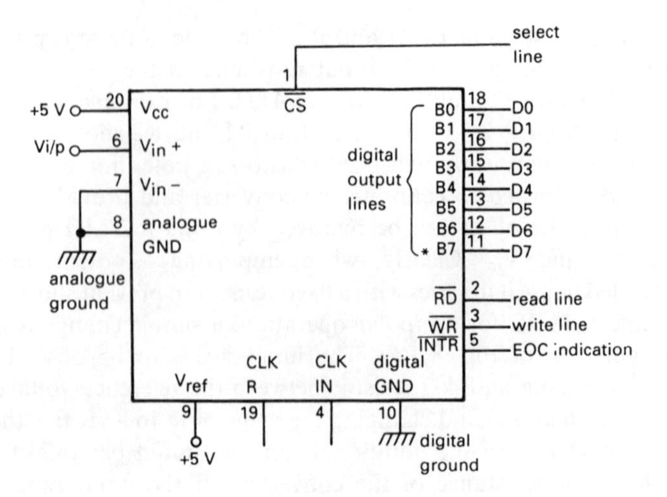

Fig. 8.5 The National Semiconductor ADC1005 10-bit A/D converter.

dedicating a line to accept an SC command, either IC begins an A/D conversion when a microcomputer supplies lines CS and WR with an active-low pulse. At the end of the pulse, the conversion begins. During conversion, the INTR line is at a logic 1 state. The converter indicates the completion of conversion by changing the state of the INTR line from high to low. When the microcomputer detects the change, it can execute a read instruction which activates the CS and the RD lines (both are active-low). This instructs the converter's 3-state output latches to deliver the digital result to the data bus.

Transferring the result from the ADC1025 converter to a microcomputer through a 16-bit data bus can be achieved in one step, whereas the ADC1005 delivers the result in two steps. The first supplies data lines D0 to D7 (through lines B0 to B7 of Fig. 8.5) with bits 2 to 9 of the result respectively (bit 9 is the MSB). The second step delivers zeros to lines D0 to D5 and supplies bits 0 and 1 to lines D6 and D7 respectively (again through lines B0 to B7 of Fig. 8.5).

8.4 Multichannel conversion ICs

It is explained in Chapters 6 and 7 that a multichannel data converter can be designed either by employing several individual data converters, or by the time-sharing of a single converter. This section considers multichannel conversion ICs. It first considers an IC containing four D/A converters generating four analogue-voltage signals. The second part of this section considers two multichannel A/D conversion ICs based on the time-sharing of an 8-bit A/D converter.

8.4.1 Multichannel D/A converters

Let us consider an example of a commercial multichannel D/A converter. It is the Analog Devices AD7226 quad 8-bit D/A converter. It is a 20-pin IC which can be directly connected to the data bus of a microcomputer. It includes four level-triggered input latches as well as an on-chip control logic to demultiplex selection signals delivered by a microcomputer. This feature allows the latter to select any of the four matched D/A converters by supplying a binary address to two selection lines (lines A0 and A1 in Fig. 8.6). The converter provides a maximum differential non-linearity of ± 1 LSB and can be operated from a single or dual supply. The former is adopted in Fig. 8.6, where the supply voltage is $+5$ V and the reference voltage delivered to pin 4 is 1 V.

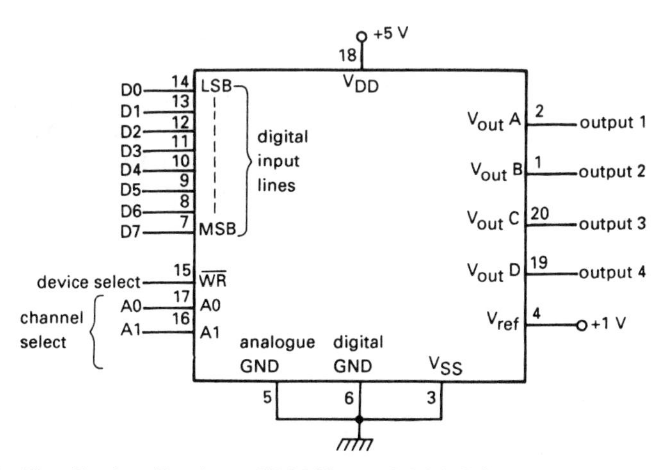

Fig. 8.6 The Analog Devices AD7226 quad 8-bit D/A converter.

Line WR permits a microcomputer to control the instant of transferring an 8-bit value to the converter and latching it. The transfer is accomplished through the data bus. When WR is set to a logic 0 state, a channel identified by lines A0 and A1 is selected and its input latch becomes transparent and accepts data. Changing the state of WR to a logic 1 latches the data within the IC and an equivalent analogue voltage is generated. The IC includes on-chip output amplifiers through which four analogue-voltage signals are produced.

8.4.2 Multichannel A/D converters

Data-converter manufacturers have produced many multichannel A/D conversion ICs; let us consider the National Semiconductor ADC0808 and ADC0816. These are 8-channel and 16-channel 8-bit A/D conversion ICs, respectively, which perform conversions with no missing codes. Each of these converters contains a 3-state output latch and an on-chip multiplexer (the multiplexer possesses eight channels in the ADC0808, and 16 channels in the ADC0816). The operation of this type of converter is the same as that explained in Section 7.7. The multiplexer is employed to permit a controlling device, such as a microcomputer, to direct any of the analogue-input signals to the input of an 8-bit successive-approximation A/D converter. Neither IC includes an internal S&H device. The conversion commences when a microcomputer supplies an active-high SC pulse. The device performs an A/D conversion in 100 μs with a clock frequency of 640 kHz. During digitization, the EOC signal is at a logic state 0. It changes to a logic 1 when the conversion is complete and can therefore be utilized to inform the microcomputer of the availability of the result. This can be achieved by using the EOC line either as an input to the microcomputer or as an interrupt-request line. The microcomputer can read the result after sending an active-high output-enable (OE) signal to the converter.

The ADC0808 converter is a 28-pin IC. It is shown in Fig. 8.7, where the address selection is implemented via three input lines: ADD A, ADD B, and ADD C (the most significant bit is delivered to ADD C). The ADC0816 converter, on the other hand, provides four lines to perform the 16-channel selection. The connection of the ADC0816 is very similar to that of the

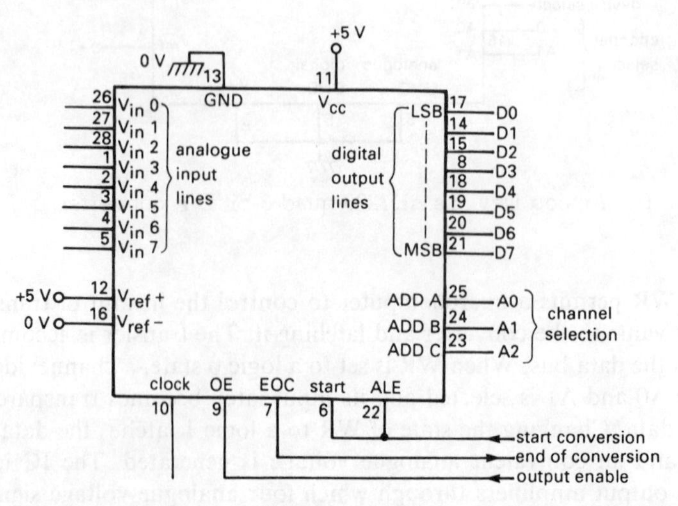

Fig. 8.7 The National Semiconductor ADC0808 8-channel 8-bit A/D converter.

ADC0808. The former, however, is a 40-pin IC, as it possesses additional lines to accommodate the 16-channel conversion and to provide other features. The output of its multiplexer is supplied to pin 15, while the input of its A/D conversion block is supplied from pin 18. This provides the designer with the ability to include a S&H and/or an amplifier in between these two pins. If this is the case, then the input of the add-on circuit is supplied from pin 15 and its output is given to pin 18. This provides a configuration similar to that of Fig. 7.9. If the add-ons are not required, then pins 15 and 18 need to be joined together in order to supply the output of the multiplexer to the input of the A/D conversion block. With the ADC0808, however, such add-ons can be placed at the input of individual channels requiring high accuracy, i.e. before the multiplexer. Further information on applying the ADC0808 is given in Example 8.3.

8.5 Examples of interfacing to microcomputers

Many examples of interfacing data converters to microcomputers are presented in Chapters 6 and 7. Further examples are given here.

Example 8.1: **A ZN425E D/A converter is interfaced to a microcomputer through a digital-output port. The microcomputer is programmed to continuously supply digital values to the inputs of the D/A converter. A system operator must be provided with the ability to manually control the supply of signals to the converter. How can this requirement be met, assuming that the microcomputer neither possesses a keyboard, nor provides input lines for accepting external commands?**

Since there are no means of supplying commands to the microcomputer, hardware can be utilized to permit an operator to control the supply of information to the D/A converter. The circuit given in Fig. 8.8 employs an 8-bit latch for this purpose. It is the 74LS373 IC, which is also used in Chapters 1 and 2 (see Figs. 1.2 and 2.15). The latch is introduced in between the output port and the D/A converter. The output-control line (OC) of that IC is connected to the 0 V, thereby enabling the output permanently. The operation of the IC can thus be controlled through the enable line (En). When that line is supplied with a logic 1, the output lines of the IC will be in the same state as their corresponding input lines. Changing the state of the En line to a logic 0 latches the output. Consequently, by connecting line En to a single-pole single-throw switch an operator can enable and disable the update of the input lines of the D/A converter. The microcomputer will continue to update the digital-output lines, but the information will only be

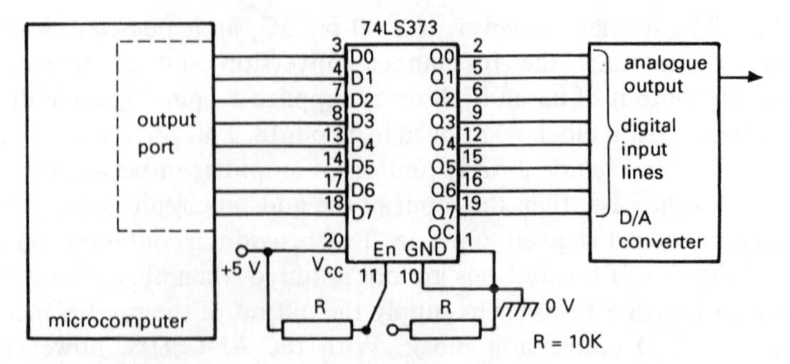

Fig. 8.8 Manual-control operation.

transferred to the input lines of the D/A converter when the output of the latch is enabled.

Example 8.2: **Consider an application where a microcomputer is required to monitor the change in amplitude of an analogue signal only when it is greater than 3 V. How can this be achieved by the use of a microcomputer-controlled 8-bit A/D converter?**

Let us employ the ZN427E converter described in Section 8.3.1. The microcomputer supplies the SC signal and reads the digitized value after the conversion is complete. Since the conversion time of this converter is short, the microcomputer can be programmed to wait until the end of conversion.

A decision is needed on how to arrange the procedure. Let us consider the following solutions.

Solution 1

The maximum rate at which the microcomputer can command the A/D converter to read an analogue input depends on whether the former needs to execute other tasks in addition to monitoring the signal, or not. The conversion can therefore be arranged to occur continuously or at regular intervals. After each conversion, the microcomputer needs to examine the received digital value to determine whether its magnitude is greater than 3 V or not. The result is discarded if it is not more than 3 V. This implies that with this approach, the microcomputer spends time converting signals which it may not need to use. Let us therefore consider a different solution.

Solution 2

An alternative solution is to use hardware to reduce the time spent on the monitoring of the analogue signal. A comparator, such as that of Fig. 4.2, can be utilized to remove the need for unnecessary conversions. The negative input terminal of the comparator is supplied from the analogue-input signal which is also delivered to the input of the A/D converter. The comparator compares the amplitude of the input signal with the voltage drop across R3 (which is 3 V in this case). If the analogue-input signal is less than 3 V, then the output of the comparator is a logic 1. This output changes to a logic 0 only when the analogue signal exceeds 3 V. This signal can, for example, be inverted by a logic gate and used to activate a rising-edge-triggered interrupt of the microcomputer. The signal requests from the microcomputer the commencement of a digitization process. The microcomputer responds by sending a SC signal and reads the result as soon as it is available. The drawback of this solution is that the magnitude of the input signal may change as a result of the time spent between responding to the interrupt-request signal and the SC instant.

This problem can be solved if the output of the comparator is also applied to command a S&H device to capture the signal. In this way, the signal is captured as soon as its amplitude exceeds the 3-V level. The microcomputer can then read the value through the A/D converter.

Solution 2 saves computation time, since the A/D conversion only takes place when needed. However, it requires additional hardware.

Example 8.3: **Design a circuit to permit a microcomputer to monitor each of eight unipolar analogue-input signals. The resolution requirement is 8 bits and each input signal is in the range 0 to +5 V.**
The requirements can be satisfied by using the ADC0808 converter presented in Section 8.4.2. Besides the three channel-selection lines, two output command lines are needed from the microcomputer. These are the SC and OE command lines. They can be introduced by using a circuit similar to that of Fig. 2.3. The converter, however, requires these signals to be active-high. Therefore, if the OR gates of Fig. 2.3 are replaced by NOR gates, then output CS2 can be used to provide the SC command, and CS1 supplies the OE command.

Note that both outputs of Fig. 2.3 correspond to the same address defined by line CS. This implies that conversion is initiated when the microcomputer writes to that address, and the output is enabled when the microcomputer reads from that address. If the device is to be memory-mapped, then the RD and the WR lines will be replaced by MEMR and MEMW lines respectively (see Chapter 2). Similarly, if input/output mapping is used then these lines will be replaced by IOR and IOW lines respectively. The microcomputer can select an input channel by supplying its address to the converter through three selection lines. This can be accomplished by connecting the three lower lines of the address bus (A0, A1 and A2) to the channel-addressing lines of the IC. The selection address of the SC and the OE lines is defined by the decoding circuit which produces the CS signal of Fig. 2.3. Section 2.2.3 gives examples of generating such a signal and takes into consideration the addressing area reserved for such addressing. Keep in mind that address lines A0 to A2 are utilized for selecting the conversion channel. This implies that address decoding can be implemented by decoding some, or all, of lines A3 to A15. This depends on the available addressing area in the microcomputer. They can, for example, be supplied to a circuit similar to that of Fig. 2.7(b) to generate signal X1 of that figure. This signal can be used as the addressing signal (CS) in the circuit of Fig. 2.3. This results in locating the converter at the addressing area 9FF8H to 9FFFH. This implies that channel 0 of the converter is located at addresses 9FF8H, channel 1 at 9FF9H ..., channel 7 at 8FFFH, i.e. one location is reserved for each channel.

8.6 Microcomputers with analogue I/O

The operation of numerous microcomputer-controlled systems is based on employing A/D and D/A converters within the same system. For example, a bipolar analogue signal can be supplied from an external source to a microcomputer-controlled robotic mechanism requesting motion. The sign and magnitude of this signal specify the direction of motion and the distance of travel respectively. An A/D converter can therefore be employed to read this signal. The same system can have one or more D/A converters through which the microcomputer commands the mechanism to perform the requested motion. Assume that the demand signal is supplied by an operator through the use of a potentiometer. This implies that the rate of change in the setting of the input potentiometer is not fast and an instantaneous response is not necessary. It is therfore acceptable to program the microcomputer to examine that signal at a low rate, say once every 50 ms.

This is not usually the case when the A/D converter is utilized to monitor a fast signal in a real-time microcomputer-controlled application. For instance, an A/D converter can be employed to supply a microcomputer with consecutive digital samples corresponding to a rapid change in the magnitude of a signal. Reading the digitized value may be one of several tasks handled by the microcomputer. After receiving each value, the microcomputer may need to process it and supply the result to an external device via a D/A converter. The use of data converters introduces delays and quantization errors. In a microcomputer-controlled system, the delay consists of the time needed for converting the signals and the time required for their processing. These delays can set an upper limit to the rate of collecting and producing data.

8.6.1 Examining the effect of using data converters

Let us carry out three tests to observe the effects of the delays and the quantization errors in a microcomputer using input and output channels.

(1) Analogue comparison without a microcomputer

We begin with a simple test using the configuration shown in Fig. 8.9(a), where an analogue signal is digitized and then reconverted to an analogue form immediately. The microcomputer is only used to supply the timing signals to the A/D converter, but does not read the resulting digital value. It monitors the EOC line and commands a start of conversion immediately after the completion of another. The EOC signal is used to enable the output of the A/D converter which will be latched until being updated after the completion of the next conversion. Both the analogue output produced by the D/A converter and the analogue-input signal can be observed and compared by the use of an oscilloscope.

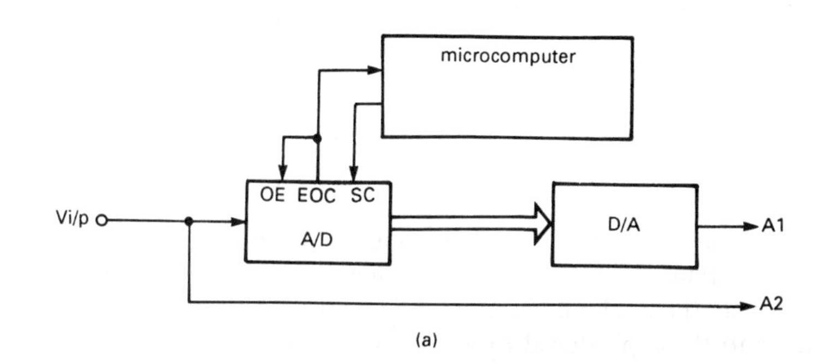

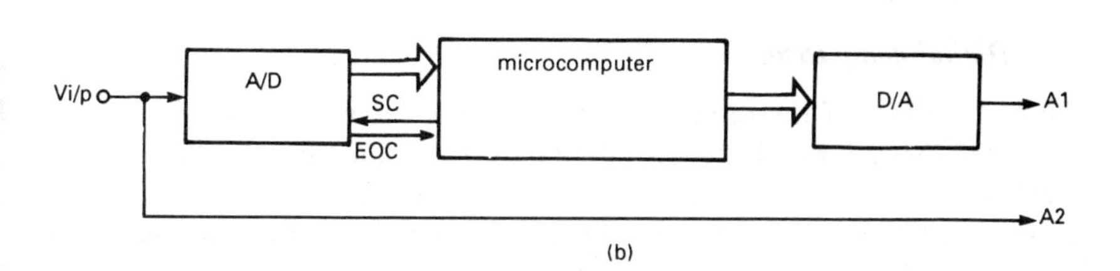

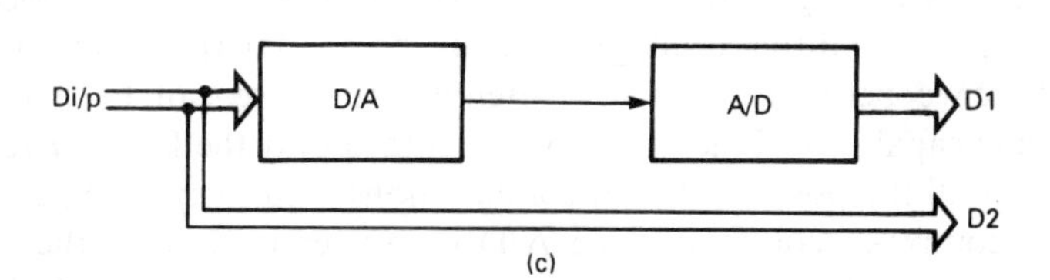

Fig. 8.9 Reconstruction of signals: (a) analogue signals; (b) effect of a microcomputer; (c) digital signals.

When observing the output signal on an oscilloscope, one will recognize the effect of quantization as it appears in the form of steps representing the quantization levels of the digital value. The number of steps and the amplitude of each step depends on the resolution of both the A/D and D/A converters. Examining the two signals also reveals the phase shift between them. This is a function of the conversion time and is best observed when the analogue input is a time-varying signal, e.g. a sine wave.

(2) Analogue comparison when a microcomputer is used

We now extend the test to determine the effect of the processing time of the microcomputer on the output signal. The test uses the configuration of Fig. 8.9(b), where the microcomputer performs the following steps repeatedly:

- Executes a short software-timing routine (a delay)
- Commands an A/D conversion
- Reads the digital value produced by the A/D converter
- Supplies the digital value to a D/A converter

The software-timing routine is introduced to represent the time taken by a microcomputer to process the input signal and attend to other tasks. When comparing the analogue-input and analogue-output signals, they should be similar to those produced by the configuration of Fig. 8.9(a) except for an increase in phase shift. The increase is proportional to the delay introduced by executing the timing routine in the microcomputer.

(3) Digital comparison

The comparison of the input and output signals is not necessarily carried out in analogue form. Instead, a digital comparison can be performed by using the configuration shown in Fig. 8.9(c). With this configuration, a digital value is supplied to the input of a D/A converter either from switches (Fig. 6.3) or from a microcomputer (Examples 6.2 and 6.3). The resulting analogue signal is supplied to the input of an A/D converter, which in turn produces a digital value. Since the value of the digital input is known, then one only needs to know the value of the output. Consequently, the digital output can be displayed by using LEDs (as in Fig. 7.5). Alternatively, the microcomputer can be used to read the resulting value and to display the input, output and the difference between them in a digital form. Use can be made of the processing capability of the microcomputer to supply the D/A converter with a series of all the possible digital combinations. After supplying each value, the mircomputer commands the A/D converter to digitize the produced analogue signal. The microcomputer reads the digital result and places it in a table in memory. In this way, the microcomputer generates a list of the input and output values, highlighting areas of high deviation.

8.6.2 Producing the optimum analogue I/O

Examples are given in the previous chapters of producing and operating the most suitable data converter for a given application. It is explained that such tasks are dependent on many factors, such as the type of signal, cost, execution time, development time, etc. The examples ilustrate how to design the optimum input and output analogue interface by using individual devices, one for A/D conversion and another for D/A conversion. If required, the operation of such a system may be arranged to update the inputs of the D/A converter while waiting for a digitization process to complete, thus saving execution time.

This approach, however, is not necessarily the optimum design for a system requiring the use of analogue input and output channels. The requirements may be satisfied by adopting an alternative approach. It is based on the fact that an A/D converter can be designed by using a D/A converter and associated circuit. Thus, an analogue I/O device can be designed by the time-sharing of a single D/A converter. It is used as an A/D converter for part of the time and as a D/A converter for the rest of the time. Obviously, the maximum sampling rate is less than when using two separate converters, but it is of a lower cost and is suitable for applications having no requirement for high-rate sampling. An example of such a converter is shown in Fig. 8.10, where the A/D conversion can be based on the ramp, tracking, or successive-approximation method. The latter is normally more attractive although it needs an S&H device at the input in order to achieve high accuracy (as explained in Section 7.3). The analogue output is supplied to the outside world through the use of an analogue switch which is closed by the microcomputer when it aims to supply a value to the analogue-output channel. The capacitor is used to store the produced analogue value and will therefore need refreshing as described in Section 5.6. It is therefore best if the D/A converter supplies the analogue-output channel all the time except when an A/D conversion is needed.

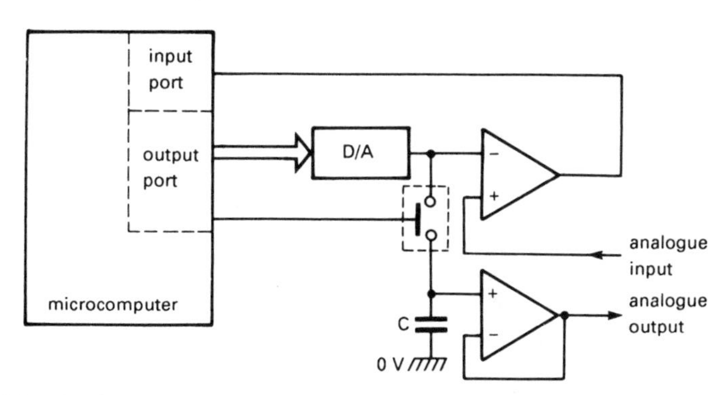

Fig. 8.10 A microcomputer-controlled low-cost analogue I/O circuit.

Each of the above-discussed solutions is therefore best suited for certain types of application. High resolution and fast conversion rate do not always form the optimum solution. Instead, the best solution to a data-conversion problem depends on the application requirements and is often based on obtaining the optimum performance/cost ratio.

Chapter 9

Counters and timers

9.1 Introduction

Counters and timers are incorporated in a variety of applications, such as digital pulse timing, frequency changing and duty-cycle control. They are also widely used in counting the number of occurrences of events, measuring the duration between events, generating waveforms, introducing time delays and controlling the real-time activities of microcomputers through the generation of interrupt signals at regular intervals. This chapter considers the use of such devices and methods of their interfacing to microcomputers.

9.2 Counters

Generally, the two main input lines of a counter are the 'reset' and the 'clock'. When the former is activated, the count produced by the counter will become 0. A counter will not be ready to count until the reset signal is removed. The clock input line, on the other hand, provides the timebase for the operation of the counter. Depending on the type of counter, toggling the clock input line causes the counter to increment or decrement its count by 1 bit.

By controlling the flow of pulses to the clock input line of a counter, an external device (e.g. a microcomputer) can start and stop the operation of the counter without clearing the count. Most counters are designed to deliver all the bits of their count through digital-output lines, but there are exceptions, as will be shown. By using a latch, an external device can freeze the count delivered by a counter at any instant of time without affecting the counting process. This is of interest in applications where a microcomputer needs to find out the number of pulses counted between two events. In such applications, the counter is likely to be operating continuously, without an external device disabling its operation or resetting it regularly. When operating in such a mode, the counter is referred to as 'free-running'.

9.2.1 Classification of counters

Counters can be classified according to the direction of changing their count as a result of receiving a clock pulse.

- An *up-counter* is a counter that increments its count with every clock pulse. When such a counter is incremented beyond its maximum value, it will overflow to set the count to 0.
- A *down-counter* decrements its count by 1 with every clock pulse. Allowing the count to be decremented after it reaches 0 will lead to setting the count to its maximum value and then continuing to count downwards.
- An *up/down counter* permits an external source to set the direction of the count. This is achieved through a dedicated input line whose digital state determines the direction of the count (such a counter is used in Example 12.3).

An alternative way of classifying counters is according to the relationship between their count and the clock signal:

- A *synchronous counter* changes its count with the clock signal.
- An *asynchronous counter* can alter its count according to its control lines and independently of the state of the clock.

Some counters are *presettable*. This implies that an external device can load a count to the counter, thereby controlling its counting range. Depending on the type of counter, the loading of the count can either be performed synchronously or asynchronously with the clock.

9.2.2 Code of counting

Counting is normally carried out in a binary fashion. The code is altered as shown in the binary column of Table 9.1, where 16 increments are performed in four stages. An example of such a counter is either of the two devices in Fig. 7.1. Since most numbers are presented to users in the decimal form, the requirements of many applications are best satisfied by employing counters to count in decimal instead of binary. The relationship between binary and decimal is given in Table 9.1, where a further column illustrates the presentation of decimal numbers in binary form. This type of code is referred to as BCD (Binary Coded Decimal). The table shows that four stages of BCD are sufficient to represent numbers 0 to 9. A fifth stage is therefore needed to display the 16 numbers of Table 9.1. With the use of gates, a 4-stage binary counter can be modified to reset when the count progresses beyond 9. This is not necessary, since the demand for decimal counters led manufacturers to produce several types of BCD counters. Figure 7.4 shows how two such counters are cascaded to form a counter which will count

Table 9.1 *Code conversion*

Decimal	*Binary*	*BCD*	*Hexadecimal*
0	0000	0000 0000	0
1	0001	0000 0001	1
2	0010	0000 0010	2
3	0011	0000 0011	3
4	0100	0000 0100	4
5	0101	0000 0101	5
6	0110	0000 0110	6
7	0111	0000 0111	7
8	1000	0000 1000	8
9	1001	0000 1001	9
10	1010	0001 0000	A
11	1011	0001 0001	B
12	1100	0001 0010	C
13	1101	0001 0011	D
14	1110	0001 0100	E
15	1111	0001 0101	F

from 0 to 99 before resetting to 0. A different code is presented in the last column of Table 9.1. It is referred to as the hexadecimal code. It is easier to use and read than a binary count, since it uses a single digit to represent up to 16 numbers. As a comparison, the decimal number 255 can be presented as 11111111 in binary, as 001001010101 in BCD and as FF in hexadecimal.

9.2.3 Resolution of counters

The resolution of a counter defines the number of bits it can count before resetting to an initial count. Generally, the number of output lines available in a counter depends on its resolution, although there are exceptions. Cascading counters will increase the resolution. For example, a 4-stage binary counter resets at 16, while an 8-stage binary counter resets at 256. Two 4-stage counters are shown cascaded in Fig. 7.1 to form an 8-stage counter. If the microcomputer needs to read the resulting count, then the output of the counter should be interfaced to the digital-input lines of the microcomputer. The number of these lines is defined by the resolution of the counter.

For a given resolution, the frequency of the clock signal determines how long a counter can count before resetting to 0. In some applications, e.g. measurement of time between events, a designer has to decide what clock frequency to choose. A low clock frequency allows the counter to count long intervals without overflowing. A higher frequency, on the other hand, permits achieving more accurate measurements, but will reduce the counting interval unless the resolution of the counter is increased. The latter may lead to an increase in the number of components and the number of lines needed to interface the counter to the microcomputer.

9.2.4 The counting process

As explained above, the arrival of a pulse at the clock input line of a counter leads to the alteration of the count by 1. The counting process of a binary counter, such as the one presented in Fig. 7.1, can be explained with the aid of Fig. 9.1. Only two output lines (Q0 and Q1) are shown. The change of the state of line Q0 coincides with a falling edge of the clock signal, while the change of the state of line Q1 occurs when the state of line Q0 changes from high to low. The change of the states of a 4-stage counter can be seen by examining the binary code of Table 9.1, where the four bits represent outputs Q3, Q2, Q1 and Q0 of a 4-stage counter. It shows that line Q3 is at a logic 0 for the first eight counts and then changes to a logic 1 for the next eight counts. In other words, it only changes once within the table. Line Q2, on the other hand, changes twice, line Q1 changes four times and line Q0 changes eight times. This shows that the frequency of each output line is halved as we move from one column to the next from right to left.

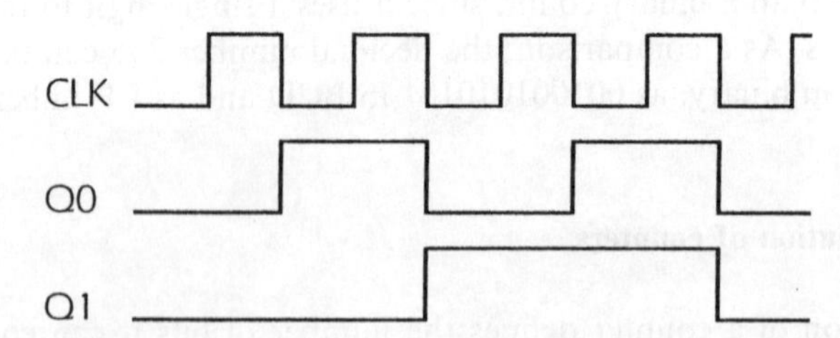

Fig. 9.1 Signals in a binary counter.

A binary counter can be formed by the use of D-type latches. Figure 9.2 illustrates the formation of a 2-stage binary counter by two such latches available in a single 74LS74 IC. The inverted output of each stage is fed back to the D input line. The state of that output will be effective with the arrival of each clock pulse, causing the output of the latch to change state. The relationship between the clock input line and the two outputs is the same as that given in Fig. 9.1.

Example 9.1: **Modify the counter presented in Fig. 9.2 to provide a microcomputer with the ability of resetting the count as well as enabling and disabling the input and output signal.**

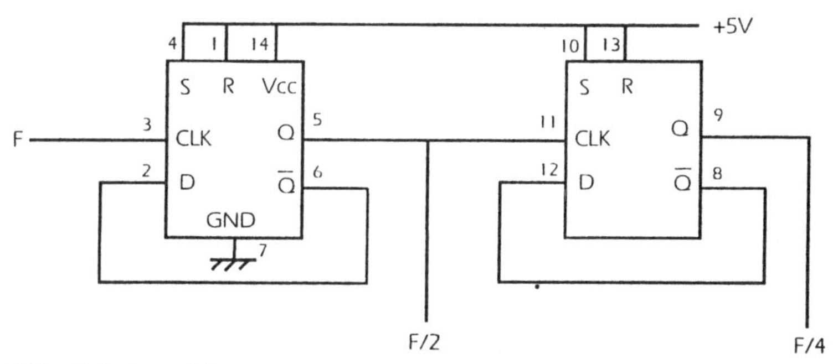

Fig. 9.2 A 2-stage binary counter.

In Fig. 9.2, the reset line of each of the two latches is connected to the +5-V line. If that line is connected to a digital-output line of the microcomputer, the latter can reset the latches by setting the state of that line to logic low. Otherwise, it will supply a logic-high signal to permit the counter to operate.

The ability to enable and disable the input and output signals can be introduced by the use of three AND gates, as shown in Fig. 9.3. The simple circuit receives clock pulses through line X and includes two control lines (IE and OE) to give the microcomputer the ability to enable or disable each of the input and output lines independently. Each control line can be driven from a digital-output line of the microcomputer, and the corresponding signal is enabled only when the microcomputer supplies a logic-high signal.

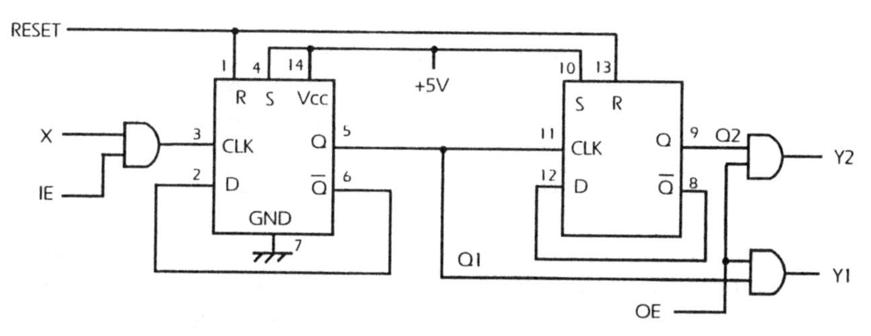

Fig. 9.3 Microcomputer-controlled counter.

Example 9.2: **Draw the count-to-time relationship of an 8-bit up/down counter when used as a free-running up-counter, showing the effect of reset and the frequency of the clock. Starting with the counter as an up-counter, show the effect of disabling the clock, and then illustrate the effect of reversing the direction of the count.**

A free-running 8-bit up-counter will count from 00000000 (00 hex.) to 11111111 (FF hex.) and then overflows to 00000000 as it repeats the count. This is shown in Fig. 9.4(a), where the count changes from 0 at instant t1 and overflows at t2. An external-reset signal is applied at instant t3 to clear the count. The counter then repeats its count between instants t3 and t4. The effect of changing the frequency of the clock can be seen by assuming that the frequency is increased at instant t4 and a reset is applied at t6. The

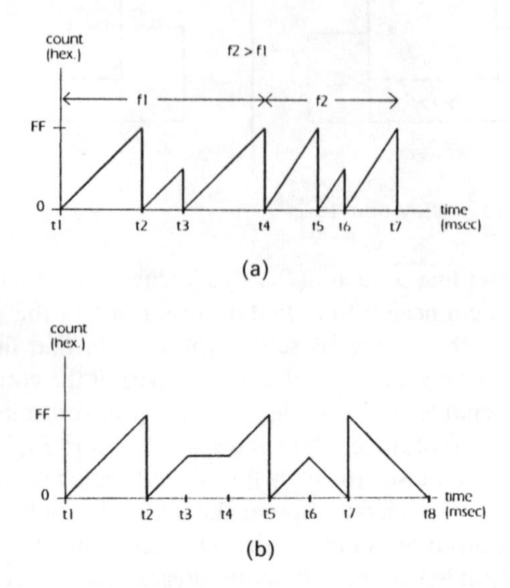

Fig. 9.4 The counting process: (a) effect of changing frequency and applying reset; (b) effect of disabling clock and reversing direction.

count will repeat in the same way but at a faster rate.

The second requirement is to show the effect of disabling the clock of the 8-bit up-counter. This is given in Fig. 9.4(b), where the counter is assumed to be free-running between instants t1 and t3 where the clock pulses are disabled. The counting process will stop and the output of the counter stay constant until the clock is re-enabled at instant t4. The counter will continue to count in the free-running mode until instant t6 when the counter is commanded to change the direction of the count. The count will start to decrease and the counter will continue to operate in the free-running mode.

Note that if a D/A converter is connected to the output of the counter, then the relationship presented in Fig. 9.4 will be given as an analogue voltage and can be observed on an oscilloscope.

9.2.5 Counters in practical circuits

Counters are used in a vast number of interfacing circuits. Commercially available counters, however, do not always fulfil the counting requirements of electronic circuits. They may not provide the required maximum count or the exact number of stages. They can be cascaded, or reset at specified intervals, in order to meet the requirements.

Table 9.1 represents the count of a 4-stage binary counter. It can also be used to observe the count of a lower-resolution counter. For example, the three least significant bits represent the count of a 3-stage counter. By ignoring the most significant bit, one can see that the count progresses from 000 to 111 and then repeats. This implies that a higher-resolution counter can perform the function of another of a lower resolution, as shown in the following example.

Example 9.3: **As a part of a general-purpose add-on board to a microcomputer, a counter needs to be included. Choose a counter so that the resolution can be varied according to the application and does not exceed 12 bits.**

A 12-bit counter IC, such as the 74HCT4040, can be included on the board. It is shown in Fig. 9.5, with the reset line connected to the 0-V line. This will allow the counter to operate in the free-running mode. It will count from 0 to 4095 and then resets to 0 to repeat the count, which is delivered through the 12 output lines Q0 to Q11. These lines can be connected to a suitable connector for external use.

Since the 12-bit resolution is not needed for all the applications, the requirement for a lower number of stages can be satisfied by utilizing output

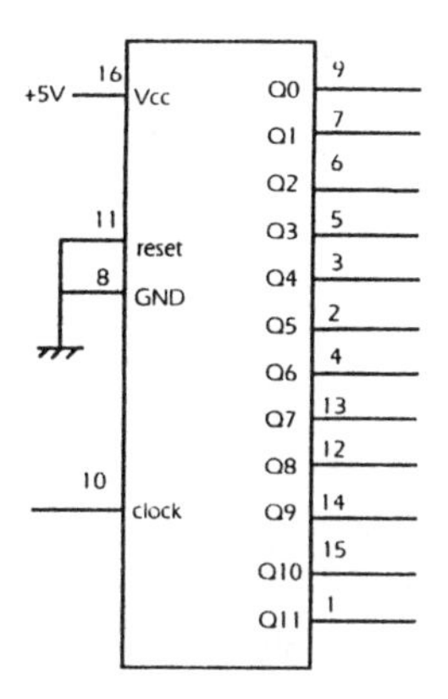

Fig. 9.5 Twelve-bit counter.

lines Q0 to Qx (where x is less than 11). The counter will continue to operate as a 12-bit counter without the need for applying a reset signal when the required count is reached. Reset will only be required if the desired count is not a multiple of 2, as shown in Example 9.10.

Example 9.4: **How can an interface circuit be modified to interrupt a microcomputer once every millisecond? Assume that the circuit includes a 6-kHz signal.**

A 1-ms interval corresponds to a frequency of 1 kHz. The requirement can be satisfied by dividing the 6-kHz signal by 6. A counter can be employed to count from 0 to 5 repeatedly and give a pulse once in every counting cycle. One solution is to use a binary counter plus logic gates to generate a reset signal at the sixth count. An alternative solution is to utilize the 74HCT4017 IC, which is a decade counter/divider with 10 decoded outputs (lines Q0 to Q9). These outputs are set high sequentially as they advance 1 with each clock pulse, starting with Q0 and continuing to Q9 before repeating the process. In other words, each of the output lines is at a logic-high state for only one clock pulse. Figure 9.6 illustrates how this IC can be employed to count from 0 to 5. Output line Q0 can be utilized to deliver the required output to an interrupt line of the microcomputer.

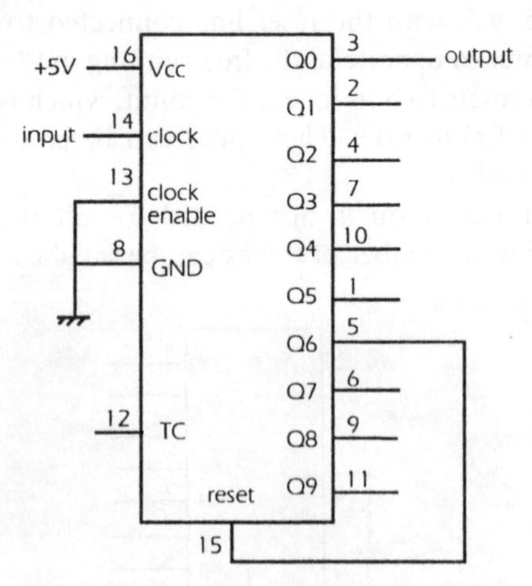

Fig. 9.6 Divide-by-6 counter.

When using devices to alter the frequency, one should consider their effect on the shape of the output waveform. For instance, the output line of the 74HCT4017 IC in the last example will be high for the duration of a single clock cycle. This implies that the output will be high for only one out of six stages, giving a mark-to-space ratio of 1:5. If the application requirement specifies an output of a mark-to-space ratio of 1, then a 3-input OR gate can be utilized to form an output signal from ORing lines Q0, Q1 and Q2 (the 74HCT4075 is a triple 3-input OR gate IC).

In many counter-based applications, a microcomputer reads a counter at regular intervals. If the available clock signal possesses a high frequency, then the counter will overflow (and starts counting from 0) before the microcomputer has the chance to read the count. Such a problem can be avoided if the frequency of the clock is reduced by the use of a divide-by-n counter. Rather than incorporating additional components, a counter of a higher resolution may be used to satisfy the requirements, as shown in the following example.

Example 9.5: **The operation of an interface circuit is based on delivering an 8-bit count regularly to a microcomputer at a rate of 1 kHz. How can the count be produced if the available clock signal has a frequency of 2 MHz?**
A 1-kHz rate means that the microcomputer will read the count once every millisecond. Since the clock has a frequency of a 2 MHz, each clock pulse has a duration of 0.5 μs. An 8-bit counter will overflow after 256 pulses, i.e. after 128 μs. This, however, is approximately $\frac{1}{8}$th the interval between successive readings of the counter by the microcomputer. Consequently, the clock frequency needs to be divided by 8 before being delivered to an 8-bit counter.

Rather than using additional components, the 12-bit counter employed in Example 9.3 can be used here. When operating in the free-running mode, such a counter counts 4096 clock pulses before resetting to 0. Since each clock pulse is 0.5 μs, the counter will overflow after 2.048 ms. Since the microcomputer will read the count every 1 ms, output lines Q3 to Q10 can be utilized to deliver the required 8-bit count.

This implies that the three least significant stages (whose count is delivered through output lines Q0, Q1 and Q2) perform a divide-by-8 action. Stages 4 to 11 behave like an 8-bit counter which receives clock pulses at the reduced frequency.

In the above example, the 74HCT4040 IC has been used to perform the function of a divide-by-8 device as well as an 8-bit counter. The requirement was satisfied without the need to use stage 12 of the counter. This implies that the same counter can be employed to satisfy additional requirements, e.g.

- If the need arises to alter the resolution of the counter from 8 bits to 9 bits. This can be achieved by taking the output from lines Q3 to Q11 without changing anything else.
- If the sampling rate of the microcomputer is altered to 500 Hz. The division factor will then need to be changed from 8 to 16. This implies taking the 8-bit count from output lines Q4 to Q11 without altering anything else.

If the ratio between the available clock frequency and the one required to achieve the count without an overflow is higher than what a 12-bit counter can provide, then a 14-bit counter such as the 74HCT4020 can be incorporated to satisfy the requirement. This is shown in Fig. 9.7, where it can be seen that the output lines of only 12 out of the 14 stages are connected to output pins of the IC. This permits the use of a 16-bit DIL package for the IC, but means that the output of stages 2 and 3 is not accessible. The IC is therefore very suitable for applications such as that of Example 9.5, where a few of the least significant bits only are utilized to provide a divide-by-*n* function.

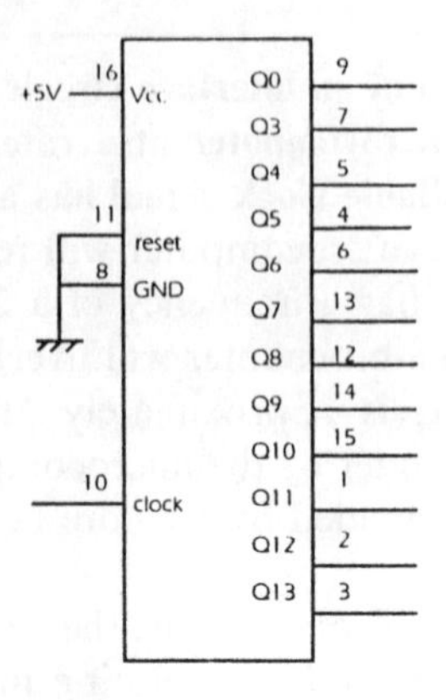

Fig. 9.7 Fourteen-bit counter.

9.3 Timers

Timers are employed to coordinate events and are extremely important components in interfacing applications, especially for real-time control. Using one or more timers within an interface circuit helps to achieve accurate timing and can replace the need for software delay loops. Timers may also be incorporated to carry out functions which will remove the need for interrupts or, at least, reduce the required rate of their occurrence.

9.3.1 Classification of timers

Timers can be classified according to the way their timing activities are defined.

Non-programmable

The timing activities of such timers are configured by hardware. The mode of operation and timing of the signals produced are normally set according to the value of components (resistors and capacitors) connected externally. Once the value of the components is selected and the timer is configured, the behaviour of the timer is defined. This, however, does not mean that a microcomputer cannot be provided with the ability to alter the mode of operation of the timer. Examples are introduced in this chapter where the operation of such a timer is controlled by a microcomputer and the output signals it produces are altered in real time.

Programmable

Software is used to define the mode of operation and the timing activities of such a timer. It is available in the form of a DIL IC, or integrated within a microcontroller IC (some microcontrollers include a few programmable timers). A microcomputer can, for instance, load a programmable timer with a digital number corresponding to the rate of turning a switching device on and off. After loading the timer, the microcomputer continues with its other activities, leaving the timer to generate the timing signals required for the operation of the switching device. The microcomputer only needs to send information to the timer if the switching rate is to be altered. More information is given in Section 9.4.6.

9.3.2 Examples of timer ICs

Let us consider three different types of timers.

555

This low-cost widely used timer is suitable for inclusion in a wide range of circuits. Books and many articles have been written describing its use in various applications. Its ability to source or sink load currents of up to 200 mA makes it attractive to directly drive relays and lamps. It is shown in Fig. 9.8(a) and Fig. 3.1.

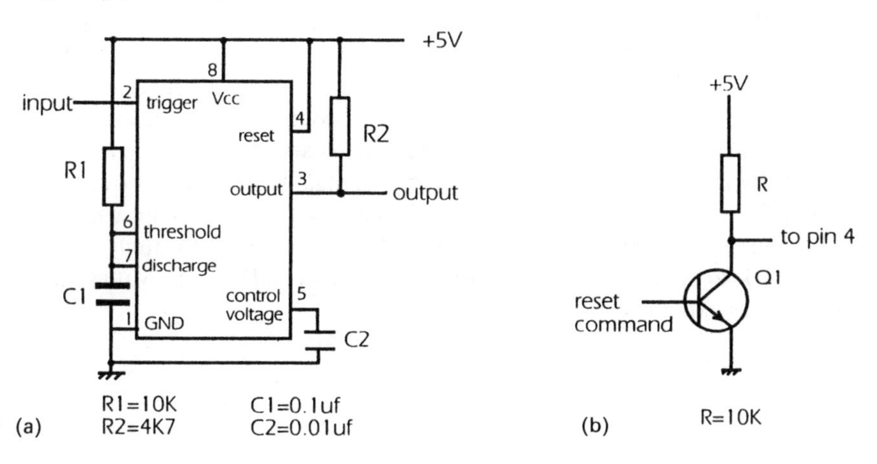

Fig. 9.8 Using a 555 timer: (a) monostable; (b) controlled operation.

It contains an SR flip-flop which will be set when the voltage supplied to its trigger input line (pin 2) falls to below one-third of the supply voltage. It will reset when the voltage delivered to its threshold input line (pin 6) exceeds two-thirds of the supply voltage. When the flip-flop is set, the output line (pin 3) is set to a logic-high state and the timing capacitor C1 starts to charge through resistor R1. The charging continues until the voltage at the threshold line (which is equal to the capacitor's voltage in the configuration of Fig. 9.8(a)) exceeds two-thirds of the supply voltage. When this happens, the flip-flop will reset, causing C1 to discharge through the discharge line (pin 7) and, at the same time, sends the output line (pin 3) to a logic-low state.

This implies that the configuration of Fig. 9.8(a) forms a monostable which is triggered when the signal on the trigger line changes state from logic high to logic low. This sends the trigger input below $V_{cc}/3$. The output line will then go high and C1 starts to charge. The duration of the output pulse is therefore set by the time needed to charge C1 through R1 and is given by $t = 1.1(R1\,C1)$. A designer can therefore set the pulsewidth by choosing the value of both R1 and C1. Replacing R1 by a resistor R3 in series with a potentiometer provides a manual setting of the monostable. The value of R3 defines the minimum pulsewidth of the output, while the sum of that resistor and the potentiometer's maximum setting defines the maximum pulsewidth.

Pin 4 is the reset line and is connected to the supply voltage in Fig. 9.8(a). If the need arises, the operation of the timer can be controlled by an external device, such as a microcomputer. The output is forced low when the reset line is taken low. This can be accomplished by replacing the link between pin 4 and the supply voltage by the circuit of Fig. 9.8(b). The transistor can be driven by an external device, e.g. a digital-output port of a microcomputer (as described in Chapter 10). When the transistor is turned OFF, pin 4 will be connected to the supply voltage via resistor R. Switching the transistor ON will activate a reset condition, forcing the output pin to a logic-low state. In this way, the microcomputer will use a single digital-output line to control the output of the timer and consequently controls the operation of other circuits driven by the timer. The microcomputer will, however, only control the duration of operation, but not the pulsewidth or frequency. These are set by the value of the external components, as explained above.

The 555 timer can also be configured as an astable. This is shown in Fig. 3.1, where C1 charges through R1 and R2, but discharges through R2 only. The mark-to-space ratio of the output signal depends on the values of R1 and R2, as explained in Example 3.2.

LM122

The National Semiconductor LM122 is an example of a high-precision timer which provides timing periods from microseconds to hours. It can be controlled from an output port of a microcomputer and can be configured in one of several ways to meet a variety of timing applications.

Figure 9.9 shows the timer with a few connections left to be made manually (e.g. by links) according to the application. The trigger input line (pin 2) allows an external device, such as a microcomputer, to initiate the timing interval which is determined by the time needed to charge capacitor C through resistor R to 2 V. The circuit connects R to the reference voltage V_{ref}, which is 3.15 V. Once the values of R and C are chosen, the timing interval will be fixed and the links can be used to configure what happens at the end of the timing interval. The output of the timer is a transistor whose collector and emitter are connected to pins 9 and 10 respectively. One can choose to connect the load to either terminal. One can, for example, connect the load between the collector and the supply voltage. A link can then be placed to connect the emitter to the 0-V line (pin 5). The state of the transistor during and after the interval of charging C is determined by a signal delivered through the logic input terminal (pin 1). A signal of a logic-high state at that line turns off the output transistor while the capacitor is charging and then turns it on when the capacitor is charged. A signal of a logic-low state at that line will achieve the opposite. This timer is employed in Example 14.6 to operate a relay in a temperature-control application (the connection is shown in Fig. 14.11).

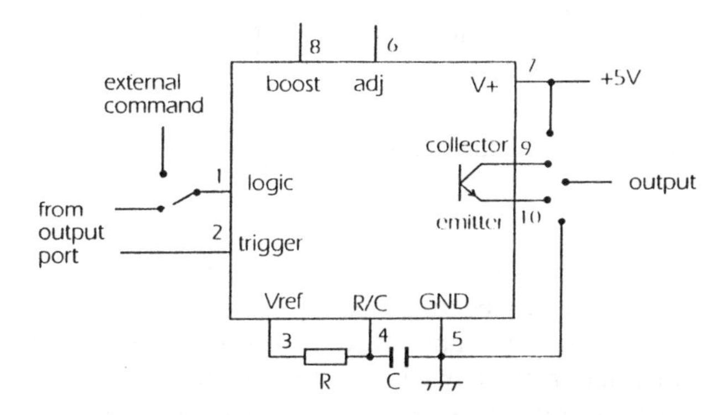

Fig. 9.9 Using the National Semiconductor LM122 timer.

uA2240c

The Texas Instruments™ uA2240c programmable timer/counter is shown in Fig. 9.10. It contains an internal oscillator whose frequency is set by an external resistor R1 and a capacitor C1 connected to pin 13. The output of the oscillator is connected internally to pin 14 to permit external use of the oscillator's frequency, which equals 1/(R1 C1). The IC permits operation over a wide frequency range, as the allowable value of R1 ranges from 1 kΩ to 10 MΩ, and the value of C1 from 100 nF to 1000 μF. The uA2240c is a 16-pin IC containing an 8-bit counter to count the pulses generated by the oscillator. The count appears on eight open-collector output lines (Q0 to Q7). The IC includes a reset line (pin 10) which, when activated by a logic-high signal, will set output lines Q0 to Q7 to a logic-high state.

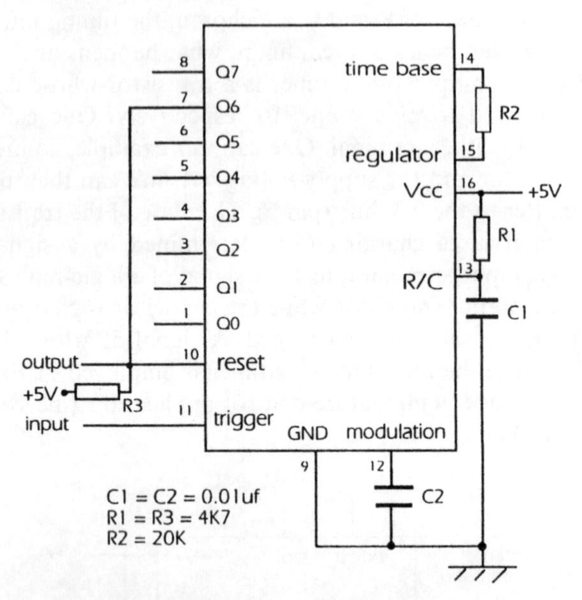

Fig. 9.10 Using the Texas Instruments uA2240c timer.

After resetting the counter, the reset line can be driven to a logic-low state, making the IC ready to be triggered by an active-high external signal (minimum duration of 2 μs) supplied to the trigger input line (pin 11). This will start the oscillator, thereby allowing the counter to start counting from 0. The IC can be operated as an astable with an active-high triggering pulse and continues to operate until a reset signal is applied to stop the count. The reset signal will only be effective if the trigger input is low. Between triggering and resetting the uA2240c IC, additional triggering pulses have no effect. In order to use the IC as a delay device or a monostable, one or more of the counter's output lines (Q0 to Q7) can be connected to the reset line. The circuit of Fig. 9.10 connects line Q6 to the reset line. Consequently, counting will be started by a logic-high triggering pulse and continues until the IC is forced to reset after 64 pulses.

Other delay factors can be easily obtained. Two or more of the open-collector output lines can be connected together to the point marked 'output' in Fig.

9.10. Reset will be activated when all these lines are at a logic-high state. For example, if output lines Q0, Q1 and Q7 are connected to the reset line together, then the reset will take place after 131 pulses (represented by the binary number 10000011). The reset line can therefore be used as the output line providing the delayed signal. It will be at a logic-high state for the duration of a single pulse only and for a logic-low state for the rest of the pulses until the next reset. Example 9.7 uses the uA2240c as a programmable divide-by-N device.

9.3.3 Timers in practical circuits

Let us consider examples using the timers described above.

Example 9.6: **In a microcomputer-controlled sequential process, the frequency of operating an industrial unit needs to be changed in 10 sequential steps. How can this be achieved with minimum intervention from the microcomputer?**
In order to minimize intervention from the microcomputer, the hardware approach can be adopted. The requirement can be satisfied by designing an astable whose frequency can be changed in 10 sequential stages. The circuit of Fig. 9.11 employs the 74HCT4017 decade counter/divider IC to divide an input frequency into 10 different channels (this IC is also used in Example 9.4). Each output channel drives a resistance connected to a single common point. Each of these resistors is a charging resistor for capacitor C1 of a 555 timer.

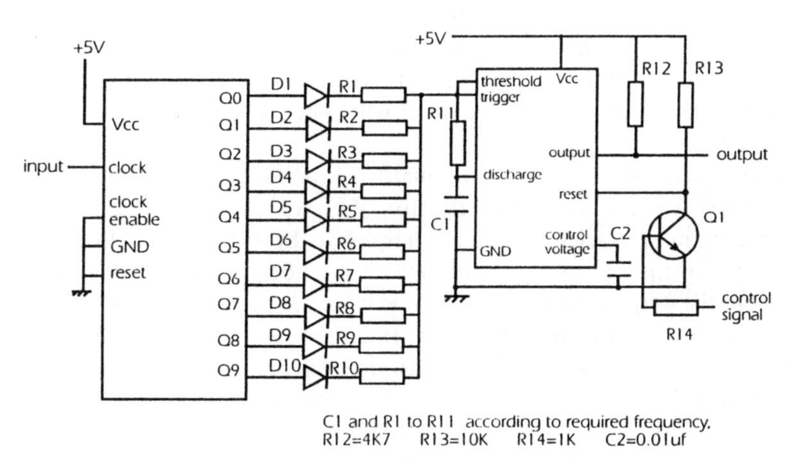

Fig. 9.11 Sequential frequency changer.

The 74HCT4017 IC is driven by external clock pulses. When these pulses arrive at the clock input line of the 74HCT4017, the output is advanced from one channel to the next, thereby choosing a different resistor to charge capacitor C1. This changes the frequency at the output of the 555 timer. The 10 required frequencies can be set by the value of each of the 10 resistors (R1 to R10) as well as capacitor C1.

The clock signal supplying the 74HCT4017 IC can be produced by the microcomputer, by an external circuit, or by a timer (e.g. another 555). The frequency of that clock defines the duration of generating each of the 10 frequencies. Transistor Q1 is introduced to provide the microcomputer with the ability to switch the output of the circuit on and off. Turning on the transistor resets the 555 timer and sets the output line to 0. No output will be produced until Q1 is turned off by the microcomputer (the operation of transistors is explained in the next chapter).

If the requirement of the above example was to produce less sequential steps (e.g. five), then one of the output lines (e.g. Q5) could be connected to the reset pin of the 74HCT4017 in a similar way to Example 9.4.

Example 9.7: **Design a timer-based frequency divider where the division factor of 1 to 255 can be programmed by a microcomputer through an output port.** An 8-bit output port can supply the required division factor. Let us choose the uA2240c IC. It possesses eight output lines which can be connected to the reset line to determine the division factor. The requirement here is for a circuit to combine the 8-bit command from the microcomputer with the eight output lines from the timer in order to form a signal to reset the timer.

The circuit of Fig. 9.12 incorporates analogue switches. Each of IC1 and IC2 is a CD4066A quad bilateral switch (the operation of this IC is explained in Section 5.2.3). The control lines of these switches are connected to the output port of the microcomputer. In this way, the microcomputer determines which of the output lines of the timer will contribute to the formation of the reset signal.

Use is made of the internal oscillator of the uA2240c. Its base frequency is defined by the values of resistor R1 and capacitor C1 leaving the microcomputer to select the output frequency through its eight digital-output lines. Choosing R1 as a potentiometer provides a manual adjustment to increase the frequency range obtainable from the circuit. Note that the output frequency is in the form of a pulse with a mark-to-space ratio dependent on the selected count (see Section 9.3.2). If required, the output waveform can be modified by the use of a D-type latch connected as a divide-by-2 counter (as shown in Fig. 9.2). Although the latch will halve the output frequency, it will produce a waveform with an equal mark and space.

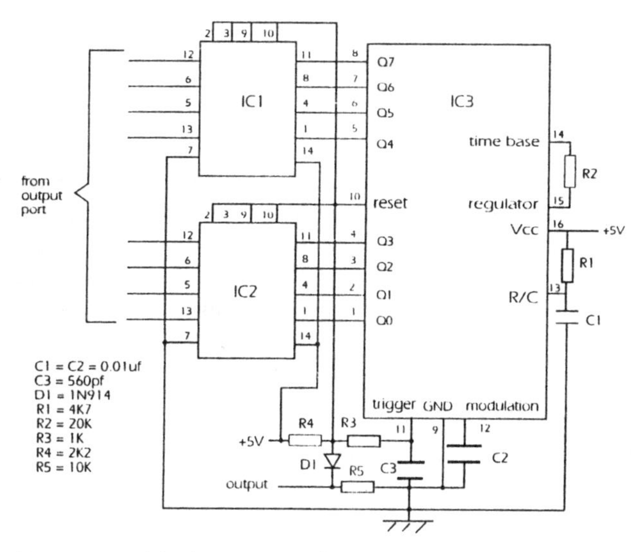

Fig. 9.12 Programmable frequency changer.

9.3.4 Watchdog timers

This is a delay circuit which will generate an output signal if not reset within the delay period. One of its main uses is to detect an improper operation of a microcomputer, e.g. if the microcomputer is stuck in an endless software loop. In such a situation, the output of the watchdog timer will reset the microcomputer and can be employed to block the transmission of signals from the microcomputer to the outside world. It can also be employed to sound an alarm or activate a back-up circuit.

A watchdog timer consists of a clock and a delay device. The delay is set by the designer according to the rate at which the microcomputer can send the reset signal. The microcomputer can be programmed to send a regular reset signal to the delay device of the watchdog timer. This signal can be a part of a regularly occurring interrupt within the microcomputer. The clock can be derived from that of a microcomputer, but this may not be the best choice, since the clock itself may become faulty. Using a separate clock source and a delay device will make the watchdog timer an independent circuit. The delay device can be a shift register or a counter.

Using a shift register

Supplying a clock signal to an 8-bit shift register will generate a time interval of eight clock cycles before the output of the last shift stage is activated. A microcomputer can be programmed to reset the shift register regularly. If it fails to do so, then the last stage of the shift register will generate a signal which will reset the microcomputer and take any other necessary action.

Using a counter

An N-bit counter can be incorporated as the delay device. The number of required bits depends on the relationship between the frequency of the clock and the delay interval after which the watchdog activates a reset. For instance, using the output of the most-significant stage of an 8-bit counter as the source of reset implies that a reset will be activated after 128 clock pulses (i.e. a count of 10000000).

Comparison

When comparing a shift register and a counter, one can conclude that a counter is attractive to use when the required delay corresponds to a large number of clock cycles. A shift register is thus suitable if the number of clock cycles is low, e.g. not more than eight. In this way, the delay function is introduced by the use of a single IC.

9.4 Changing frequency

One of the main uses of counters and timers is in frequency-changing applications. Each of the output lines of a free-running counter possesses a frequency which is a fraction of that of the signal supplied to the clock input line. The ratio of output and input frequencies, however, depends on the type of counter and on which of the output lines is used to generate the output signal. Producing the exact division factor is not always a straightforward process. A designer may spend some time comparing counters in order to find the best combination to satisfy the requirements with the minimum number of ICs and/or at the lowest cost.

9.4.1 Using a binary counter

It is shown in Section 9.2.4 that a binary counter can be utilized as a frequency divider. The division factor is equal to 2 to the power n, where n is the number of the stage whose output line is selected. For instance, the 8-bit binary counter presented in Fig. 7.1 is made from cascading two 4-bit binary counters. Line QA of the left counter changes state at half the rate of the clock frequency and hence divides the clock frequency by 2. Similarly, line QB of that counter gives a divide by 4. In other words, the eight output lines give the following division factors: 2, 4, 8, 16, 32, 64, 128 and 256. The division is obtained from using the outputs of a counter during its normal operation. Based on this principle, Example 9.5 uses a single 12-bit counter as both a divide-by-8 device and an 8-bit counter.

9.4.2 Using a BCD counter

If a binary counter is not suitable to provide the right division factor, then a BCD counter may satisfy the requirement. An example of such a counter is shown in Fig. 7.4, where its output lines provide the following division factors: 2, 4, 8, 10, 20, 40, 80, 100. Difficult-to-achieve division factors (e.g. 1280 or 5120) can be introduced by cascading BCD and binary counters.

9.4.3 Using special counters

The 74HCT4017 IC is utilized as a divide-by-6 counter in Example 9.4. The division factor can be changed to any value between (and including) 2 to 10 depending on which of the output lines is utilized to reset the counter, if any. A very wide range of division factors can be produced by cascading a few such ICs. For instance, 18, 21, 36, 49, 72 and 89 are some of the division factors produced from cascading two such ICs. Note that the mark-to-space ratio of the produced waveform may differ according to the sequence of the two ICs. For instance, a division factor of 27 is formed by cascading two such ICs, one of which is set to divide by 3 and the other to divide by 9. If the divide-by-3 operation is performed by the second IC, then the output signal will have a mark-to-space ratio of 1:2. If the ICs are swapped, then the output signal will possess a mark-to-space ratio of 1:8. One can therefore adapt this approach to define the ratio of input and output frequencies and, at the same time, set the relationship between the mark and space of the generated waveform.

An alternative IC is given in the following example.

Example 9.8: **Design a frequency divider to derive a 50-Hz signal from a 10-MHz clock.**

The requirement can be satisfied by designing a divide-by-200 000 counter. One solution can be based on using a D-type latch as a divide-by-2 counter (as shown in Fig. 9.2) cascaded with five BCD counters (e.g. the 74LS90 used in Fig. 7.4). This will satisfy the requirements, but a solution which uses six ICs is not always favourable.

An alternative solution can be based on the use of the 74LS390 dual decade ripple counter IC. It contains four counters. Each of lines 1 and 15 is an input to a divide-by-2 counter, while each of lines 4 and 12 is the input to a separate divide-by-5 counter. Using an output of one counter to feed another, the four counters perform a divide-by-100 function. Three such ICs can be cascaded to provide the division by 200 000. They should be configured as two divide-by-100 counters cascaded with a divide-by-20 counter. Configuring the 74LS390 IC to fulfil the requirement is shown in Fig. 9.13. Since there is no need to reset the resulting divider, then the active-high reset lines (1MR and 2MR) are connected to the 0-V line.

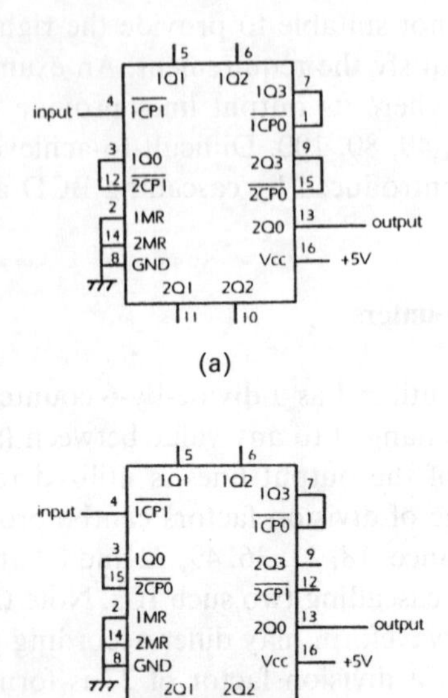

Fig. 9.13 A frequency divider: (a) divide-by-100; (b) divide-by-20.

9.4.4 Using additional circuits

Satisfying the requirements of various applications is not always a straightforward task, especially when the involvement of the microcomputer with other tasks leaves little execution power to handle the frequency-division task. Adding more hardware will reduce the involvement of the microcomputer, but some control functions still need to be performed by the microcomputer.

Example 9.9: **Design a microcomputer-controlled frequency divider where the microcomputer is capable of selecting between the following eight division factors: 2, 4, 8, 16, 32, 64, 128 and 256.**

The division factors can be obtained from the outputs of an 8-bit binary counter. The main task is then to design an interface circuit which allows the microcomputer to select any of the eight output lines of the counter. An 8-channel digital multiplexer can be incorporated to fulfil the needs. An example of such a multiplexer is the 74LS151 IC described in Section 5.2.2 and shown in Fig. 9.14. The microcomputer uses one of its output ports to supply a binary number through lines X0, X1 and X2, where the last is the most significant bit. The line whose number is the same as the supplied binary number will be connected to the output. In this way, the microcomputer can select the required frequency by the use of three digital-output lines. The number it delivers will be latched by the output port and the frequency will continue to be produced without any further action from the microcomputer.

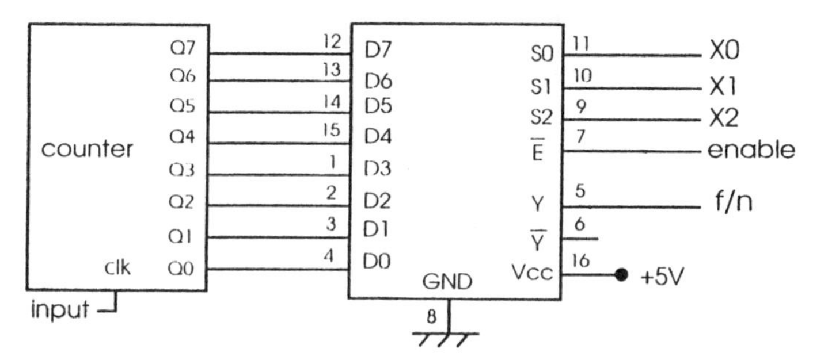

Fig. 9.14 Microcomputer-controlled frequency divider.

The active-low enable line of the IC (pin 7) can be connected to another digital-output line of the microcomputer. The connection gives the latter the ability to disable the supply of the clock pulses by changing the state of that line from low to high. During the disable condition, the state of output line Y is low.

Certain division factors cannot be obtained by the use of counter ICs on their own. An additional circuit may need to be included outside the counter to activate a reset when the required count is reached.

Example 9.10: **Design a divide-by-129 counter.**

Solution 1

The 129 is represented in a binary form as 10000001. The count is represented by 8-bits, but a free-running 8-bit binary counter will count 256 pulses before overflowing to 0. The division factor of such a counter can be reduced by the use of logic gates. The counter should be allowed to count from 0 and reset when the count is 129. This implies generating a reset signal when output lines Q7 and Q0 are at a logic-high state and all the other output lines are at a logic-low state. Four gates are used in Fig. 9.15(a) to satisfy the requirement. The 2-input AND gate (4 in a 74LS08 IC, see Appendix) produces a logic-high output when its inputs are set to logic-high, while the NOR gates (3 in a 74LS27 IC) will generate a logic-high signal when all of its inputs are at a logic-low state. Consequently, the 3-input AND gate (3 in a 74LS11 IC) will generate a logic-high signal to reset the counter only when the count reaches 129. If, however, the reset line of the counter is activated by a logic-low signal, then the 3-input AND gate will be replaced with a 3-input NAND gate (3 in a 74LS10 IC).

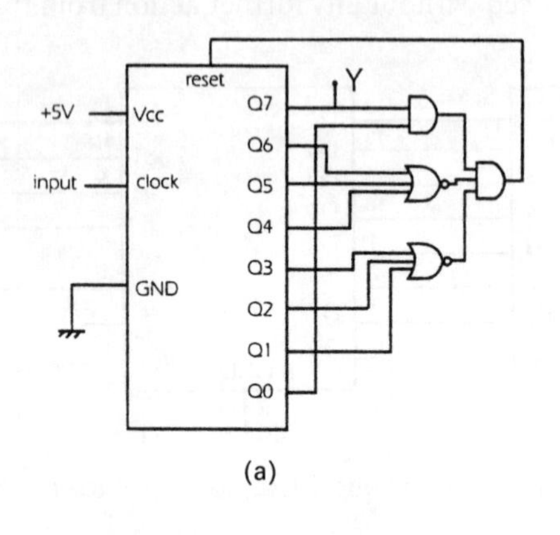

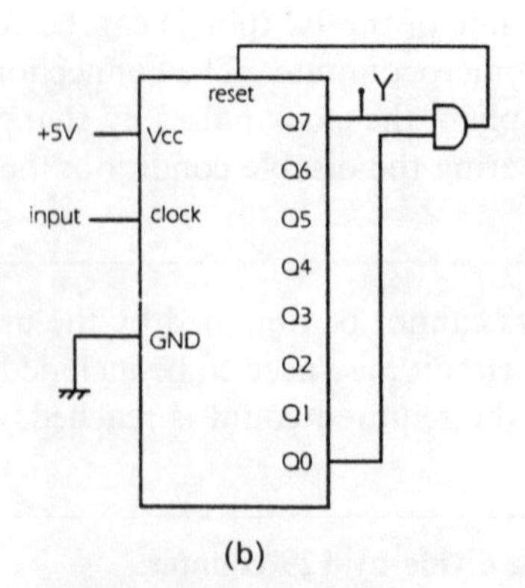

Fig. 9.15 Divide-by-129 counter: (a) using four gates; (b) reducing number of gates.

Solution 2

In many applications, designers can use their knowledge of the operation of the devices, or the behaviour of the circuit, to reduce the number of components. A divide-by-129 is a good example. From understanding the operation of up-counters, one concludes that bit Q7 will not be at a logic-high state until the count exceeds 127 (which is 01111111 in binary). This will be followed by a count of 10000000 and then 10000001, which is 129. This implies that when counting from 0 to 129, the only count where both Q7 and Q0 are at a logic-high state is 129. Consequently, the circuit of Fig. 9.15(a) can be simplified to that of Fig. 9.15(b) and perform the same task.

Example 9.11: **Design a circuit to generate a ramp waveform whose frequency is set by a microcomputer through a digital-output port.**
One way of generating a ramp signal is to connect the output lines of a free-running counter to the input lines of a D/A converter (an IC combining both an 8-bit counter and an 8-bit D/A converter is presented in Section 8.2.1). The generated waveform is given in Fig. 9.16(a). Its frequency can be set by the provision of a signal to reset the counter at regular intervals.

In the centre of the circuit of Fig. 9.16(b) is an 8-channel digital comparator. It is the 74HCT688 IC, which is supplied with two 8-bit numbers, one from the eight output lines of a counter and the other from the digital-output port of the microcomputer. As the counter counts from 0, its count is compared with the number delivered from the microcomputer. When the two numbers are equal, the digital comparator changes the state of its output line Y from logic high to low. This signal resets the counter, which will start counting from 0, and the cycle is repeated. During the counting period, the output of the D/A converter is continuously increasing, thereby producing the required ramp signal.

This shows that the ramp signal is produced with little effort from the microcomputer. One has to remember, however, that the number given by the microcomputer represents the period of the generated waveform (which is the inverse of the frequency). This implies that increasing the frequency is achieved by decreasing the digital number.

9.4.5 Using timers

Timers are incorporated as frequency dividers in various interfacing circuits. Examples 9.6 and 9.7 illustrate how two different types of timers are employed as frequency changers. Programmable timers are very widely used for this purpose, as explained in the following section.

9.4.6 Using programmable counters/timers

Such devices are very attractive for use as frequency dividers, especially in real-time control applications. Many microcontrollers include one or more

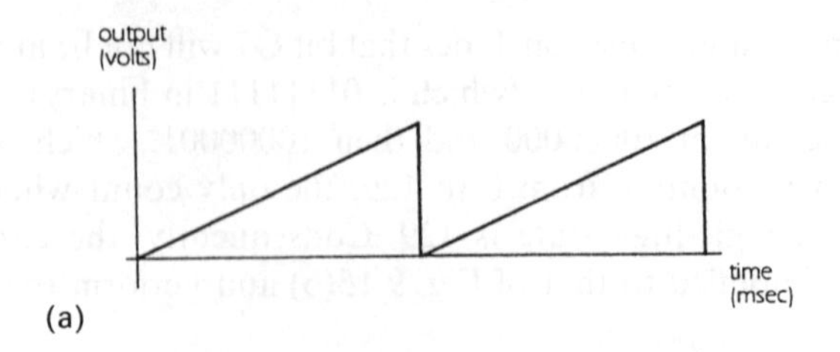

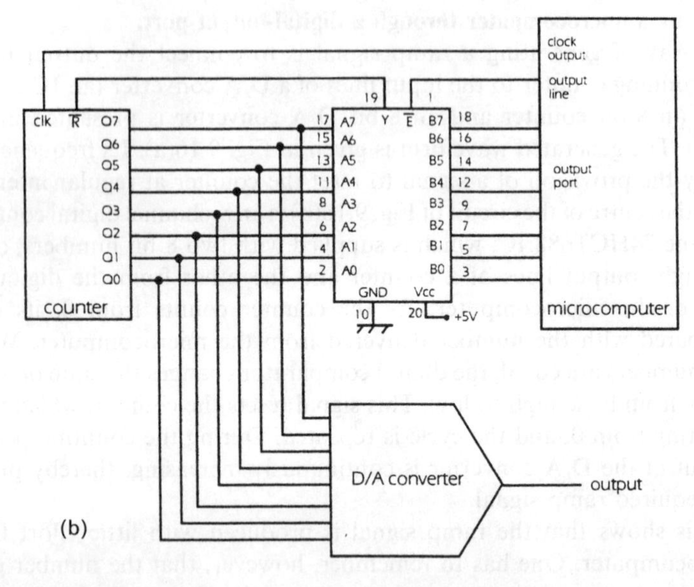

Fig. 9.16 Microcomputer-based frequency changer: (a) waveform; (b) circuit.

such devices plus other peripherals within the microcontroller IC. Programmable timers/counters are also available in IC form, e.g. an IC containing three 16-bit programmable timers/counters.

Such devices can be easily run under the control of software and, generally, they can be operated in one of a few modes. When configured as free-running counters, they count from 0 to maximum count and then overflow to 0 to repeat the counting process. The microcomputer writes to an 8-bit or a 16-bit register to identify the count at which the output needs to change state. When that count is reached, an interrupt can be generated to instruct the microcomputer to perform a certain task and update the register with a count corresponding to the next required instant of changing the state of the output. Depending on the type of application, processing power can be saved if the timer/counter has two registers. One indicates the count where the output line needs to change to a logic-low state, while the other register identifies the instant when that line should change to a high state. Processing power is further saved if the count in the second register resets the timer/counter. This will give repeatability, and the output of the timer/counter will toggle between high and low without the need to interrupt the microcomputer. In this way, the microcomputer only writes to the two registers once. The required frequency and the mark-to-space ratio will be generated without further intervention from the microcomputer, unless a change is required.

9.5 Setting time intervals

Examples are presented in this book illustrating uses of microcomputers and interface circuits in the formation of delay functions. This may be the interval between the repeated occurrence of an event, or the interval between two different events. The source of the first event may be external to the microcomputer, which will generate the second event after a preprogrammed interval. For instance, the first event may be the instant where the mains voltage crosses the 0-V line, while the second event can be the instant of turning-on a power-switching device (Section 10.8). The interval may be defined by a binary counter (Example 13.7) or by a timer (Example 14.7). The solution to most microcomputer-based time-related problems relies on the introduction of regularly occurring interrupts. Setting the duration between interrupts is the subject of Section 9.5.4.

One of the main areas of microcomputer use in time-related applications is in the generation of pulses, or in the control of their width. If the pulses occur regularly, then they can be referred to as rectangular waves. When setting the specification of a rectangular wave, it is important to clarify whether the requirement is to set the pulsewidth or the mark-to-space ratio. Although the two may seem the same, but there is an important difference.

- The pulsewidth is independent of the frequency of the waveform. The state of a digital line will, for instance, be set to logic high at the beginning of a cycle and will be changed to logic low after 1 ms, regardless of when the cycle terminates.
- The mark-to-space ratio is the relationship between the interval of the high and low states within a digital cycle. If, for instance, a mark-to-space ratio of 1 is required from a 500-Hz signal (corresponds to a period of 2 ms), then the waveform will be at a logic-high state for 1 ms and at a logic-low state for the other millisecond. If the frequency is increased to 510 Hz, then the duration of both the high and low states need to be adjusted to keep their ratio constant.

This is further explained in the following two sections.

9.5.1 Setting the width of pulses

This section illustrates two ways of satisfying the requirement of keeping the pulsewidth constant.

Using a monostable

A monostable can be employed as shown in Fig. 9.17. The 74HCT4538 IC is selected. It is a dual retriggerable precision monostable multivibrator. The duration of the output pulse is set by the value of the externally connected resistor and capacitor and is given by T = 0.7RC. A potentiometer can be employed in series with each of the resistors of Fig. 9.17 to allow setting of the required pulsewidth manually.

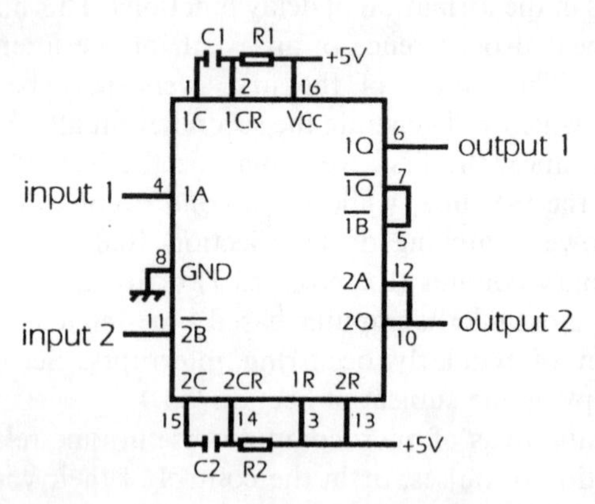

Fig. 9.17 Using a monostable.

The above-mentioned monostable can be used in either of two modes:

- In a non-retriggerable mode, the output pulsewidth (T) of the monostable starts from the edge of the first pulse applied to the monostable. It will not be affected by other pulses applied before it is complete.
- A retriggerable mode of operation allows the duration of the output signal to extend if additional input signals are applied before the end of the output pulse. This implies that the output pulse generated by such a monostable starts from the application of the first pulse and lasts for a period (T) from the last triggering pulse.

Both monostables of Fig. 9.17 are connected in the non-retriggerable mode and will be used in this configuration later in this chapter. The first monostable is configured to be triggered by a rising edge (leading edge) of the input signal. This is achieved by using line 1A (pin 4) as the triggering input. Line 1B (pin 5) is connected to the inverting output line of this monostable (pin

7). The second monostable is configured to trigger by a falling edge of the input signal. This is achieved by connecting input line 2A (pin 12) to the non-inverted output line 2Q (pin 10). Line 2B (pin 11) is then used as the input line of this monostable.

Using a counter-based circuit

Counters can be easily incorporated as pulsewidth-setting devices where the width is defined as a count of the clock pulses. The higher the frequency of the clock, the lower is the error between the required and the generated pulsewidths. On the other hand, a higher frequency means a higher count within a given interval. This creates the need for a counter with a higher resolution.

9.5.2 Setting the mark-to-space ratio

Many variable-frequency applications impose the requirement of setting the mark-to-space ratio of the signal delivered by a frequency divider. The operation of a large number of control applications relies on achieving the best mark-to-space ratio to fulfil a given requirement. For instance, the frequency in a.c.-motor drives (Chapter 15) is adjusted to set the speed of the motor, while the mark-to-space ratio is controlled to maintain the shape of the waveform and to control the number of harmonics. The power is delivered to the motor through power-switching devices where the mark-to-space ratio defines the time for turning each device on and off.

9.5.3 Application examples

The utilization of microcomputers in defining the required time interval is best explained by a few examples.

Example 9.12: **Modify the circuit of Example 9.11 to generate a rectangular wave. The microcomputer should be provided with the ability to set the mark-to-space ratio as well as the frequency.**
The requirement of Example 9.11 is satisfied by the use of a digital comparator connected as shown in Fig. 9.16. This can be employed to set the frequency here, except that the D/A converter can be replaced by a lower-cost digital circuit. Setting the mark-to-space ratio can also be performed by the use of a second digital comparator, as shown in Fig. 9.18. The microcomputer supplies two digital numbers through two output ports. Port 1 supplies a count equivalent to the required period of each cycle. The counter starts

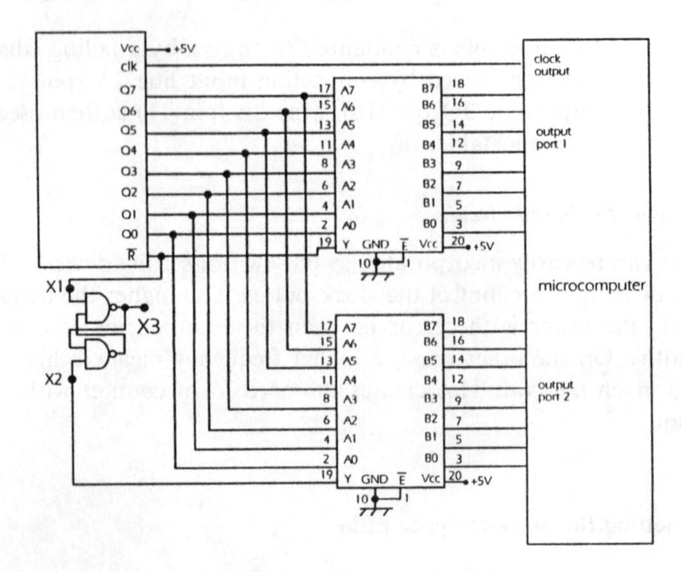

Fig. 9.18 Setting frequency and mark-to-space ratio.

counting from 0 until its count (which is delivered through lines Q0 to Q7) reaches the number set by the microcomputer. The digital comparator will change the state of its output line (Y). This alters the state of line X1 from high to low and resets the counter, which will start counting from 0 again.

Output port 2 of the microcomputer supplies a number to define the required width of the pulses. This number represents the 'mark' of the wave and is obviously less than the number delivered by port 1. Figure 9.19 shows the output of ports 1 and 2 as constant values Pt1 and Pt2 respectively. The figure also shows the output of the counter increasing with time before resetting to 0. During the counting process both lines X1 and X2 are at a logic-high state. When the count becomes the same as Pt1, the state of X1 changes to low. This resets the counter and changes its output to 0. The result is changing the state of line X1 to logic-high again, and the counter continues to count from 0. In the same way, the state of line X2 changes to low for the duration of a single pulse, when the count becomes the same as the digital number delivered by port 2.

Both lines X1 and X2 are connected to the inputs of a flip-flop made from two NAND gates. The output of the flip-flop (line X3) is set to a logic-high state when line X1 changes state to low. When line X2 is pulsed, it resets the output of the flip-flop to a low state. The output continues to be in that state until the end of the cycle, as shown in Fig. 9.19. The result is a waveform of a mark-to-space ratio and a frequency set independently by the microcomputer through two 8-bit digital-output ports.

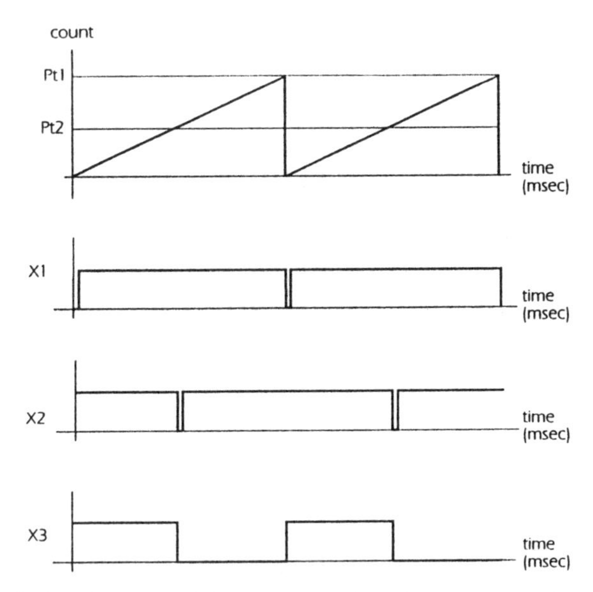

Fig. 9.19 Timing diagram of frequency and mark-to-space ratio setting circuit.

Example 9.13: **Design an interface circuit to produce a pulse every 25 ms where the pulse should be at a logic-high state for 300 μs and should be derived from a 10-kHz clock. Allow for future requirements where the 10-kHz signal will be replaced by another of a different frequency. The output frequency should change in proportion to the input signal, but the mark-to-space ratio should not change.**

The requirement is for an output signal with a period of 25 ms. In other words, the signal should have a frequency of 40 Hz. This frequency can be derived from the 10-kHz input signal by using a divide-by-250 counter. In this way, the present requirement is satisfied and future changes of the frequency of the clock will allow the output to change in proportion. A divide-by-250 can be formed by cascading two divide-by-5 counters (available in a single 74LS390 IC used in Example 9.8) with a divide-by-10 counter (the 74LS90 IC of Fig. 7.4).

Having satisfied the frequency requirements, we can now derive the required pulsewidth to satisfy the first requirement. This can be achieved by the use of a monostable at the output of the frequency-dividing counters. Although this approach will satisfy the present needs, it does not take into account those of the future (since it will continue to produce pulses with a width set by hardware regardless of the frequency of the input signal). This solution is therefore not acceptable.

An alternative solution is shown in Fig. 9.20. It incorporates an 8-stage up-counter, such as the one used in Fig. 7.1. When free-running, this counter

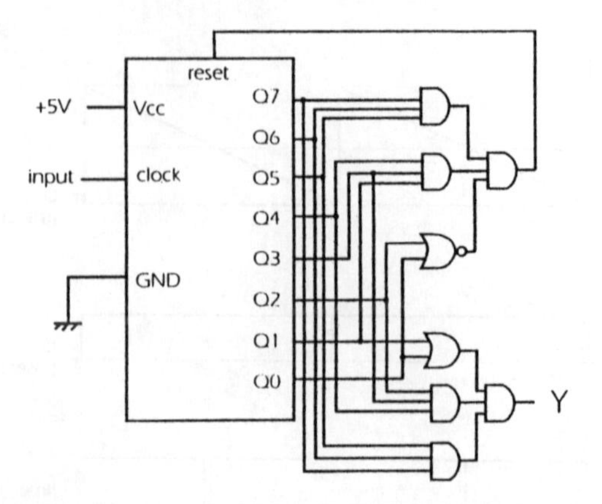

Fig. 9.20 A frequency divider with a mark-to-space setting.

has a counting cycle of 256 steps. Three AND gates and a single NOR gate are combined to produce a reset signal. It will be at a logic-high state when all the output lines of the counter are at a logic-high state except Q0 and Q2. This corresponds to the binary number 11111010, which is equivalent to the decimal number 250. The combination will therefore provide the division factor needed to satisfy the frequency requirement.

The next step is to set the required duration of the pulses. This is achieved by the use of an additional four gates to deliver an output at line Y. This line will be at a logic-high state when output lines Q7, Q6, Q5, Q4, Q3 and Q2 are high, while Q1 and Q0 are not at a logic-low state simultaneously. This covers three states of the count; these are 11111101, 11111110 and 11111111. This implies that the output line will be at a logic-high state for three successive cycles of the input clock. For the first requirement, the frequency of the clock is 10 kHz. Since the duration of each cycle of the input clock is 100 μs ($\frac{1}{10}$ kHz), the output will be high for 300 μs. The second requirement is also satisfied, since the generated pulse will be at a logic-high state for three clock pulses regardless of its frequency.

A designer can choose from many ways of setting the mark-to-space ratio of signals. The complexity of the solution depends on the required ratio. Try to solve the following example before reading the solution.

Example 9.14: **In a variable-frequency application, the mark-to-space ratio is needed to be kept constant.**

(a) Design a circuit where the mark of the waveform covers 10% of a cycle.

(b) Alter the circuit to allow the mark to cover 30%.

(c) Modify the circuit to deliver a mark-to-space ratio of $\frac{1}{2}$.

A simple way of fulfilling the requirements of some mark-to-space ratio applications is to use the 74HCT4017 IC, since each of its output lines goes high for 10% of the time. Using the IC in the free-running mode (i.e. with the reset line connected to the 0-V line), the output of any line will be high for 10% of the cycle regardless of the clock frequency. This satisfies the requirement of part (a) of this example.

The requirement of part (b) is to increase the mark to 30%. Since this is a multiple of 10%, then three output lines of the 74HCT4017 IC can be combined by an OR gate to satisfy the requirement. This is shown in Fig. 9.21(a) where, again, the IC is operated in the free-running mode.

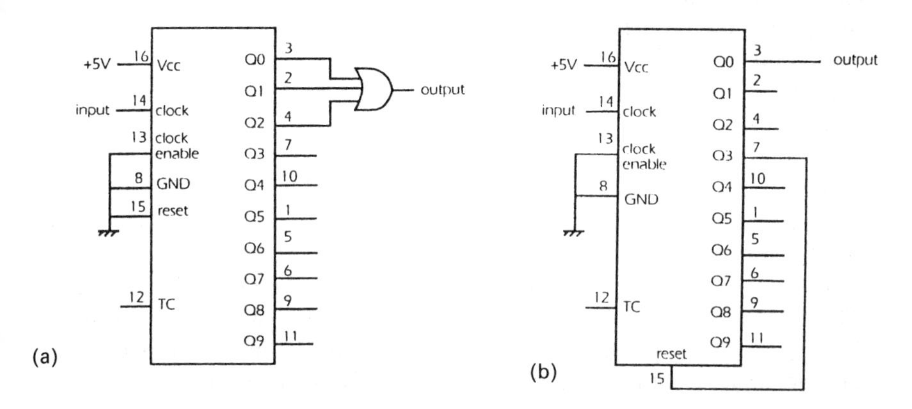

Fig. 9.21 Pulsewidth setting: (a) 30% mark; (b) 33% mark.

The requirement of part (c) may seem more difficult to achieve, since we need the mark to cover one-third of the cycle. In fact, the 74HCT4017 IC can be used again to provide the required mark-to-space ratio. Figure 9.21(b) shows the IC with its output line Q3 connected to the reset line. In this way, the circuit will count to 3 before resetting. Taking the output of the circuit from line Q0 implies that the output will be high for one-third of the cycle, thereby satisfying thc requirement.

9.5.4 Setting duration between interrupts

The operation of the majority of microcomputer-based real-time control systems relies on the use of regularly occurring interrupts. The method of deriving the interrupting signal depends on the requirements. For instance:

- A fixed rate of interrupts can be achieved by hardware, e.g. a timer whose delay is set by external components, or a counter which will interrupt the microcomputer when a specified count is reached.

- Some flexibility can be added by allowing a user to place a link or a microcomputer to close an analogue switch in order to select any of a few interruption rates.
- Further flexibility can be provided if the microcomputer is allowed to choose the rate by supplying a digital number. This number can be utilized as a division factor to reduce the frequency of a fixed clock signal to one of many rates of interruption, as shown in the following example.

Example 9.15: **A microcomputer supplies a 50-kHz clock signal. Design a circuit to allow the microcomputer to derive the rate of occurrence of interrupts by dividing the 50-kHz signal by an 8-bit number.**

The 8-bit number can be supplied by the microcomputer through a digital-output port. This will be used as a division factor of 1 to 255. The resulting interruption rate will vary between 50 kHz (i.e. once every 20 μs) and 196 Hz (i.e. once every 5.1 ms).

A counter can be incorporated to count the 50-kHz clock pulses. The main step in this design is to define a method of using the digital number supplied by the microcomputer to generate the interrupt and reset the counter. An attractive solution can be based on the use of a digital magnitude comparator. The circuit of Fig. 9.22 utilizes the 74HCT688 IC (which is also used in Example 9.11). This IC compares the 8-bit digital number supplied by the microcomputer with the output lines of an 8-bit counter. The output of the comparator (line Y) is at a logic-high state and will only change to a logic-low state when the two values are equal. The change of state of this signal is employed to interrupt the microcomputer and reset the counter to repeat the counting process. In this way, interrupts are generated regularly without further interference from the microcomputer (unless a change of their rate is needed). One should keep in mind, however, that the duration of the signal delivered to the interrupt line is low. Depending on the specification of the required interruption signal, the pulse may need to be widened by the use of a monostable.

9.6 Counting frequency

There are several ways of designing a frequency counter. An interrupt-based method makes a microcomputer read a counter when interrupted regularly. This is fine if the microcomputer can respond to the interrupt and read the counter before the count is incremented by the next clock pulse. If this is not guaranteed, then the signal which creates the interrupt can also be utilized to stop the clock pulses from reaching the counter until the microcomputer reads and resets the counter. A microcomputer can be programmed not to

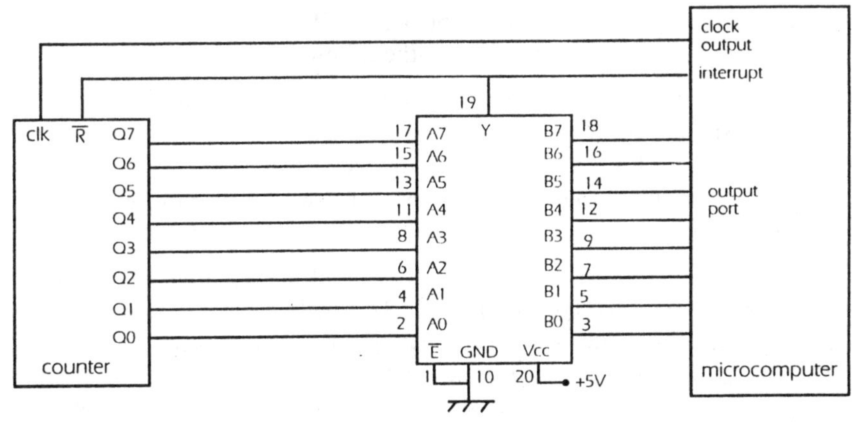

Fig. 9.22 Setting the rate of interrupts.

use the result until adding several successive readings. This can help to obtain a more accurate result. The drawback of this approach is the time taken to obtain and add the readings, leading to a reduced rate of generating results. There are ways around this problem. One way is to store the last 16 readings in a table. Whenever a new reading is obtained, it will be stored instead of the oldest reading in the table. This is easy to apply, since all the microcomputer needs to do is to use a pointer to identify the oldest reading. That pointer will be incremented every time a new reading is stored. With this approach the microcomputer can produce a new average every time a new reading is obtained, as all it needs to do is add the contents of the table and scale the result accordingly.

Example 9.16: **The operation of a microcomputer is based on a regular real-time event controlled by a high-priority interrupt. How can that microcomputer be employed as a frequency-measuring device? Include a means to permit the microcomputer to select the frequency range.**

The requirement to allow the microcomputer to alter the frequency range automatically increases the complexity of the design. It needs an out-of-range detector as well as a mean of changing the range. Let us divide the solution into a few sections.

Method of measurement

Generally, the measurement of frequency can be performed by counting the number of cycles of the applied signal within a fixed time interval. The frequency is normally measured over a relatively long time. Use can be made of the available regular high-priority interrupt. If the rate of its occurrence is high, then a software counter is required to be loaded with a value Y and decremented every time the interrupt occurs. When the count equals to 0, the interrupt-service routine performs the following actions:

1 Reads the number of cycles counted by a hardware counter.
2 Resets the hardware counter.
3 Reloads the software counter with the value Y.
4 Scales the count to give the number of cycles per second (which is the frequency).

The circuit of Fig. 9.23 can be employed to fulfil the requirement. The main part of the circuit is the counter (IC2) which performs the frequency measurement. Since the input frequency is unknown, then it is possible to have an input frequency which is too high to measure without the counter overflowing. On the other hand, the input frequency may be too low to be measured accurately. A means of changing the range of the measured frequency can, in this case, help to achieve a more accurate result. Allowing the microcomputer to select the range will make the frequency-measuring process simpler for the operator, since it removes the need for manual examination of the measured frequency.

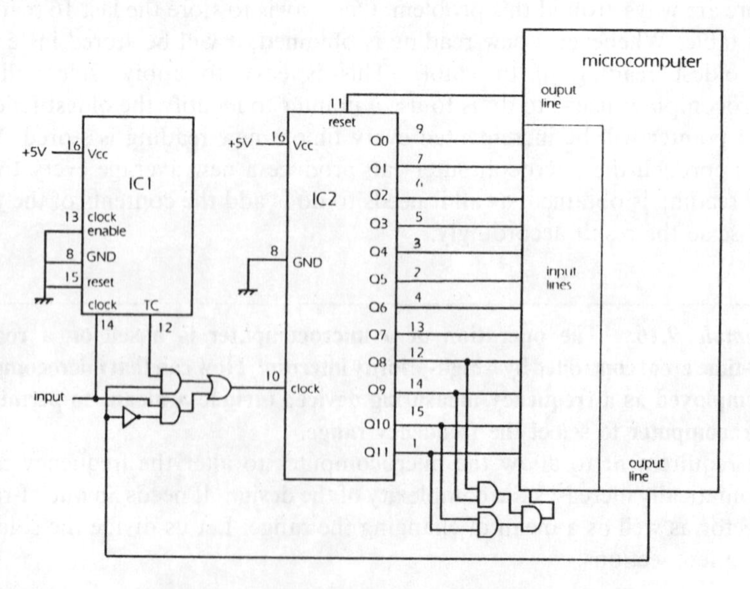

Fig. 9.23 A frequency counter with range selection.

Method of range selection

This is based on the use of logic gates. The microcomputer needs to decide whether to count the number of cycles directly, or to pass it through a divide-by-10 counter in order to reduce its frequency before counting. The

74HCT4017 IC is chosen as the frequency divider (IC1 in Fig. 9.23) and the 74HCT4040 IC is selected as the counter (IC2) to provide a 12-bit resolution. The microcomputer selects the input path of the signal through a single output line. Setting the line low at start permits the input signal to pass directly through the AND and OR gates to the clock input line of the counter. This does not stop the 74HCT4017 IC from dividing the input frequency by 10, but its output is blocked by the other AND gate. Operation will start when the microcomputer uses another of its digital-output lines to pulse the reset line of the counter, which starts counting from 0. The microcomputer then reads the count regularly and resets the counter.

Out-of-range detection

A designer has a choice of several methods of preventing the microcomputer from reading the wrong value as a result of a counter overflow. One method is to use a hardware out-of-range detector, formed by the use of three AND gates in Fig. 9.23. The result of ANDing output lines Q8, Q9, Q10 and Q11 is the delivery of an out-of-range indication some time before the counter overflows. This gives the microcomputer the chance to respond before the overflow takes place. A decision is required on the best way of interfacing the indication to the microcomputer. The use of an interrupt line may seem attractive, but if such a line is not available, or if it is preferred not to interrupt the operation of the microcomputer unnecessarily, then the out-of-range indication can be connected to a digital-input line of the microcomputer. Reading that line can be performed within the servicing routine of the regularly occurring interrupt.

With such an approach, the designer should choose the out-of-range count to be sufficiently far away from the overflow point. The circuit of Fig. 9.23 gives the indication when the count reaches, or exceeds, 1111 0000 0000. In other words, the counter will count up to 3840 cycles before indicating an overflow. Note that there is no need to latch the indication, since the continuation of the count will only alter the lowest eight bits of the count. This is true as long as the microcomputer will respond within the next 256 clock cycles.

On-line range selection

Having defined a way of generating the out-of-range indication, let us decide on how to implement the range-setting task. A suitable method will be to program the microcomputer to examine the out-of-range indication before attempting to read the count. If that line is at a logic-high state, then the microcomputer will alter the range of measurement, by setting the range-selection line to a logic-high level. This stops the input pulses from flowing directly to the counter and enables the transfer of the lower-frequency pulses from the output of the 74HCT4017 IC to the input of the counter through the

AND and OR gates. The microcomputer will then reset the counter and repeat the measurement. This action supplies the counter with pulses at one-tenth of the input frequency. Software will obviously take care of scaling the measured frequency.

While working on the high-frequency scale, the need to change to the low-frequency scale will be realized by the microcomputer when it reads the count and compares it to a preprogrammed limit. If the count is less than that limit, then the microcomputer can command a change of range, by setting its range-selection line to a low state. It can then repeat the measurement after resetting the counter.

For simplicity, only one additional divide-by-10 range is included in Fig. 9.23. Additional ranges can be introduced by using a different type of counter, or by cascading two or more 74HCT4017 ICs.

Error in measurement

The presented solution may lead to a small error due to the time needed for the microcomputer to respond to its regularly occurring interrupt. The microcomputer needs to complete the software instruction it is handling, and then calls the interrupt-service routine. The error depends on the difference between the time needed to execute the shortest and longest instructions of the processor used. Let us assume that the difference is 2 μs. The effect of the error will then depend on the time interval allowed for the measurement. This time is normally selected according to the allowed range of measuring frequency. If, for instance, the measuring interval is chosen to be 1 s, then the error is 2/1 000 000, but if that interval was 100 μs, then the error is 2/100.

The following example presents an alternative method of measurement.

Example 9.17: **Design a microcomputer interface circuit to count the frequency of a square wave and inform the microcomputer when the count is ready. Assume that there are no regular interrupts within the operation of the microcomputer, and no access to interrupt lines.**

The circuit of Fig. 9.24 employs two counters to count the number of pulses within a fixed time interval before informing the microcomputer. Counter A counts the number of cycles of the input signal, while counter B controls the counting interval and is driven from a stable clock which can be external or derived from the clock of the microcomputer. The required resolution of each counter depends on the required frequency range of the input signal as well as the duration of each measurement (allowing the microcomputer to select the optimum frequency range can be introduced in the same way as in the previous example).

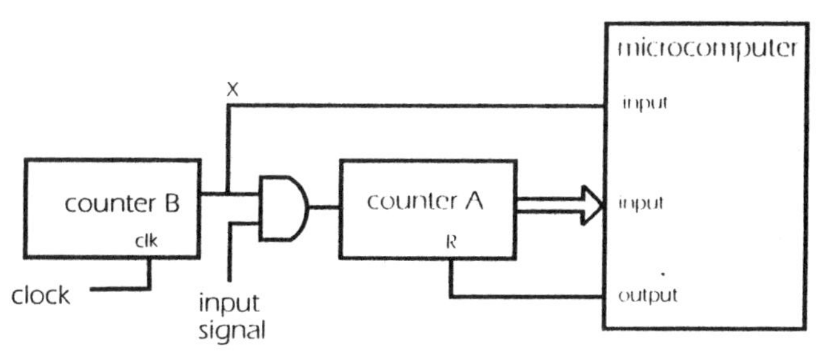

Fig. 9.24 Externally controlled frequency counter.

With the help of an AND gate, one of the output lines of counter B (call it line Qn) is employed to define the duration of supplying pulses to counter A. When line Qn is at a logic-high state, counter A counts the input pulses. The counting will stop as soon as the logic state of line Qn changes to low. If there was access to an interrupt line of the microcomputer, then this transition would have been used to interrupt the microcomputer to read counter A. But an access to such a line is not available and the indication from counter B has already stopped counter A from counting. Counter A will continue to be in this state until the state of line Qn changes to high again. The time between changes in the state of line Qn can be set relatively long, e.g. once every second. This is very likely to be long enough for a microcomputer to examine line X while carrying out other activities. In other words, reading line X is dealt with as one of the tasks of the microcomputer without the need for a dedicated interrupt.

Without synchronization, there is a possibility of a small error which depends on the instant at which line Qn changes state relative to the measured clock pulses. If synchronization is required, then a D-type latch can be employed, as shown in Fig. 9.25. The change of state of line Qn will only affect the operation at the beginning of a clock cycle. This implies that the counter will only count the number of full clock pulses.

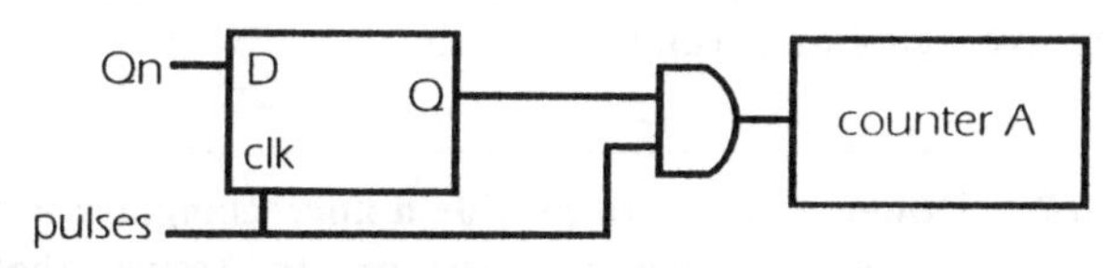

Fig. 9.25 Synchronized frequency counter.

If the frequency to be measured is of a low value, then taking several readings and averaging the results will take a very long time, which may not be acceptable in some applications. Rather than counting pulses, the duration of each cycle can be measured, as shown in the following section.

9.7 Measuring time intervals

There are several ways of measuring time intervals. The choice between them is greatly influenced by the type of interface circuit, the duration of the interval and the acceptable error in measurement. One needs to choose between performing the microcomputer-based measurement by analogue or digital means.

- An analogue approach can be based on the use of a ramp generating circuit. The slope of the ramp is either fixed and programmed into the microcomputer, or is adjustable by the microcomputer. Hardware can be used to permit the operation of the ramp generator only during the time of interest. The hardware can then inform the microcomputer when the interval is complete. The latter reads the amplitude of the ramp signal through an A/D converter (note that no fast response from the microcomputer is necessary if the hardware has stopped the ramp before informing the microcomputer). The ramp starts from 0 after being reset at the beginning of the interval and continues until its end. The time is calculated by the microcomputer after it reads the instantaneous value of the ramp at the end of the interval. Further measurements can take place within the interval (while the ramp is being generated) by additional digitization of the ramp through the A/D converter. Knowing the slope, the microcomputer can calculate the time since the start of the ramp. Allowing the microcomputer to control the slope of the ramp, gives it the ability to extend the range of measurement while covering the highest possible proportion of the allowable range of the A/D converter. This helps the performance of a wide range of measurements at the best resolution.
- A digital approach can be based on the use of a counter which is reset at the beginning of the interval, and then instructed to count until its end, as shown in Example 9.18.

With either approach, a more accurate measurement can be obtained if the interval to be measured corresponds to one or more regularly repeated events. The microcomputer can be programmed to repeat the measurement several times before averaging and using the results.

Example 9.18: **Design a circuit to provide a microcomputer with the ability to measure the time between two successive events. Assume that there is no access to the microcomputer's interrupt lines.**

The design can be based on employing a counter to start counting clock pulses when the first event occurs and terminate the count as soon as the second event begins. The circuit of Fig. 9.26 can be used. It employs an SR latch as a means of controlling the counting interval. Event 1 triggers the latch to set its output line Q to a logic-high state. This allows the AND gate to conduct clock pulses to the input of the counter. The counting will stop when event 2 resets the latch sending line Q to a logic-low state. The microcomputer can read the count, and then use a digital-output line to reset the counter while line Q is low.

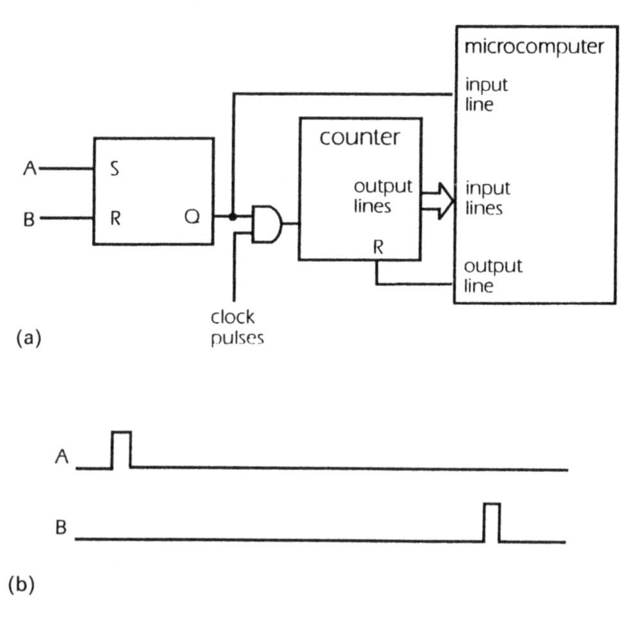

Fig. 9.26 Measuring time intervals: (a) circuit; (b) input signals.

In order to ensure that the count will always start from 0, a rising-edge triggered monostable (such as the 74HCT4538 IC shown in Fig. 9.17) can be employed to reset the counter when event 1 occurs. This approach is also useful in case the microcomputer was involved with time-consuming activities and did not read the counter prior to the recurrence of event 1. If this happens, then the monostable will reset the counter and a reading will be lost (which is better than obtaining a wrong reading). An alternative approach is to store the result in a latching buffer which is enabled by event 2. This implies using additional components in order to prevent losing a reading. The decision depends on the application and on the significance of each reading.

A decision is also needed on what frequency the counter should be clocked at. A 12-stage binary counter (e.g. the 74HCT4040 IC) counts up to 4095 pulses before resetting to 0. The maximum permitted count for the circuit of Fig. 9.26 should be less than this, say 3500. The designer needs to set a limit to the maximum interval that this circuit can measure, say 10 ms. This means that the maximum clock frequency is 350 kHz. If, for example, the microcomputer supplies a frequency of 1 MHz, then a divide-by-3 counter can be formed by using the 74HCT4017 IC as shown in Fig. 9.21(b). This gives a clock frequency of 333.3 kHz, which is close to, but less than, the maximum selected frequency.

The counter will therefore count 3333 pulses in 10 ms. The error in the measurement is the period of one clock pulse, which is 3 μs in an interval of 10 ms; this is considered low for most applications, even when added to the narrow pulsewidth generated by the monostable to reset the counter. The latter should obviously be reduced by choosing small values of R and C.

Note that the error is small, because of using a high-resolution counter. Interfacing such a counter with a microcomputer requires many input lines which, in some applications, is not possible. If this is the case, then the designer may be forced to reduce the number of lines at the expense of a higher error.

Exercise

Modify the circuit of Fig. 9.26 to allow the microcomputer to change the frequency of the clock, thereby extending the range at which the circuit can operate successfully.

Hint: The solution can be similar to that of Example 9.16.

The SR flip-flop in the circuit of Fig. 9.26 produces a pulse whose width equals the interval between the two events. Removing that flip-flop will alter the function of the circuit to measure pulsewidth rather than an interval between two events. The signal will be supplied directly to the input of the AND gate. A reset signal will still be needed from the microcomputer, or from a monostable to clear the counter.

Example 9.19: **How can interrupts be used to allow a microcomputer to measure the duration of a pulse? Assume that the microcomputer possesses an internal counter.**

Two monostables can be utilized to fulfil the requirement. One monostable is configured to detect the rising edge of the input pulse and produce a narrow output pulse to interrupt the microcomputer. The need for a fast response dictates the use of a high-priority interrupt. The interrupt-service routine starts the internal counter. The second monostable is configured to generate an output pulse when a falling edge is detected. This will interrupt the microcomputer, which will stop the counter and read the count. The microcomputer will then reset the counter and has it ready to count during the next pulse. As explained earlier, the error in an interrupt-based measurement depends on the time difference between requesting and activating an interrupt. The longer the measured interval, the lower is the percentage error.

The connection of the monostables is shown in Fig. 9.17, where one monostable is configured for triggering from a rising edge of the input signal (input 1 of Fig. 9.17), while the second monostable is configured for triggering by a falling edge of a signal applied to input 2 of Fig. 9.17. The pulse to be measured can be connected to both inputs. Each of its edges will trigger one of the monostables. Output 1 will generate the first interrupt, while output 2 will generate the second interrupt.

Example 9.20: **Design a circuit to allow a microcomputer to detect the magnitude and direction of the phase shift between two square waves.**
The first task is to detect the magnitude of the phase shift between the two waveforms. This can be achieved by the use of an Ex-OR gate, as shown in Fig. 9.27. The two square waves are applied to input lines A and B. The output of an Ex-OR gate is at a logic-high state only when its two input lines are at different logic states. This implies that the signal conducted through line C will be at a logic-high state for the duration of the phase shift between the two waveforms. The width of the pulse can be measured by the use of a counter (as explained earlier in this section). The measurement is allowed by connecting line C to the input of an AND gate. This gate permits the flow of clock pulses to the counter during the duration of the phase shift.

The next task is to detect which of the two waveforms is leading the other. This is accomplished by the use of a D-type latch, as shown in Fig. 9.27. When waveform A changes state from low to high, it clocks the latch, which will capture and latch the state of waveform B at that instant. This will be delivered to an input port of a microcomputer through line E. A logic-high state at that line implies that waveform B is leading waveform A by an angle represented by the pulsewidth delivered by line C and represented by the count of the counter. Similarly, a logic-low state at line E means that waveform A is leading B.

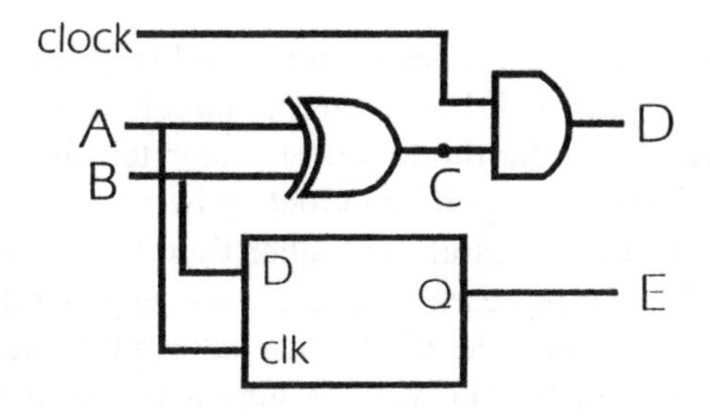

Fig. 9.27 Measuring phase shift.

Chapter 10

Switching devices

10.1 Introduction

A 'switching device' is a term identifying a device which conducts power when commanded by a suitable controller to do so. Switching devices, such as transistors, FETs, MOSFETs, IGBTs, thyristors, triacs and relays, are incorporated in a variety of applications, ranging from very low to extremely high powers. The interface between a low-power circuit and a high-power switching device often requires the inclusion of an amplifying circuit which incorporates an isolating device, e.g. an optocoupler or a pulse transformer. During their operation, switching devices generate heat due to switching and conduction losses. The generation and dissipation of heat is the subject of Section 14.7. This chapter explains the operation of a few switching devices and illustrates methods of interfacing them to microcomputers. It starts by describing a device which is not a controlled switching device, but which is employed in a vast number of switching circuits. This device is the diode.

10.2 Diodes

A diode is a two-terminal device designed to conduct power in one direction. Figure 10.1(a) shows diode D1 in a simple circuit where an a.c. signal is supplied to the anode of the diode whose other terminal (the cathode) is connected to a resistor. The diode conducts power only when it is forward biased, i.e. when the anode voltage is higher than the cathode voltage. The waveform of Fig. 10.1(b) represents the a.c. voltage generated by the power supply. Conduction through the diode only takes place during the positive half-cycle of the a.c. signal. The voltage across the load is shown in Fig. 10.1(d). During conduction, there is a small voltage drop across the diode, normally referred to as V_f. Its magnitude depends on the current conducted

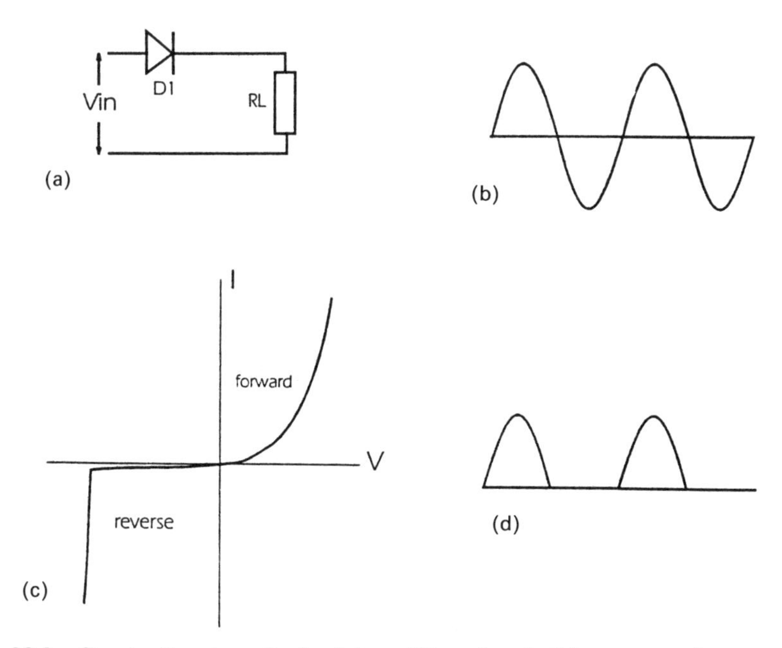

Fig. 10.1 Conduction in a diode: (a) rectifier circuit; (b) a.c. waveform; (c) diode *V*/*I* characteristics; (d) rectified signal.

through the diode, as shown in Fig. 10.1(c). This figure also shows the effect of the reverse voltage across the diode. It causes an extremely low current to flow until the breakdown voltage is reached (operation should be away from the breakdown voltage). If the direction of the diode of Fig. 10.1(a) is reversed, then conduction will take place during the negative half-cycle of the a.c. signal.

10.2.1 Packages and configurations

There is a wide choice of diodes. They are available in different packages and ratings. The increase in the use of diodes in rectifier circuits led to the availability of packages containing a few individual diodes, or packaged in a special configuration. One of the most common configurations is the full-wave bridge rectifier, represented in Fig. 10.2(a). It is a 4-terminal package whose physical size, price and shape depend on its rating. Its use removes the need for using four individual diodes. During the positive half-cycle of the a.c. input voltage (V_{in}), current flows from the supply through diode D2 to the load and then back to the supply through D3. In the negative half-cycle, current flows to the load through diode D4 and returns to the supply through D1. The shape of the output waveform is shown in Fig. 10.2(b) (a full-wave

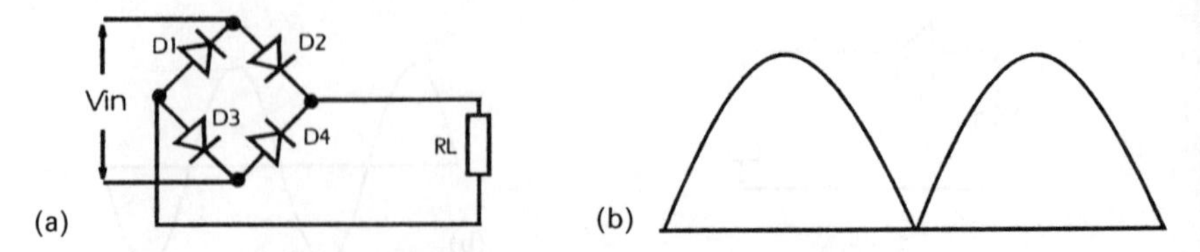

Fig. 10.2 Full-wave rectifier: (a) circuit; (b) waveform.

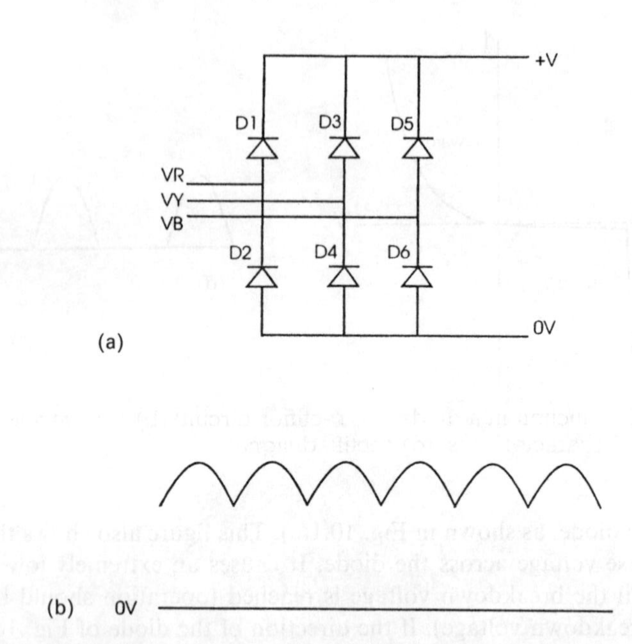

Fig. 10.3 Three-phase rectifier: (a) circuit; (b) waveform.

rectifier is incorporated in the power supply circuit of Fig. 16.6). A different use of this type of rectifier is shown in Fig. 15.8, where the diodes can be replaced by thyristors to allow a microcomputer to control the power delivered to a d.c. motor.

Rectifying a 3-phase a.c. signal needs six diodes connected as shown in Fig. 10.3(a). Current conduction through the diodes depends on the potential difference between the three input voltage signals. The sequence of the instances of starting a conduction through the diodes is D1, D6, D3, D2, D5, D4, and will repeat sequentially to generate the waveform of Fig. 10.3(b). This configuration is known as an uncontrolled 3-phase bridge, since current will be conducted according to the voltage difference and not according to commands from an external controller. A controlled version of this type of rectifier can be constructed by replacing half or all of the diodes with thyristors, as explained in Section 15.2.2.

10.2.2 Flywheeling diode

This is a normal diode connected across an inductor in a power-switching circuit, as shown in Fig. 10.4. When the transistor (Q) is commanded to turn off, the current will change. The inductor will use its stored energy in an attempt to maintain the flow of current. The result is an increase in voltage at point B according to the relationship: $V = L\ (di/dt)$. This indicates that the faster the switching process, the higher is the generated voltage. The result is an increase in the voltage at point B above that of point A. The use of the diode provides a path for the flow of current to dissipate the stored energy and stop the voltage at point B from rising to an excessive value.

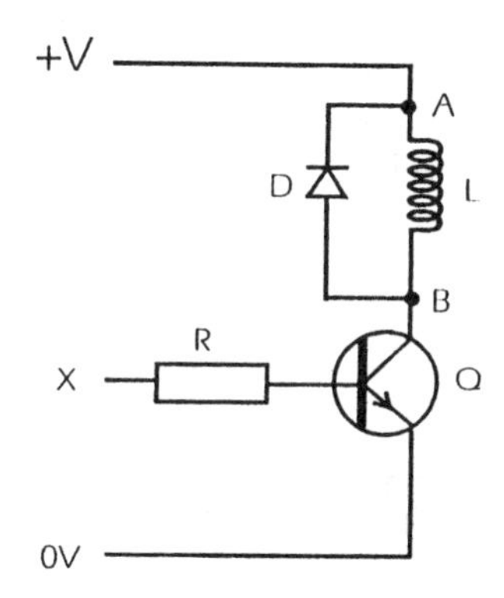

Fig. 10.4 Using a flywheeling diode.

10.2.3 Zener diodes

This is a special type of diode. In addition to permitting conduction in the forward direction, it allows conduction in the reverse direction (i.e. when reversed biased) only when the applied reverse voltage exceeds a fixed specified value. An example of using a zener diode is given in Fig. 10.5(a), where the maximum voltage at point Y is defined by the reverse value of the zener diode Z. Zener diodes are available in a wide range of reverse voltage and power values, enabling one to select a particular zener diode that will closely satisfy a design requirement (the selection should also take into consideration the power rating of the zener diode). Zener diodes are of low cost and are employed to protect against overvoltage or to provide a voltage reference. The voltage is, however, not exact, due to tolerance, and it is affected by temperature as well as the value of current.

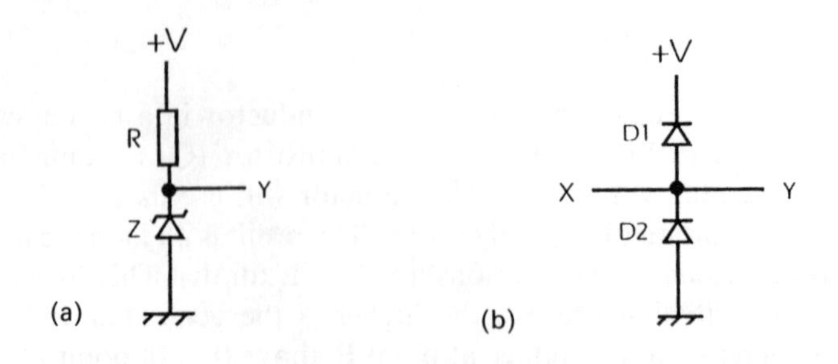

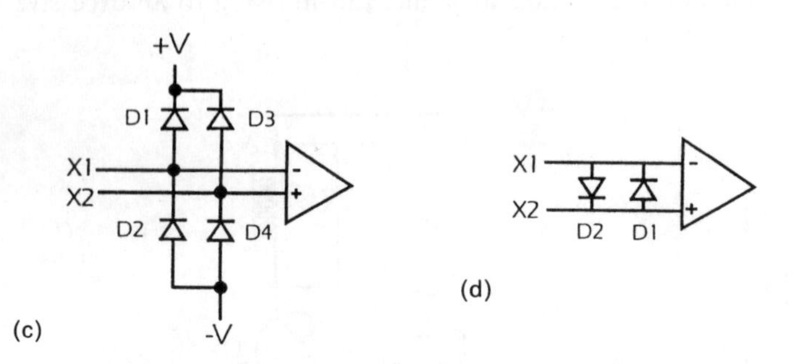

Fig. 10.5 Using diodes: (a) zener diode; (b) protection with diodes; (c) limiting the input voltage; (d) limiting the voltage between terminals.

10.2.4 Examples of using diodes

Diodes are extremely useful devices employed in such applications as rectification, protection and voltage referencing. They find a wide use in power-electronics applications and are included in several circuits throughout this book.

Example 10.1: **How can diodes be used to protect an input line of a microcomputer and ensure that its voltage is within the allowable limits?**
The circuit given in Fig. 10.5(b) uses two diodes to protect the input line. Diode D1 is connected between the input line X and the d.c. supply, while diode D2 is connected between the 0-V line and input line. If the input voltage becomes high, then D1 starts to conduct, keeping the voltage at line Y slightly higher than the supply voltage (by the voltage drop across the diode). Similarly, if the input voltage becomes negative, then diode D2 conducts and keeps the voltage of that line close to 0 V. Zener diodes will also be useful, since they will prevent high voltages from damaging the circuit.

The method of using diodes to protect input lines is used in Fig. 10.5(c) to protect the two input lines of an operational amplifier. Assuming that the operational amplifier is operated from a dual supply, the anodes of diodes D2 and D4 are connected to the −V supply, rather than to the 0-V line (compare with Fig. 10.5(b)). Current will only flow through any of the four diodes if the instantaneous voltage of one or both of the input signals exceeds that of the d.c. power supply of the operational amplifier. Diodes D1 and D2 of Fig. 10.5(d) are given a different protection task. They ensure that the voltage between the two terminals of the comparator will not increase to more than the voltage drop across a single diode. During normal operation, the diodes will not conduct and have no effect on the operation.

Example 10.2: **In many interface circuits, individual logic gates of different types are required. Rather than using an IC for a single gate, how can diodes be employed to form a 2-input AND gate and a 2-input OR gate?**
Two diodes are incorporated in the configuration presented in Fig. 10.6(a) to form an AND gate. Lines X1 and X2 are the inputs and line Y is the output. If any of the inputs is at a logic-low state, then the corresponding diode starts to conduct, forcing the state of output line Y to a logic low regardless of the state of the other line. The output can only be at a logic-high state when both diodes are reverse biased and not conducting. This is the case when both X1 and X2 are at a logic-high state. Similarly, diodes D1 and D2 of Fig. 10.6(b) are connected to form an OR gate. Output line Y is at a logic-low state only if both of the input lines (X1 and X2) are at a logic-low state. Otherwise, a logic-high signal at any of the input lines leads to conduction through the corresponding diode. This changes the state of the output line to a logic high.

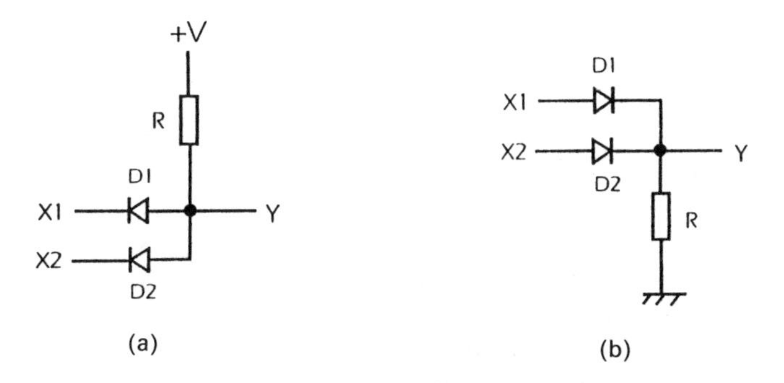

Fig. 10.6 Replacing logic gates with diodes: (a) 2-input AND gate; (b) 2-input OR gate.

10.3 Transistors

Many microcomputer-based instruments and controllers utilize transistors as controlled switches for various levels of power. Transistors are also incorporated as amplifiers in audio circuits, or to provide a microcomputer with the ability to drive other switching devices, e.g. relays and thyristors. They are also employed to interface low-voltage microcomputer-based circuits to devices or circuits operating at other voltage levels.

10.3.1 NPN and PNP transistors

A transistor has three terminals: they are the base B, collector C and emitter E. The conduction of current through a transistor is controlled from its base. An NPN transistor, such as the one shown in Fig. 10.7(a), can be operated by supplying a signal of a suitable level to its base when the voltage applied to the collector is higher than that of the emitter. The junction between the base and emitter is equivalent to a diode and needs to be forward biased to start conduction. A signal from an output port of a microcomputer causes current to flow into the base of the transistor. This turns on the transistor, allowing current to flow from both the collector and the base to the emitter. The collector current is higher than that of the base by a factor h_{fe}, which represents the gain of the transistor and is not the same value for all transistors. The relationship between the transistor currents is represented by the following equations, where I_e is the emitter current, I_b is the base current, I_c is the collector current, and h_{fe} is the current gain.

$$I_e = I_b + I_c \tag{10.1}$$

and

$$I_c = I_b h_{fe} \tag{10.2}$$

Hence

$$I_e = I_b(1 + h_{fe}) \tag{10.3}$$

If $h_{fe} \gg 1$, then Equation 10.3 can be approximated to:

$$I_e = h_{fe} I_b \tag{10.4}$$

In the circuit of Fig. 10.7(a), the load resistor R2 is connected between the d.c. supply and the collector of the transistor. The transistor is turned on when a high-level logic signal is delivered from an output port of a microcomputer. The duration of passing current between the collector and emitter is determined by the period of supplying the base of the transistor with current. If the transistor has an h_{fe} of 100, then (according to Equation

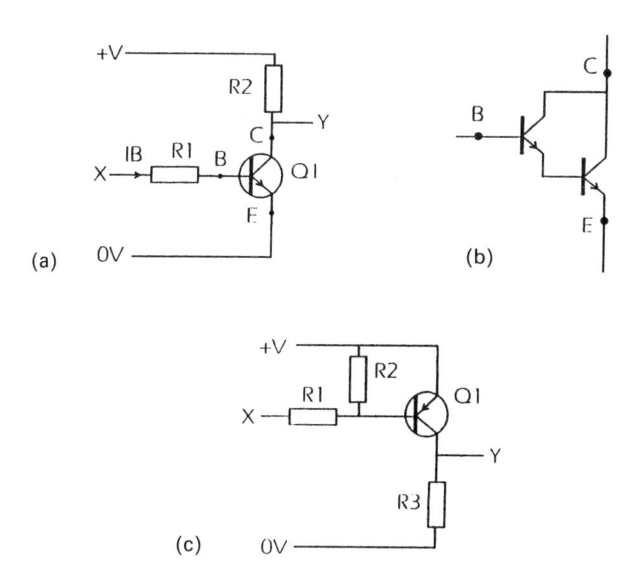

Fig. 10.7 Transistors: (a) NPN; (b) Darlington; (c) PNP.

10.2) the current passing through the load will be 100 times the current supplied by the output port of the microcomputer. The transistor is therefore acting as both a controlled switch and a current amplifier. A few transistors can be cascaded to provide a much higher amplification which will satisfy the requirements of high-power loads, such as motors and heaters. Transistors are available from several manufacturers, offering the user a very wide choice from small devices for low-power applications to power transistors capable of handling very high currents and voltages. Rather than cascading transistors, the load requirement may be satisfied by the use of a very high gain transistor. It is known as the Darlington transistor and consists of a pair of transistors connected as shown in Fig. 10.7(b). The gain of the device is the product of the h_{fe} of the two internal transistors. Its base-to-emitter voltage is higher than that of a single transistor (equal to the sum of V_{be} of both transistors).

Current gain is not all that a transistor can achieve. The voltage of the d.c. supply in Fig. 10.7(a) is shown as +V, because it does not have to be the same as that of the microcomputer. This means that a transistor can be employed to interface a microcomputer to a circuit, or to a load, operating at a different voltage. One should obviously select a transistor capable of withstanding the voltages and capable of conducting the required current. There are limits specific to each transistor which should not be exceeded. These are the maximum collector current ($I_{c\,max}$), the maximum collector-to-emitter voltage ($V_{ce\,max}$), the maximum base-to-emitter voltage ($V_{be\,max}$), and the maximum transistor power (P_{max}). These limits are given in catalogues and data sheets and should be observed at all times. Section 10.3.2 describes the characteristics of a transistor and explains how to define the operating point and its effect on the operation.

So far, this section has only considered NPN and Darlington transistors. Another very useful transistor type is the PNP. It is similar to the NPN transistor, except that all the polarities are reversed. For instance, the circuit given in Fig. 10.7(a) allows conduction through the load when the input signal is at a high logic level. If the opposite is required, then a PNP transistor can be connected, as shown in Fig. 10.7(c). In this circuit, conduction through the transistor and hence through the load resistor R3 takes place as long as the input signal is at a low logic level. Therefore, the choice between the two circuits depends on the required relationship between input and output.

10.3.2 Transistor characteristics

When the transistor of Fig. 10.7(a) is turned on, there is a voltage drop between the collector and emitter which limits the magnitude of the current flowing through resistor R2. The output of the circuit is at point Y. When the transistor is conducting, the voltage at that point is equal to the voltage drop between the collector and the emitter. The output voltage becomes the same as the supply voltage when the transistor is turned off, since there is no current flowing through R2 and hence there is no voltage drop.

The voltage between the collector and emitter is referred to as V_{ce}. Its value depends on the value of the collector current I_c and the characteristic of both the transistor and the load. The relationship between V_{ce} and I_c is shown in Fig. 10.8. A few curves are drawn, each corresponding to a value

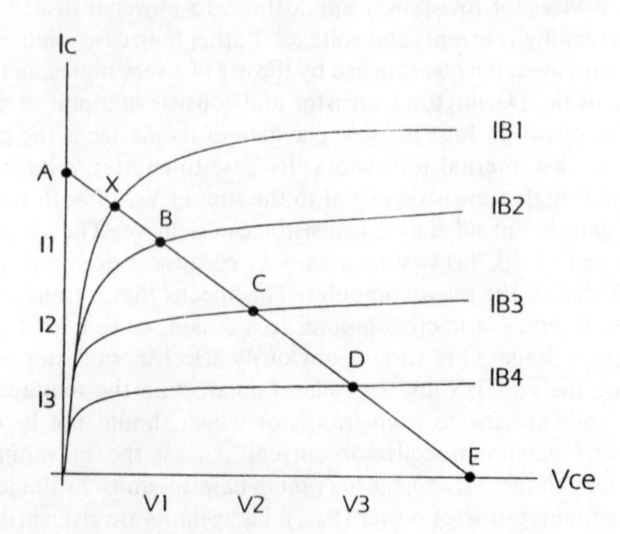

Fig. 10.8 Transistor characteristics.

of the base current I_b, where:

$$I_{b1} > I_{b2} > I_{b3} > I_{b4}$$

The value of the load resistor (call it R_L) limits the magnitude of the collector current. The line between points A and E represents the effect of R_L, where point E corresponds to a situation where the transistor is off. In other words:

$$I_c = 0 \quad \text{and} \quad V_{ce} = V_{cc}$$

where V_{cc} is the supply voltage. Point A, on the other hand, corresponds to:

$$I_c = V_{cc}/R_L \quad \text{and} \quad V_{ce} = 0$$

An operating point of the transistor will be along line AE at a point where it intersects one of the curves. Note that only four curves are drawn in Fig. 10.8. In reality, there is an infinite number of curves depending on the value of I_b.

A value of $I_b = I_{b3}$ will result in operation at point C where I_c and V_{ce} are represented by points I2 and V2 respectively. Increasing I_b to I_{b2} will shift the operating point to B, thereby increasing I_c to I1 and reducing V_{ce} to V1. Increasing I_b further will move the operating point to X, which is the highest value of I_c that can flow in the transistor for a load line AE. Increasing I_b further will not change I_c. The transistor is referred to as 'saturated' where the operating point is moved from the 'active' region to the 'saturation' region. The collector-to-emitter voltage V_{ce} is at its lowest value, which is referred to as $V_{ce\,sat}$.

Let us consider a situation where a transistor is operating at point C and supplied with a time-varying signal, e.g. a sine-wave, at its base. The result is a variation of I_b which leads to a corresponding variation of both I_c and V_{ce}. The time-varying signal will add a time-varying current to I_b leading to a swing between, say, I_{b2} and I_{b4}. This leads to a swing of I_c between I1 and I3 and a corresponding swing of V_{ce} between V1 and V3. Note that as I_b and I_c increase, V_{ce} reduces. This causes the output signal at point Y of Fig. 10.7(a) to be the inverse of the time-varying signal applied to the base. The transistor is therefore acting as an analogue inverter which is also amplifying the input signal. If the amplitude of the input signal is large, or if the amplification of the transistor is high, then the output waveform may be clamped. This is because the value of the output is limited by the parameters of both the transistor and the load. In some situations, however, the operating point (also known as the quiescent point) may not be in the middle of the operating region. This leads to an unequal clamp on large signals which can be corrected if the operating point is moved to the middle.

Transistors are widely used as switches and are driven with sufficient power to operate in the saturation region when turned on and in the cut-off region ($I_c = 0$) when off. This is attractive, because it reduces the voltage between the collector and emitter during conduction, leading to lower conduction losses. These losses are represented by:

$$P = V_{ce} I_c \tag{10.5}$$

If operation was at point C, then V_{ce} is at half the supply voltage. This leaves the other half of the supply voltage across R_L to generate I_c. One can then replace the last equation with:

$$P = (V_{cc}/2) \times V_{cc}/(2R_L)$$

Hence,

$$P = V_{cc} \times V_{cc}/(4R_L) \tag{10.6}$$

If, for example, V_{cc} is 5 V, and R_L is 1 kΩ, then P = 6.25 mW.

If, on the other hand, the transistor is used as a switch, then it will be at saturation when turned on. This can be represented by:

$$V_{ce} = V_{ce\,sat} \tag{10.7}$$

and

$$I_c = (V_{cc} - V_{ce\,sat})/R_L \tag{10.8}$$

Using the same values as above and assuming $V_{ce\,sat}$ equals 0.2 V, the value of P can be calculated by substituting Equations 10.7 and 10.8 in Equation 10.5. The result is:

$$P = [0.2(5 - 0.2)]/1\,K = 0.96\,mW$$

which is much smaller than when operating at point C.

10.4 FETs, MOSFETs and IGBTs

This section considers three devices widely used in power-switching applications, i.e. motor drives, temperature control and robotics.

10.4.1 Field-effect transistor (FET)

This is a 3-terminal device possessing a very high input impedance and is turned on by an applied voltage. This is different from a transistor, which is turned on by an applied base current. Due to its high input impedance, a FET draws an extremely low current, making it attractive in many applications. Figure 10.9 shows two types of FET. They are the n-channel and the p-channel types. They are equivalent to NPN and PNP transistors respectively. The three terminals of the devices are shown as G, D and S corresponding to gate, drain and source respectively.

The characteristics of a FET are shown in Fig. 10.9(c), where a single curve is drawn representing the applied V_{gs} (similar to the transistor

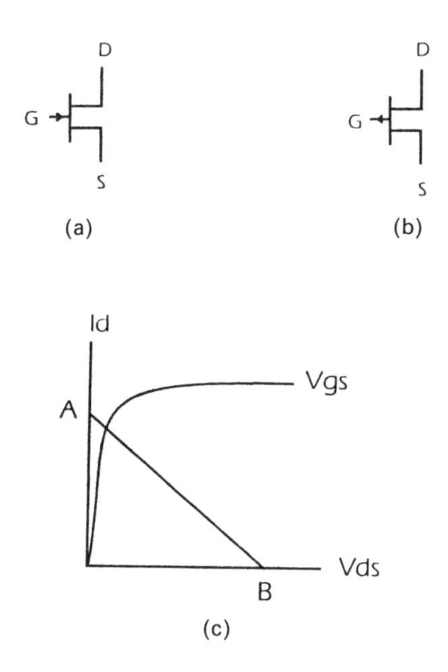

Fig. 10.9 Field-effect transistors: (a) n-channel; (b) p-channel; (c) characteristics.

characteristics, there is an infinite number of curves corresponding to the values of V_{gs}). A load line can be drawn on the device's characteristics, in a similar way to that of the transistor shown in the last section. Point A of the load line is where V_{ds} is equal to zero. The value of the drain current I_d can therefore be obtained from dividing the applied d.c. voltage by the load resistance. Point B, on the other hand, is where I_d equals zero, i.e. where V_{ds} equals the supply voltage.

FETs are used in a large number of interfacing circuits as microcomputer-controlled switches. Two such circuits are shown in Fig. 10.10. The circuit in Fig. 10.10(a) shows the use of a FET to give a microcomputer the ability to reset an integrator. When the FET is turned on, it provides a path for discharging the capacitor, thereby allowing the microcomputer to define the instant at which the integration process starts. The circuit of Fig. 10.10(b) shows the utilization of a FET in a different application. The use of this circuit provides the microcomputer with the ability to change the sign of an analogue signal through the alteration of the state of a single digital-output line. When the FET is turned on, the point joining resistors R3 and R4 is connected to the 0-V line, making the configuration an inverting amplifier with a unity gain (a FET with a low ON resistance R_{ds} is therefore preferred). Turning the FET off, on the other hand, makes the configuration a unity-gain non-inverting amplifier.

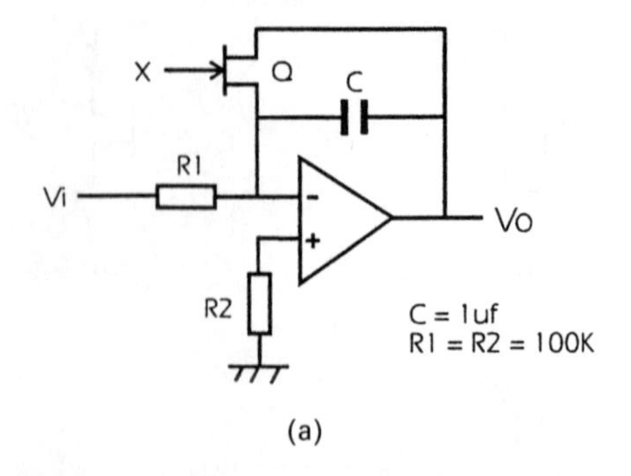

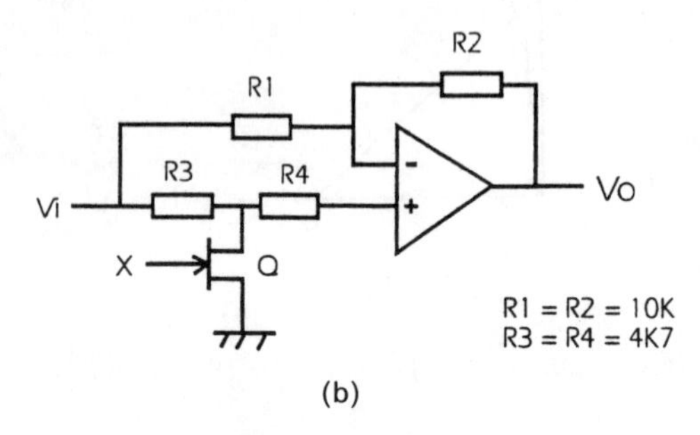

Fig. 10.10 FETs circuits: (a) integrator; (b) controlled sign-changer.

10.4.2 Metal oxide silicon field-effect transistor (MOSFET)

This is a voltage-controlled 3-terminal device. The three terminals have the same names as in a FET, but the symbol of the device is different, as shown in Fig. 10.11. A MOSFET is very attractive to use in power-switching circuits due to its high-switching speed, extremely high gain and low drive current requirement. The interface and drive circuit is relatively simple. Turning the device on is performed by applying a voltage at its gate to charge the input capacitor within the device. The value of that voltage should be higher than the threshold voltage of the MOSFET. The device will turn off as soon as the gate voltage drops below its threshold voltage. The conduction losses of the device are represented by R_{ds} which is equal to V_{ds}/I_d. Its value increases with the increase of V_{ds} which imposes a limitation in high-voltage applications. Figure 10.11(c) shows the characteristics of the device, giving three curves corresponding to:

$$V_{gs1} > V_{gs2} > V_{gs3}$$

By examining the characteristics, one can observe two regions. The first is the linear region in which the value of R_{ds} is constant (i.e. I_{ds} increases with V_{ds}). The second region is the constant-current region. Operating in the

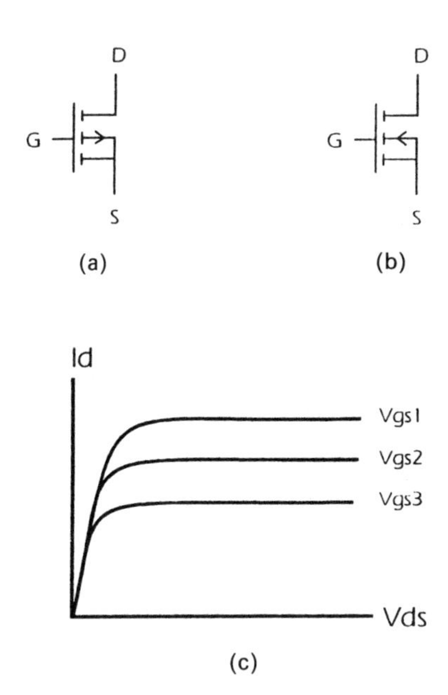

Fig. 10.11 MOSFETS: (a) p-channel; (b) n-channel; (c) characteristics.

second region increases the value of R_{ds}, since the value of V_{ds} is higher for a given value of current. One can therefore say that the value of R_{ds} can be kept low by operating in the linear region where the requirement of a higher value of I_d may be met by increasing the value of V_{gs}.

Since operating the MOSFET is based on charging and discharging the input capacitor, faster switching can be achieved if the impedance of the drive circuit is reduced. Figure 10.12(a) shows a simple drive circuit consisting of a single logic gate. Faster switching can be achieved by paralleling a few such gates. On the other hand, slower switching speeds can be achieved by inserting a resistor between the logic gate and the input terminal of the

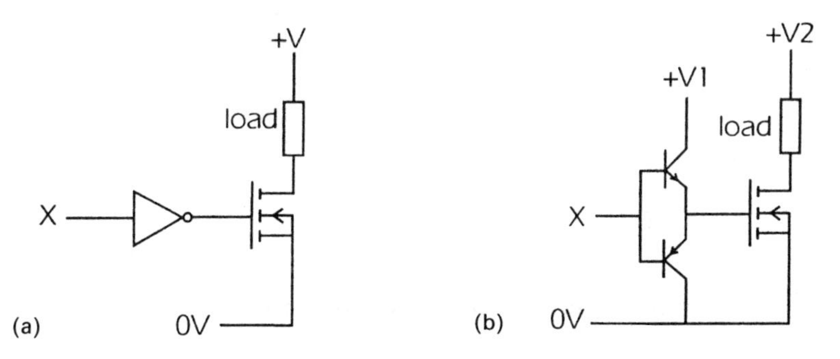

Fig. 10.12 MOSFET drive circuits: (a) logic gate as a driver; (b) transistors as drivers.

MOSFET. An alternative drive circuit is shown in Fig. 10.12(b), where two transistors are connected in a push–pull configuration to provide a fast current sourcing and sinking to the input capacitor.

10.4.3 Insulated-gate bipolar transistor (IGBT)

This 3-terminal switching device combines the benefits of a MOSFET and of a bipolar transistor. It is easy to drive, capable of very fast switching and, at the same time, offers a low saturation voltage. These features make the IGBT attractive to use, particularly in high-voltage high-frequency applications where the R_{ds} of a MOSFET is high and the switching time plus the drive requirements of a bipolar transistor are high.

Like a MOSFET, an IGBT is a voltage-controlled device. It possesses a small input capacitance and requires a relatively simple drive circuit. An example of such a circuit is given in Fig. 10.13, showing the IGBT (Q3) with a fast-recovery diode. IGBTs are widely used in inverter circuits (e.g. to replace the transistors of Fig. 15.14). When an IGBT is turned on, current starts to flow and overshoots before settling to a constant value. The peak of the overshoot and its duration are affected by the characteristics of the recovery diode. A fast diode will help to achieve a high switching speed, leading to reduced switching losses, but causes overvoltage spikes and noise. A lower switching speed means a slower recovery with a lower peak current, but it also means higher switching losses. On the other hand, turning an IGBT off leads to a voltage overshoot which depends on the switching speed and the stray inductance.

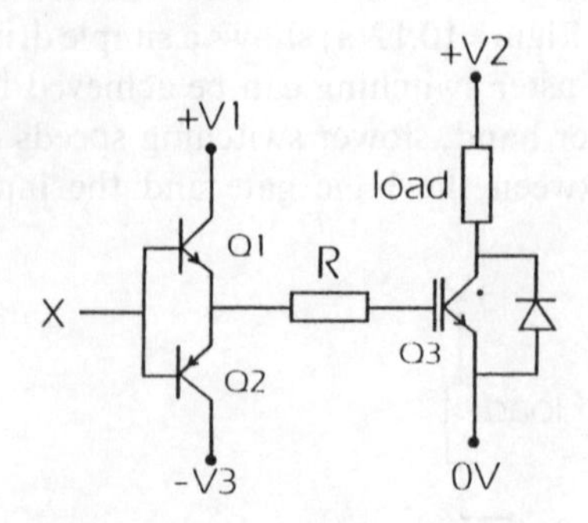

Fig. 10.13 IGBT drive circuit.

10.5 Microcomputer-controlled operations

When provided with a proper interface circuit, a microcomputer can control the conduction periods of a switching device by controlling the logic state of a single digital-output line. Since an output port latches the signal supplied by the microcomputer, a conduction commanded by a microcomputer continues until the microcomputer changes the state of the output line.

Example 10.3: **How can a microcomputer be employed to produce two successive sound effects repeatedly? The first lasts for 1 s and has a frequency of 500 Hz. The frequency of the second is 1 kHz, and it lasts for a period of 2 s.**
A simple circuit for interfacing a microcomputer to a loudspeaker is shown in Fig. 10.14. For simplicity, the speaker is supplied with a square wave produced by the microcomputer. It is delivered to the base of the transistor through the line marked 'input' in Fig. 10.14. Resistors R1 and R2 are connected in the form of a voltage divider to ensure that the transistor will turn on only when a logic-high signal is supplied from the microcomputer.

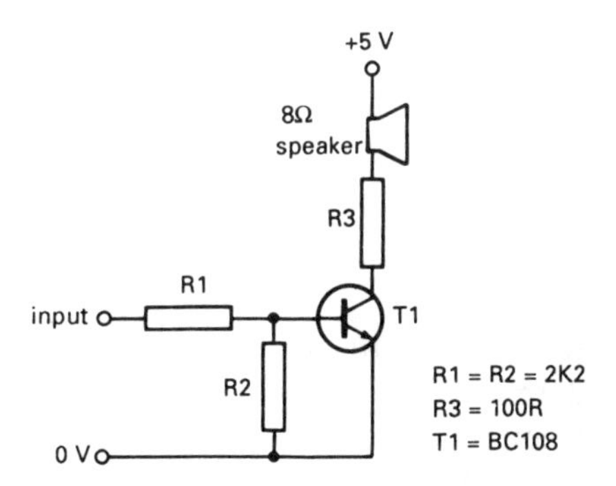

Fig. 10.14 Tone generator.

The timing requirements can be satisfied by using either a delay (the program-controlled method) or a timer-generated interrupt (the interrupt-controlled method). If the first method is adopted, then the procedure is similar to that of Example 3.1, except that here the frequency alternates between two levels. At 500 Hz, the square wave has a duration of 2 ms, i.e. the logic level of a digital-output line needs to change state once every 1 ms. The specifications demand the generation of this waveform for a duration of 1 s, i.e. complementing the state of the output line 1000 times. This is followed by generating the second waveform in the same way, except that at 1 kHz the logic level needs to be complemented once every 500 μs. Producing the second sound for 2 s can therefore be accomplished by complementing the output 4000 times before repeating the procedure and regenerating the first sound. The procedure can therefore be represented by the following steps:

1 m = 1000
2 m = m − 1
3 COMPLEMENT OUTPUT LINE
4 WAIT 995 MICROSECONDS
5 IF m > 0 THEN GO TO STEP 2
6 m = 4000
7 m = m − 1
8 COMPLEMENT OUTPUT LINE
9 WAIT 495 MICROSECONDS
10 IF m > 0 THEN GO TO STEP 7
11 GO TO STEP 1

Note that the delays used in steps 4 and 9 are 995 and 495 μs respectively instead of 1 ms and 500 μs. This is because it is assumed that the execution time of the sequential steps within each loop is 5 μs and no interrupts are generated. The microcomputer is therefore dedicated to producing the sounds.

Exercise

Write the procedure if a timer-generated interrupt is used instead of a delay. (Hint: The procedure will be similar to that of Example 3.3.)

The circuits of Fig. 10.7 and Fig. 10.14 permit unidirectional conduction of current through a load. In some situations, such as when controlling a robotic mechanism, the microcomputer needs to be provided with the ability to command motion and to alter its direction. One way of satisfying the requirements is shown in the following example.

Example 10.4: **Design a circuit to allow a microcomputer to control the direction of flow of a d.c. current through a resistor R_L.**

The circuit of Fig. 10.15 gives one way to fulfil the requirement when using a single d.c. power supply. An alternative circuit is given in Fig. 15.3(b) for operation from a split d.c. supply. The circuit of Fig. 10.15 uses transistors to control the conduction of current through the load resistor R_L in either of two paths. In the first path, current flows through R_L as a result of switching on transistors T2 and T5 while keeping T3 and T4 off. In the second path, the load current flows in the opposite direction. This is achieved by switching on T3 and T4 while T2 and T5 are off. The transistors employed in this circuit are rated for a maximum collector current of 0.1 A. This implies that for a supply voltage of + 12 V, R_L should have a value greater than 120 Ω.

Four digital-output lines of the microcomputer are reserved to control the flow of current in either of the two directions. Assume that, at the beginning, all the transistors are turned off. When the microcomputer supplies a

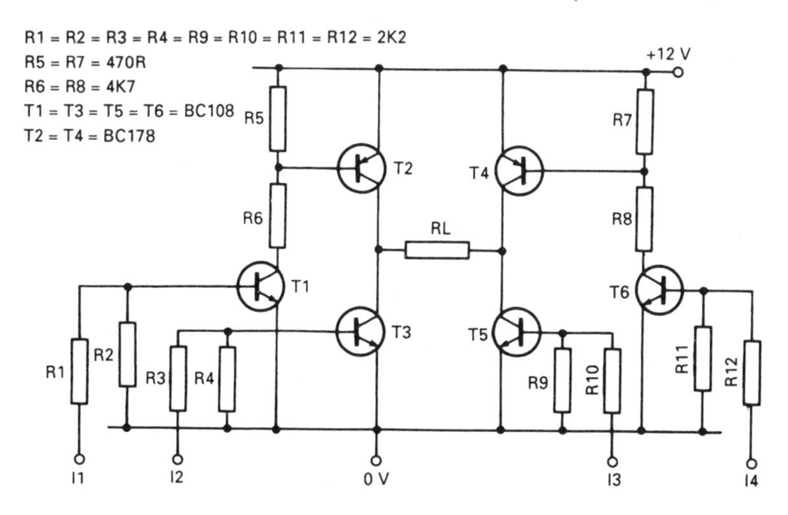

Fig. 10.15 Selecting direction of current flow.

logic-high signal to terminal I1 and I3, transistors T1, T2 and T5 turn on to commence conduction through R_L. The conduction terminates after the microcomputer supplies a logic-low signal to these terminals. Similarly, conduction in the reverse direction is performed through terminals I2 and I4.

When writing the software to operate such a circuit, one should ensure that transistors of the same leg (i.e. T2 and T3, or T4 and T5) will not conduct simultaneously. A safety interval should be introduced between commanding one transistor to switch off and instructing the other to commence conduction, thereby allowing time for the transistors to turn off. The duration of that interval should be increased if an inductive load is used. This time interval is known as the dead time and is the subject of Section 10.6.

10.6 Dead-time generating circuits

When two switching devices are connected in series across a d.c. voltage line in a configuration such as that of Fig. 10.15, only one of them should be conducting at any instant of time. This is because switching the two together will create a short-circuit across the d.c. power lines which can destroy both devices, their drive circuits, and possibly the power supply. In order to avoid turning on the two devices simultaneously, the control circuit should ensure the introduction of a short period between switching one device off and the other on. The required period is referred to as the dead time. Its duration depends on the time needed for the switching device to turn off after being commanded by the controller to do so.

When a controller decides to turn on a switching device, such as a transistor or a MOSFET, it delivers a digital pulse to a suitable drive circuit. The duration of the pulse represents the required conduction period of the switching device. The instant of terminating the pulse, however, does not correspond to the instant of turning the device off. This is because of the time interval taken for the current to decay to zero. The duration of this interval depends on many factors, such as the type of switching device and its drive circuit.

The circuit controlling the operation should be designed to provide a dead time of a suitable duration. The optimum value leads to a successful operation without introducing unnecessary delays which could degrade the response of the system or distort the shape of the output waveform. The optimum dead time varies from one circuit to another and can be introduced either by software or hardware means. The software approach is not always preferred, because the on-line microcomputer needs to be interrupted at the end of each pulse to introduce a software delay before issuing a switching pulse to the other device. The time between applying the interruption signal and the activation of the interrupt-servicing routine depends on the type of instruction being executed by the microcomputer at the instant of interruption (the microcomputer completes the execution of the instruction before responding to the interrupt). Clearly, any variation in the response time can affect the duration of conduction and can cause errors, jitter, or deviation from the desired values. The effect depends on the type of system and the rate of switching. Another limitation can be imposed if the pulses are generated at a high rate. This is because a high percentage of the processing power of the microcomputer will be devoted to responding to interruption signals and introducing pulses and their associated dead time. This is not an efficient use of the microcomputer, but can be acceptable if the microcomputer has little or nothing else to do. A timer, a counter, or a simple circuit can be incorporated to terminate one pulse, introduce a dead time and then generate a pulse to the other switching device.

Let us consider a situation where both devices need to be conducting alternately. Without a dead time, the control signal to one switching device is usually derived by inverting the control signal supplied to the other switching device (unless both devices need to be off). This can be easily achieved by an inverting logic gate. The next step is to introduce a delay between the two switching signals. This can be applied in several ways, e.g. by the use of a resistor–capacitor delay circuit. The delay will be the time of charging and discharging the capacitor (Section 4.6.2). Alternative methods are presented in the following examples.

Example 10.5: **Design a digital circuit which receives pulses from a microcomputer and generates two signals separated by a dead time.**

A widely used approach is based on incorporating a shift register as the digital dead time generator. The number of shift stages and the optimum clock frequency of driving the register depend on the required duration of the dead time. A 4-stage shift register is used in this example, as shown in Fig. 10.16(a), where the input is applied through line X and two signals are taken out. The first is from the output of the first stage (QA), while the second signal is taken from the output of the fourth stage (QD). The use of two signals synchronizes the generated outputs in order to produce a fixed delay of three clock pulses, as shown in Fig. 10.16(b). A clock frequency of 1 MHz will, for example, provide a dead time of 3 μs. Once selected, the duration of the delay will be fixed and can only be altered by varying the frequency of the clock driving the shift register.

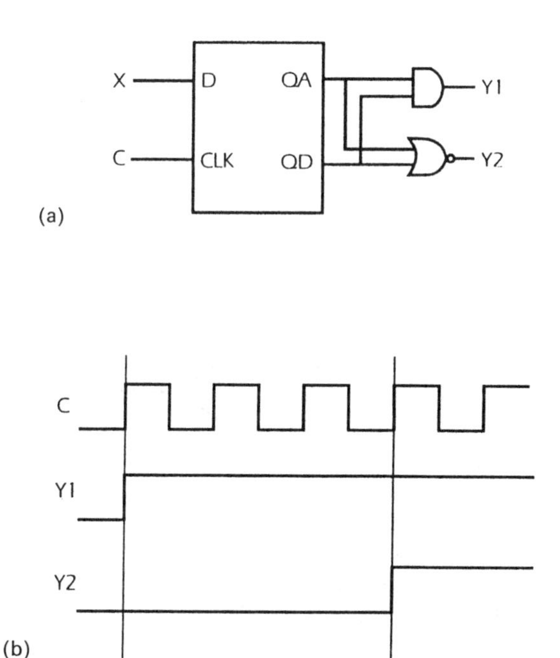

Fig. 10.16 Digital dead-time generator: (a) circuit; (b) timing signals.

The 74HCT4015 IC is a dual 4-stage static shift register. Only one of the two shift registers within this IC is needed for this example. The second register can be cascaded with the first if a longer delay is required, or if the available clock frequency was higher.

Example 10.6: **A power circuit similar to that of Fig. 15.3(b) is to be driven from a microcomputer. Design a circuit to turn the upper and lower switching devices on and off sequentially.**

One way of meeting the requirement is to use analogue means. A microcomputer-controlled triangular wave needs to be generated and delivered to the point marked 'input' in the circuit of Fig. 10.17(a). The instantaneous value of that waveform is continuously compared to the voltages at points A and B, which in turn are set by the values of resistors R1, R2 and R3. The d.c. voltage at point B is connected to the positive input line of an analogue comparator whose negative input line is connected to the input signal. The output of this comparator (line Y2) will be at a high state as long as the instantaneous value of the triangular wave is less than the voltage at point B, which is given by:

$$V_B = (V\ R3)/(R1 + R2 + R3)$$

where V is the d.c. voltage across the three resistors. The voltage at point A is higher than that at point B and is given by:

$$V_A = [V(R2 + R3)]/(R1 + R2 + R3)$$

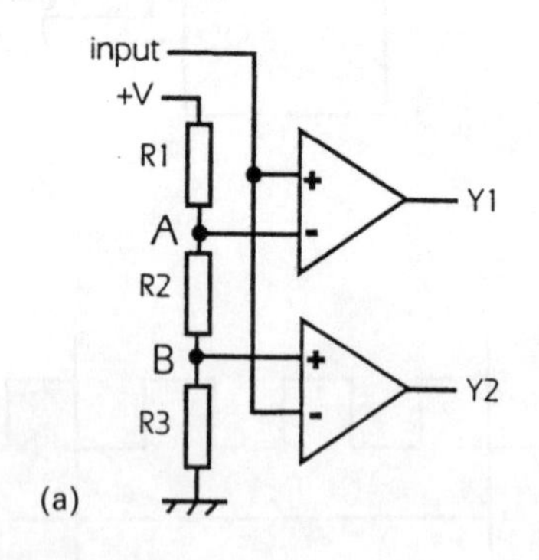

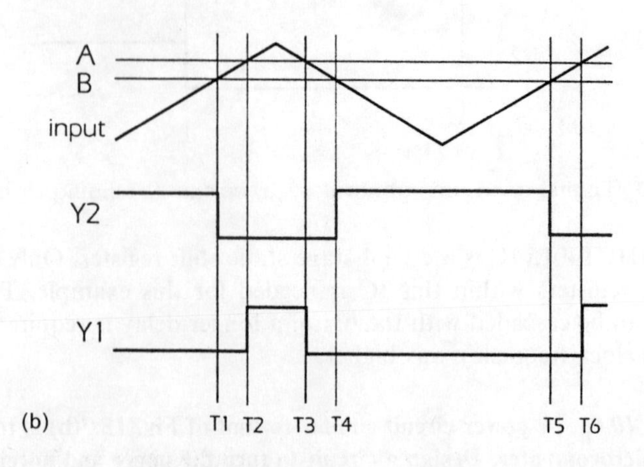

Fig. 10.17 Analogue dead-time generator: (a) circuit; (b) timing waveforms.

Note that point A is connected to the negative input line of the other analogue comparator. It is also compared to the instantaneous value of the triangular wave, which is connected to the positive input line of that comparator. This implies that the output of the comparator (line Y1) is at a high state as long as the value of the triangular wave is higher than the voltage at point A.

Figure 10.17(b) shows the triangular wave, the voltage levels at points A and B, and the two output voltages at Y1 and Y2. The value of resistor R2 sets the dead time, while the values of resistors R1 and R3 set the upper and lower levels of switching the output signal respectively. Once the resistors are selected, the levels will be fixed and the variation of the frequency of the output signal can be performed by the microcomputer through the variation of the frequency of the triangular wave. As the frequency increases, the period of the output waveform becomes shorter. This implies that the percentage time taken by the dead time increases with the increase of frequency. This sets a limit to the maximum output voltage produced by the circuit and can distort the output signal. Consequently, means of reducing the dead time become important for operation at higher frequencies. Some of these are fast switching devices and a suitable drive circuit.

Example 10.7: **In a testing application, the input and output ports of a microcomputer are taken to a connector. Any of several power-interface circuit boards can be plugged into the microcomputer through that connector. The microcomputer is required to recognize the type of the board and set the dead time accordingly. How can the circuit designed in the previous example be modified to fulfil the requirement?**

Two things are required here. The first is making the dead time adjustable and the second is to let the microcomputer examine each board and determine what value of dead time to use.

Let us start by considering the first requirement. Clearly, the microcomputer can easily alter the dead time if software is used to generate the switching pulses. But this approach can consume a great deal of the processing power and the requirement is to modify the circuit of Fig. 10.17(a). The procedure can be based on sending two d.c. levels from the microcomputer to replace the voltages of points A and B thereby eliminating the need for resistors R1, R2 and R3. This means that, regardless of the source of the triangular wave, the microcomputer supplies V_A and V_B to the comparators to produce an output in a similar way to that of the previous example. The only difference between the original and the modified circuits is the source of the voltage levels. The resistors provide fixed levels, while the microcomputer provides software-selectable levels. The dead time is set according to the difference between the two levels. The method is based on using two D/A converting channels (the interface of D/A converters to a microcomputer is described in Chapter 6).

Let us now define how to provide the microcomputer with the ability to recognize the different boards. Each power board can be provided with a few links which can be set according to the type and rating of the power-switching devices and the onboard drive circuit. Pull-up resistors can be used to set the state of the link lines to a logic high when the links are not connected. The state of a line will become low when a link is connected to join the line to the 0-V line directly, or through a low-value resistor. The microcomputer recognizes whether a link is made or not by reading the state of all the links through an input port. A table is programmed into the microcomputer relating a dead time to each combination of links. The danger of this approach is that a link in the wrong position can lead to the generation of the wrong dead time and may lead to damaging the hardware. For safety, the links can be solder links made in the factory when a board is produced for specific switching devices and drivers.

10.7 Isolation

Many systems use switching devices in high-power applications and/or in noisy environments. The microcomputer and its associated electronic circuits should be electrically isolated from noisy and high-power circuits. Isolation leads to safer operation which prevents power fluctuations and electrical noise from affecting the internal power supply and associated circuitry. Optocouplers (also called optoisolators) and transformers are examples of popular isolation devices. In the former, signals are transferred between circuits by the use of infrared light. Transformers, on the other hand, provide isolation by allowing the transfer of signals by a magnetic field. Furthermore, transformers permit the alteration of the magnitude of an electrical signal according to the turns ratio between the primary and the secondary sides of the transformer.

An optocoupler consists of a LED and a photosensitive device, e.g. a phototransistor or a phototriac. It provides good optical coupling and can withstand high voltages. It is available in IC form from several manufacturers and is described in Section 11.4. Figure 10.18 shows a circuit employing an optocoupler where the LED and the phototransistor are powered from separate power supplies. The microcomputer controls the operation of the circuit by supplying a logic-high signal from an output line to line I1 of Fig. 10.18. This signal switches on transistor T1, thereby causing current to flow through the LED, which causes the phototransistor to switch on. This leads to a conduction of current through resistors R4 and R5. Transistor T2 turns on and leads to a current conduction through the load resistor R_L.

Figure 10.19(a) shows a transistor driving a pulse transformer. The primary side of such a transformer is a winding possessing a very small resistance.

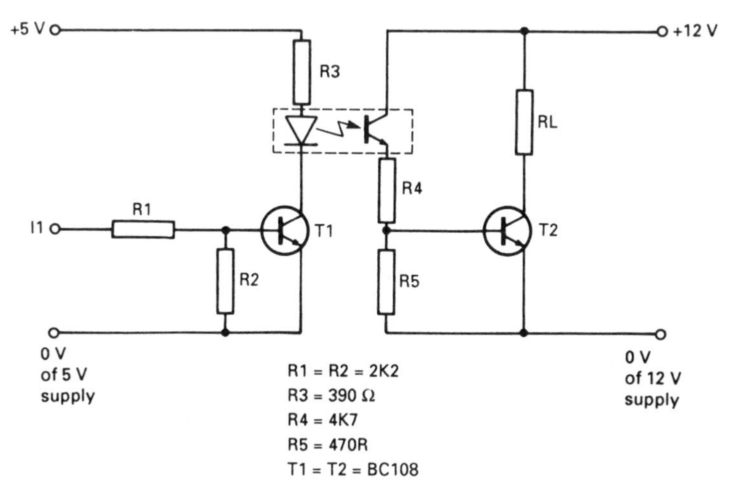

Fig. 10.18 Isolation with an optocoupler.

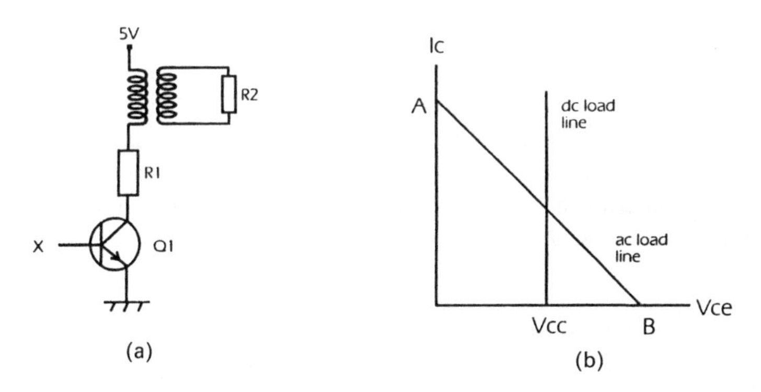

Fig. 10.19 Isolation with a pulse transformer: (a) circuit; (b) load lines.

This forms a very small d.c. load to the transistor, resulting in an almost vertical load line, as shown in Fig. 10.19(b). When a signal is applied to the base of the transistor, a corresponding current starts to flow from the supply through the primary of the transformer. The transformer will then reflect the energy to the secondary side, causing current to flow in the load R2. The time-varying signal applied to the base of the transistor gives an a.c. load line whose slope depends on the value of both R2 and the turns ratio of the transformer. It is given by:

$$R_{a.c.} = n^2 R2$$

where n is the turns ratio of the transformer. The quiescent point of the circuit of Fig. 10.19(a) is the intersection between the a.c. and the d.c. load lines.

10.8 Interfacing to SCRs

This section considers two types of SCRs (Silicon-controlled rectifiers), the thyristor and the triac. Their interface to a microcomputer opens the door to a wide range of applications, e.g. temperature and motor control, programmable light dimming, and power conversion and control.

10.8.1 Thyristor

This is a reliable unidirectional power switch with three terminals (anode, cathode and gate). Similar to a diode, conduction of current through a correctly biased thyristor is in the anode-to-cathode direction. Conduction commences when a triggering signal of low power is applied to its gate. It will latch to conduction as long as the anode-to-cathode current exceeds the latching current of the thyristor in use. Conduction is maintained until the device is forced to turn off by such methods as reverse biasing it, or reducing the magnitude of the current below the holding current. (This is the minimum current which keeps the thyristor conducting. Its value depends on the particular thyristor in use.) When conducting, the device has a very small resistance, which leads to a small forward voltage drop on the thyristor. This leads to the generation of heat, which needs to be transferred to a proper heat sink (Section 14.7).

Operation from a.c. supply

Figure 10.20(a) shows the connection of a thyristor in an a.c. circuit. Since thyristors conduct in one direction, the maximum conduction interval through the resistive load R_L is for half a cycle of the applied a.c. signal. Conduction through the thyristor commences when the controlling circuit applies a triggering pulse to its gate and terminates at the end of the half-cycle at which the pulse is applied. In this way, the controller can adjust the conduction periods, thereby defining the power delivered to the load. The control strategy is based on selecting the interval between the instant at which the a.c. voltage crosses the 0-V line and the instant of applying the triggering signal. This interval is referred to as the phase angle, and is defined in electrical degrees, where a full cycle is 360°. Once switched on, the thyristor of Fig. 10.20(a) conducts until the mains voltage crosses the 0-V line. Figure 10.20(b) shows the timing relationship between the input sine wave and the voltage delivered to the load. The conduction interval is represented by the shaded area, and starts after a triggering pulse is applied. Note that, with this configuration, the control circuit has the ability to define the instant of the start of conduction, but has no control over its end.

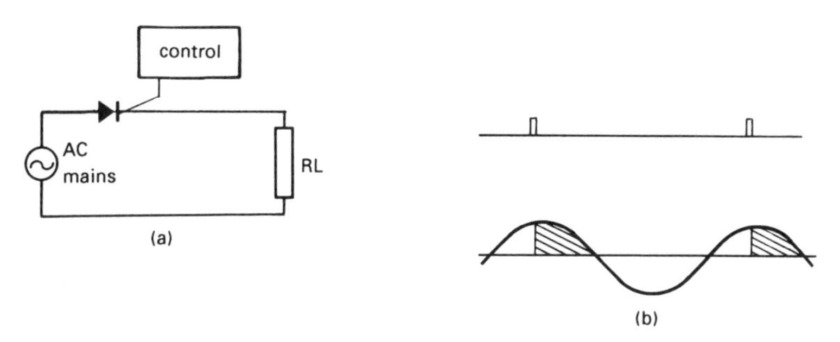

Fig. 10.20 Operation of a thyristor: (a) circuit; (b) uni-directional conduction.

Operation from d.c. supply

The operation of much power-electronics equipment is based on changing a d.c. voltage into a series of pulses; a process known as pulsewidth modulation (PWM). A designer needs to decide on which power-switching device to use. One possibility is a thyristor, which can be attractive to use, particularly when the conduction periods are long. This is because its conduction losses are relatively low. The difficulty with using thyristors to control the flow of d.c. power, however, is in turning them off. In the configuration of Fig. 10.20, a thyristor is incorporated in an a.c. circuit where conduction commences after a thyristor is triggered by a microcomputer and continues until the supply voltage crosses the 0-V line. This does not happen in a d.c. supply, creating the need for the introduction of an additional circuit to turn the thyristor off, as shown in the following example.

Example 10.8: **How can a thyristor be utilized to control the duration of conduction of d.c. power through a load?**

A commonly used configuration is shown in Fig. 10.21, where the power delivered to the load R1 is controlled through thyristor TH1. Let us start

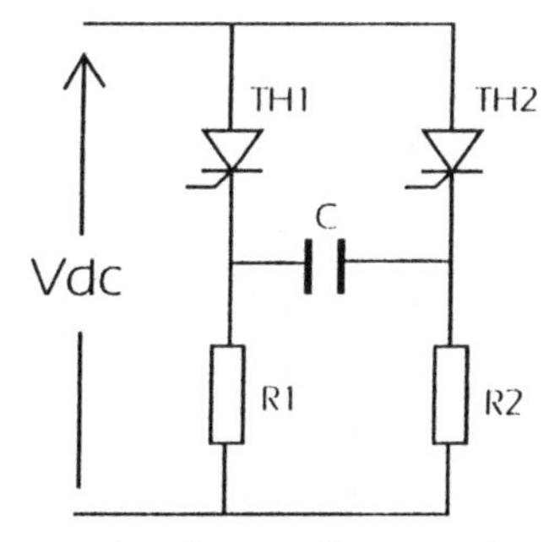

Fig. 10.21 Thyristor operation from a d.c. supply.

at the instant where TH1 is conducting and thyristor TH2 is off. The d.c. voltage is applied to R1 and capacitor C is charged to the d.c. supply voltage through TH1 and R2. In order to turn TH1 off, the microcomputer supplies a triggering pulse to TH2. This action causes TH2 to conduct through R2 and reverse biases TH1 by the voltage stored in C. The result is turning TH1 off and charging the capacitor in the opposite direction through TH2 and R1. The process continues until the microcomputer supplies a triggering pulse to TH1. This reverse biases TH2 by the voltage stored in C. Thyristor TH2 will turn off and the process is repeated. The operation of the circuit produces pulses through the load resistor R1. The amplitude of these pulses equals that of the d.c. supply. The mark-to-space ratio is set by the microcomputer and is defined by the duration between switching one thyristor and the other.

The use of a thyristor to turn off another adds to the component count and cost, since the additional thyristor requires a separate triggering circuit and a digital-output line from the microcomputer. In return, the microcomputer is provided with the ability to produce high-power pulses through a controlled switch with relatively low conduction losses. Note that conduction through TH2 and R2 only serves to turn TH1 off. Power losses can therefore be reduced if the value of R2 was high. On the other hand, the capacitor charges through R2 immediately after turning TH1 on. A high value of R2 increases the charging time and will therefore limit the upper frequency of operating the circuit.

The continuation of conduction through a thyristor is used as the basis of the protection circuit shown in Fig. 10.22 and referred to as the 'crowbar' configuration. The circuit consists of four components and can be placed between a power source (+V) and the power-input terminals of a circuit (points X and Y). The zener diode is chosen with a reverse voltage higher than the voltage of normal operation. If the input voltage increases, then the zener diode will start to conduct, causing a voltage drop across the resistor and charging the capacitor. This turns the thyristor on; it will latch to conduction and prevents the high voltage from damaging the circuit. If a

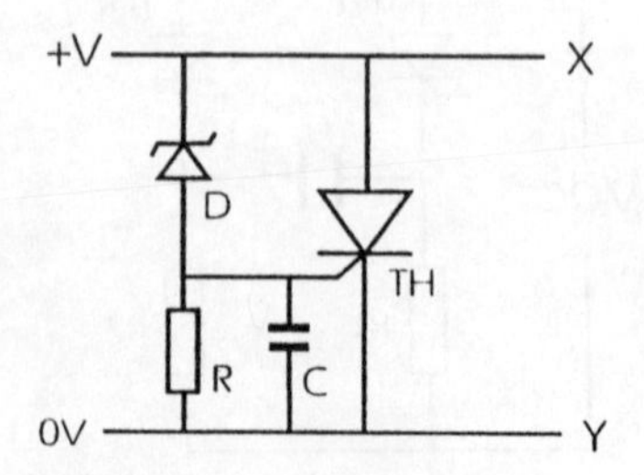

Fig. 10.22 Crowbar circuit.

fuse is connected in series with the supply, then conduction through the thyristor will cause the fuse to blow and protect against any damage.

10.8.2 Triac

Since a thyristor only conducts in one direction, conduction in the positive and the negative direction of an a.c. cycle requires the use of two thyristors connected back to back. A triggering circuit and isolation is needed for each thyristor. Alternatively, a triac can be utilized. It is similar to a thyristor except that it is a bidirectional switching device. Figure 10.23(a) shows a triac connected as a part of an a.c. circuit to achieve the same result as two back-to-back thyristors. Its interface to a microcomputer is similar to that of a single thyristor, except for the number of triggering pulses. A thyristor circuit needs a single triggering signal in a full cycle of the a.c. waveform. Two such signals are required to achieve bidirectional conduction through a triac (one signal per half-cycle). This is shown in Fig. 10.23(b), where the shaded areas correspond to the conduction periods.

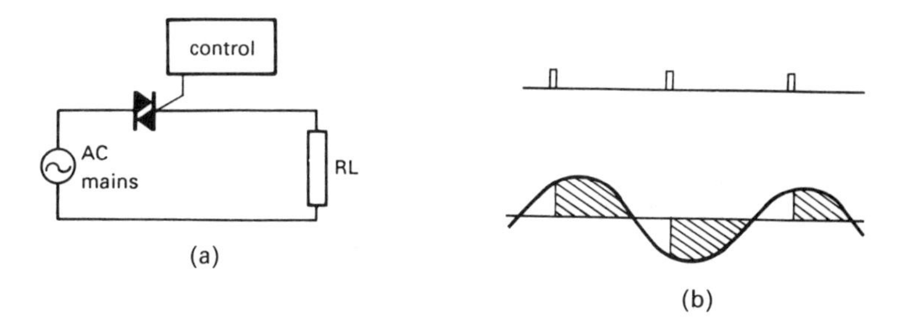

Fig. 10.23 Operation of a triac: (a) circuit; (b) bi-directional conduction.

10.8.3 Interfacing to microcomputers

There are different types of circuits for interfacing thyristors or triacs to microcomputers. The majority are isolated for safety reasons and in order to prevent noise signals from passing from the high-power circuit to the microcomputer (noise effects and sources are discussed in Section 16.7, where recommendations are given on how to reduce its effects). This section introduces a few interfacing circuits. Due to the similarity between thyristors and triacs, the presented circuits only show thyristors. Examples of controlling the phase angle are presented in other parts of this book, e.g. Examples 14.7 and 14.8.

Example 10.9: **Design a simple circuit to interface a thyristor to a microcomputer.** The circuit of Fig. 10.24(a) is a simple non-isolated circuit where the 0-V line of the low-voltage d.c. supply is directly connected to the a.c. voltage. Let us assume that the microcomputer is provided with a pulse representing the instant at which the voltage waveform crosses the 0-V line (zero-cross detection circuits are the subject of Section 10.9). The microcomputer can use counters or timers to introduce a delay before supplying a triggering pulse to the thyristor. The pulse can be supplied from a digital-output line of the microcomputer to turn on the PNP transistor Q1, allowing current to pass through resistors R3 and R4 to turn thyristor TH on.

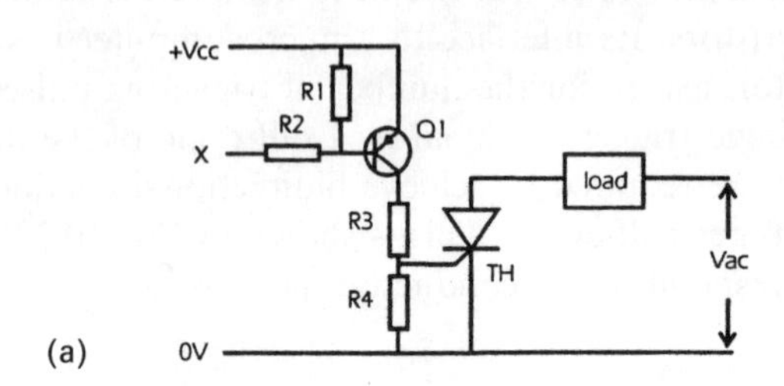

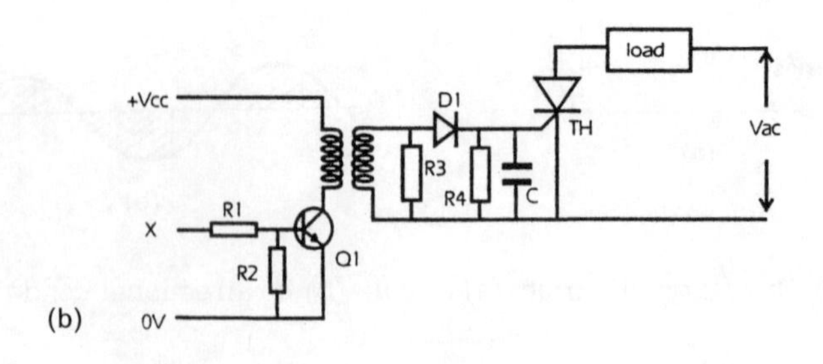

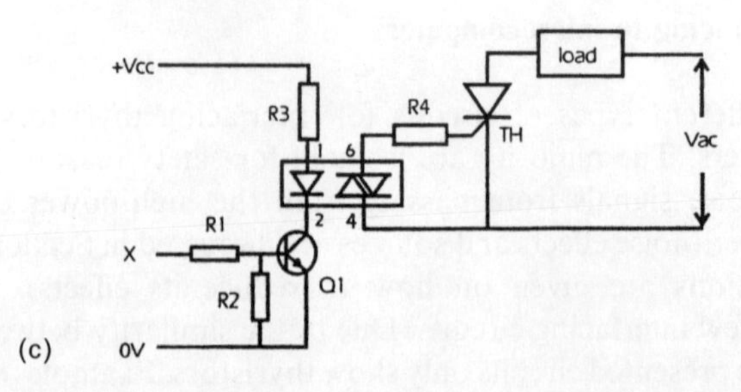

Fig. 10.24 SCR drive circuits: (a) non-isolated; (b) using a pulse transformer; (c) using an optocoupler.

Caution! Although the circuit is relatively simple, it is not recommended for direct control of power switching from a microcomputer, because of the lack of isolation. It is important to take safety precautions when using and connecting non-isolated circuits, especially if the voltages are high. If, for instance, a microcomputer is employed to provide the drive signals to the circuit of Fig. 10.24(a), then the power supply of the microcomputer is connected directly to the a.c. signal. This can upset the operation of the microcomputer, or in the case of a fault could damage the microcomputer (furthermore, some of the microcomputer parts or interface circuits could be earthed). The use of a non-isolated power circuit will also introduce the need for special precautions for any user interface. A well-isolated potentiometer should, for example, be used and the circuits can be protected with fast-acting fuses.

Isolation is recommended and is normally placed between the drive circuit and the power-switching device, as shown in the following example.

Example 10.10: **Design a circuit to allow a microcomputer to control the power delivered to a load through a thyristor. The power circuit should be isolated from the microcomputer.**

Solution 1

A pulse transformer is a very attractive device to use in thyristor (and triac) drive circuits, since it can be employed to perform the isolation task without the need for an isolated power supply at the secondary side. An example of a drive circuit is given in Fig. 10.24(b). Interfacing to the microcomputer is through a digital-output line driving transistor Q1 through line X. As in the previous example, an additional circuit is included to inform the microcomputer when the a.c. signal crosses the 0-V line. A delay is then introduced (by software or hardware means). The end of the delay interval marks the instant of sending a triggering command through line X. Changing the state of line X from low to high turns on transistor Q1 to pass current through the primary side of the pulse transformer. Energy is transferred to the secondary side of the transformer and used to activate the gate of the thyristor.

Solution 2

A different type of isolation is used in the circuit of Fig. 10.24(c). It employs an optocoupling device. This is an optically isolated triac contained in a 6-pin DIL package. When the microcomputer delivers a logic-high signal through line X, the NPN transistor (Q1) is turned on, permitting current to flow through the LED. Energy is transferred through the optocoupler to turn on the optical triac and hence the thyristor. Resistors R3 (390 Ω) and R4 (270 Ω) are employed to limit the current through the paths of conduction.

Calculations

The next step after interfacing thyristors or triacs to a microcomputer is to program the latter to control the power delivered to the load. This is performed by controlling the instants of triggering these devices, as shown in Example 14.7 for a single-phase a.c. supply, and Example 15.6 for a 3-phase supply.

10.9 Zero-cross detection

In order to be able to specify the conduction intervals of the thyristor, the microcomputer needs an indication of the instant at which the mains voltage crosses the 0-V line. A variety of circuits have been developed to perform this task.

Example 10.11: **Design a circuit to inform a microcomputer of the instant at which an a.c. signal crosses the 0-V line.**

Solution 1

Such information can be provided to the microcomputer by a circuit such as that of Fig. 10.25(a). The diode performs a half-wave rectification to the a.c. signal delivered by the transformer. During the positive half-cycle of the mains, diode D1 conducts and triggers transistor T1, causing a change of voltage at point C from +5 V to approximately 0 V. This results in changing the state of point D from logic low to logic high. During the negative half-cycle of the a.c. signal, the secondary current of the transformer flows through resistor R1. If point D is used to activate a leading-edge-triggered interrupt of a microcomputer, then the latter will be interrupted once every cycle. This approach helps the microcomputer to synchronize its internal operations with the zero-cross instant of the a.c. supply.

Solution 2

An alternative circuit is given in Fig. 10.25(b). It changes a stepped-down a.c. signal to a square wave by the use of a high-gain amplifier (a small transformer may be used to step down the voltage and to provide isolation). The square wave is then changed to short pulses by a capacitor–resistor differentiator. The diode allows only the positive pulses to pass to the rest of the circuit.

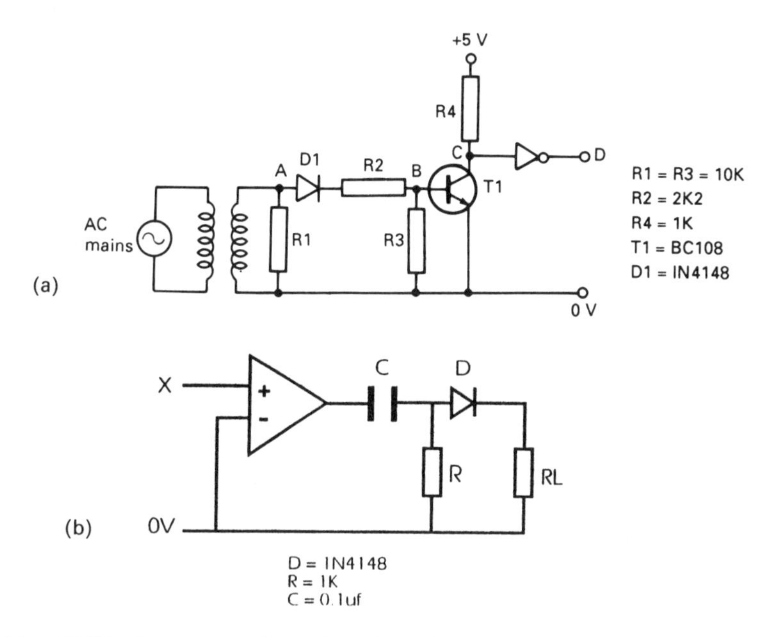

Fig. 10.25 Zero-cross detection: (a) using a diode-transistor circuit; (b) using a differentiator.

10.10 Interfacing to relays

A relay consists of two electrically isolated parts, the coil and the contacts. Energizing the coil causes the contacts to close, and de-energizing it opens the contacts. Consider the circuit of Fig. 10.26(a), where both V1 and V2 are d.c. voltages from two different power supplies. When switch S is closed, the coil will be energized and the contacts of the relay closed, causing current to flow through the load resistor R_L. The load current flows only through the contacts and not through the coil. Consequently, the contacts are capable of conducting currents of high magnitudes, whereas the coil is normally driven from a low-power circuit. When compared with transistors or SCRs, relays have a shorter switching life and are much slower to operate. They are therefore attractive for use in applications having no requirement for fast repetitive switching, e.g. turning heaters on and off.

The interface between a microcomputer and a relay can be implemented by the use of a single digital-output line from the microcomputer. The drive capability of that line can be boosted by employing a transistor to supply sufficient current to energize the coil of the relay. Figure 10.26(b) and Fig. 10.26(c) show relay-driving circuits where the load is represented by R_L.

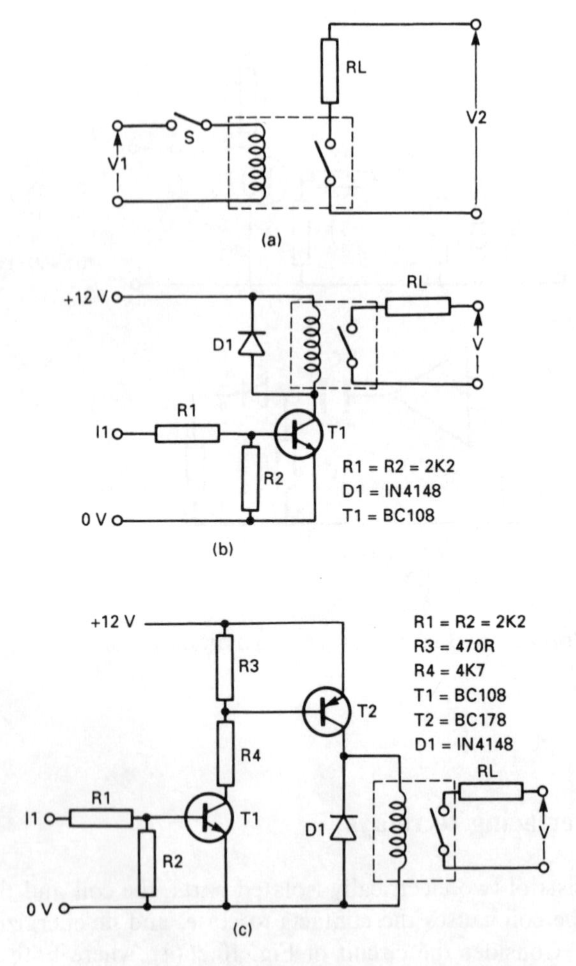

Fig. 10.26 Relay-driving circuits: (a) switch-controlled operation; (b) using an NPN transistor; (c) using a PNP transistor.

Diode D1 is employed to provide a path for the voltage generated across the coil when the transistor is switched off. Current flows through R_L when the microcomputer supplies a logic-high signal to switch the transistor on.

Clearly, if the interface is required for a specific application, then the designer chooses a relay capable of passing the required load current. If, on the other hand, the interface circuit is general purpose, then the designer may not be able to predict the load current needed for every application. Consequently, the designer can produce a relay-driving circuit. It provides an open-collector or an open-emitter terminal with a specified drive capability. It allows the user to connect a suitable relay, or a low-power load if this is desired.

Chapter 11

Optical devices

11.1 Introduction

Optoelectronic components are widely used in electronic circuits such as in light emitting and detection, optoisolation, and positioning of mechanical equipment. This chapter starts by considering light sources and detectors. It proceeds to describe optocouplers, giving a few examples of their use (further circuits are presented in later chapters). Different types of optical encoders are considered in Section 11.8 as examples of devices using light sources and detectors to measure the speed and position of moving objects. Such encoders are incorporated to generate a flexible easy-to-use measuring system in a wide range of applications, e.g. positioning elevators, forklifts, cranes, robots and motor speed controllers.

11.2 Light sources

There are many sources of light, e.g. the sun, neon lamps and LEDs. Some are designed to radiate light in a narrow beam, while others radiate in all directions. LEDs are incorporated in a vast amount of electronic equipment to indicate operating conditions or signal levels. An example is given in Section 7.4, where a group of LEDs are employed to test a data converter. Different types of LEDs are available, giving the user a choice of colours (red, orange, yellow, blue and green) and emission wavelengths in the visible and infrared range. A standard LED needs around 10 mA to supply a relatively bright light. The current requirement drops to around 2 mA if a high-efficiency LED is used. Such a LED helps to reduce the power consumption, but is more costly.

When a LED is biased in the forward direction, it conducts current (I_f). There will be a voltage drop across the device (V_f) and energy is released in the form of light plus a small amount of heat. The brightness of the LED depends on the current passing through it. Figure 11.1(a) illustrates the control of the current conduction through a LED by an NPN transistor and a resistor. The transistor is employed as a switch which can be driven to saturation by a logic signal applied to point X. The current passing through the LED is given by:

$$I_f = (V_{cc} - V_f - V_{sat})/R$$

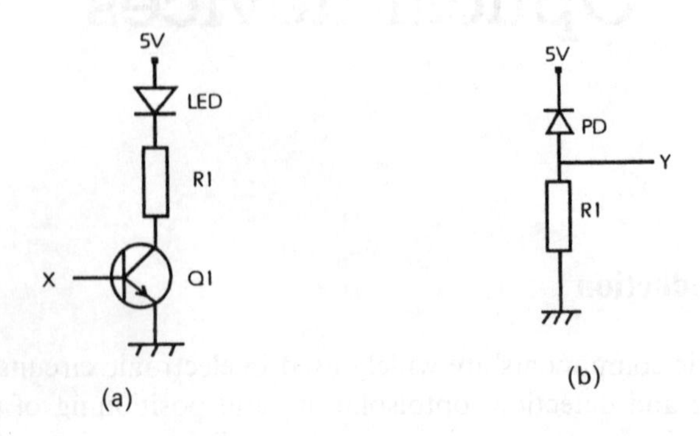

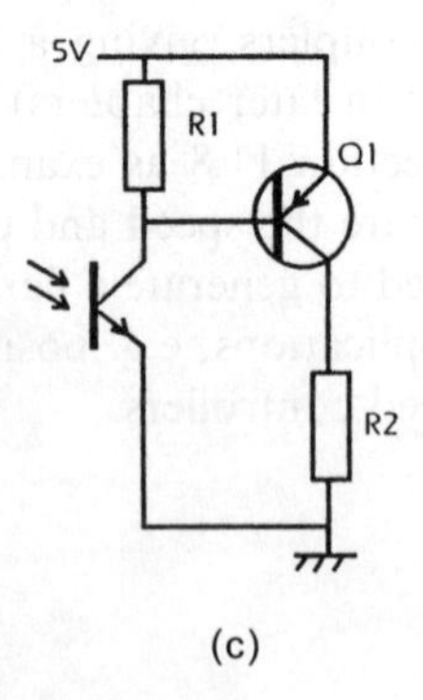

Fig. 11.1 Using optical devices: (a) LED; (b) photodiode; (c) phototransistor.

where V_{cc} is the +5-V d.c. supply voltage, V_f is the voltage across the LED, V_{sat} is the saturation voltage of the transistor, and R is the resistor. The typical value of I_f and the corresponding value of V_f are normally given in the data sheet of the LED. Assume that the LED requires a typical value of I_f of 10 mA, the corresponding V_f is 2 V, V_{sat} is 0.2 V, and V_{cc} is 5 V. The value of R can be calculated from the above equation to be:

$$R = (5 - 2 - 0.2)/0.01 = 280\,\Omega$$

A 270-Ω resistor can be used. Alternatively, a potentiometer can be employed to give an operator the ability to adjust the intensity according to the working environment. Another situation where light adjustment may be required is a unit employing a few LEDs. This is because individual LEDs may not have the same intensity, especially if they are of different colours. The resistive value of the potentiometer depends on the required dimming level. A resistor in series with the potentiometer is recommended to set the maximum limit of current allowed to pass through the LED and transistor.

11.3 Light-sensing devices

The light delivered by a light source (e.g. a LED) can be detected by a light-sensitive device. Such a device is referred to as an optical sensor. It absorbs light energy and produces an equivalent electrical signal. It is incorporated in a variety of circuits, e.g. to detect the intensity of light or the movement of objects. The latter relies on a break in the path of conduction of light between the light source and sensor. The range of wavelengths that create a response varies according to the type of device and is presented in graphical form in the data sheet of the device. When designing an interface circuit which relies on the transmission and the detection of light, it is best to select a closely matched light transmitter and sensor. Such devices are available commercially and are referred to as 'spectrally matched', where the light source may be an infrared diode and the sensor may be a phototransistor.

11.3.1 Photodiodes

This type of light sensor is similar to an ordinary diode, except that it is packaged in a transparent case and is operated in reverse bias. A leakage current flows through the device. Without light, there will be an extremely small leakage current known as the 'dark current'. As light intensity increases, energy is absorbed and the leakage current increases accordingly.

Photodiodes provide a very fast response to changes in the intensity of light. The leakage current is, however, small, which limits the use of the device in low-light-intensity applications. Its fast response, however, makes it attractive to employ for the fast detection of light signals, particularly those of high frequencies. Figure 11.1(b) shows a photodiode PD connected in series with a resistor. The light intensity will cause a proportional current to flow through the device, causing a proportional voltage drop across the resistor. This voltage drop can be measured by a microcomputer through an A/D converter, or compared to a d.c. level through an analogue comparator. The output line of the latter provides a signal to a digital-input port of a microcomputer, or to one of its interrupt lines.

11.3.2 Phototransistors

This is a transistor packaged to allow light to fall on its base, causing a current flow from the base to the emitter. This turns-on the transistor, causing a much larger current to flow from the collector to the emitter. The magnitude of the current is therefore amplified by the current gain of the transistor. This provides the device with a higher current sensitivity than a photodiode, but it slows its response. A phototransistor is therefore more suitable for lower-light-intensity applications than a photodiode.

Figure 11.1(c) shows a phototransistor connected to start conduction of current when subjected to light of a sufficient intensity. The flow of current causes a voltage drop across resistor R1 which turns on the PNP transistor Q1 to pass current through the load resistor R2. There are several applications for such a circuit, e.g. to turn a device on when an operator points a source of light towards it. The device will turn off as soon as the light source is removed. The current passing through R2 is, however, small and can be amplified by adding one or more stages of transistor amplifiers. The circuit of Fig. 14.8 is one example of a multistage transistor amplifier.

11.3.3 Photodarlingtons

A phototransistor may not be effective enough in applications requiring the detection of very low light intensities. This creates the need for a device with a higher current gain. A photodarlington is such a device. It is similar to a phototransistor, except that it includes a pair of transistors cascaded in a Darlington configuration (see Fig. 10.7) to provide a much higher current gain.

11.3.4 Slotted and reflective opto-switches

Such switches are incorporated in a large number of microcomputer-controlled systems. They are employed for such applications as the detection of a moving object, or the measurement of linear and angular motion. Figure 11.2 shows two such switches. They both contain a light source and a light sensor in the same package. The side view of a slotted opto-switch is given in Fig. 11.2(a), where the light source S (an infrared-emitting diode) and the detector R (a phototransistor, a photodarlington, or a combination of a photodiode and an amplifier) are placed facing each other. An object can be passed through the slot to break the conduction of light from the light source to the detector.

A reflective opto-switch is represented in Fig. 11.2(b). It differs from the slotted opto-switch in that its operation is based on the detection of a light signal reflected on an object. A series of reflective and non-reflective lines

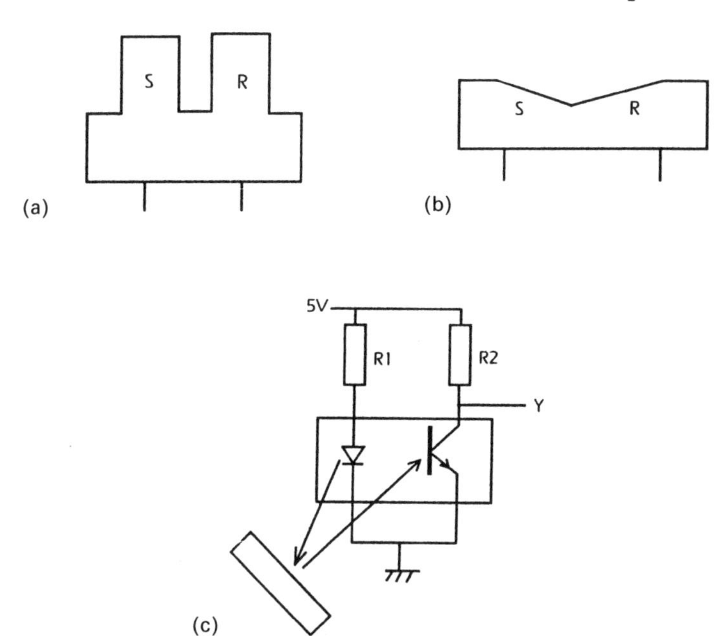

Fig. 11.2 Optical switches: (a) slotted opto-switch; (b) reflective opto-switch; (c) using reflective opto-switch.

can, for example, be attached to a strip, or a ring, and placed on the object whose movement is to be detected. Once the sensor is positioned in the right place, movement of the object causes the reflection of the light transmitted from the source back to the sensor only if it falls on a reflective line. As motion continues, the light detector produces a series of pulses corresponding to the conduction and non-conduction of light. An interface circuit can use these pulses to identify linear movements or angular displacements. The connection of such a sensor is shown in Fig. 11.2(c), where the path of light is represented by the two arrows.

The output of either sensor can be interfaced to a digital-input line of a microcomputer or to the input line of a counter. The latter can be read regularly and reset by the microcomputer.

11.3.5 Light-dependent resistors (LDR)

As the name implies, this is a resistor whose value changes with light level, making it a very useful sensor in the detection of light levels in a variety of applications. The use of this type of sensor in interfacing applications is a relatively simple task which relies on the linear reduction of the resistance of a LDR with the increase in the intensity of light.

Example 11.1: **A 16-bit microcomputer is employed as a multitask controller. One of its functions is to control the intensity of a light unit operated from a d.c. supply. How can the light be turned on and off according to the setting of a SPST (single-pole single-throw) switch and the intensity varied according to a user-selectable setting?**
The light-control task involves examining signals from three input sources: the on/off switch, the user-selectable intensity level, and a signal representing the actual light intensity. Let us consider how to meet the requirement of each task separately:

- The procedure of interfacing the SPST switch given in Example 4.1 can be employed to read the user's on/off command.
- A potentiometer can be incorporated to provide the user with the ability to set the intensity. The potentiometer, however, supplies an analogue signal representing the setting.
- An indication of the light intensity can be obtained by the use of a LDR. Since the resistance of a LDR decreases as the light intensity increases, the LDR can be connected in series with a suitable resistor to form a voltage divider. The voltage at the centre point of this divider is proportional to the intensity of light.

Having decided on the method of converting the light signal to a proportional electrical signal, the next step is to find a suitable way of interfacing. One solution is to allow the microcomputer to read the analogue signals from both the potentiometer and the LDR. Reading these analogue signals, however, introduces the need for a 2-channel A/D converter. But the microcomputer has no requirement to know the actual magnitude of the signals. All it needs is an indication of whether the light intensity has exceeded the user's setting or not. Consequently, a simpler and a lower-cost solution can be based on the use of an analogue comparator as illustrated in Fig. 11.3. Resistor R3 is a potentiometer through which the user can set the required intensity level. The circuit produces a logic-level signal which can be monitored by the microcomputer through a digital-input line. The state of this line depends on the result of comparing the voltage across the LDR with that across resistor R3. If the light intensity is lower than the setting, then the resistance of the LDR is high. The result is a logic-low signal on the output of the comparator. When the light intensity exceeds the required level, the output of the comparator changes to a logic-high state.

This approach provides the microcomputer with a single digital signal. If a fast detection of a change in the state of that line is not necessary, then the light-control task can be implemented by the use of the program-controlled approach. In many light-control applications, the requirement can be satisfied by reading the inputs at a low rate, say once every 100 ms. If the operation of the microcomputer-based multitask controller is performed by the use of

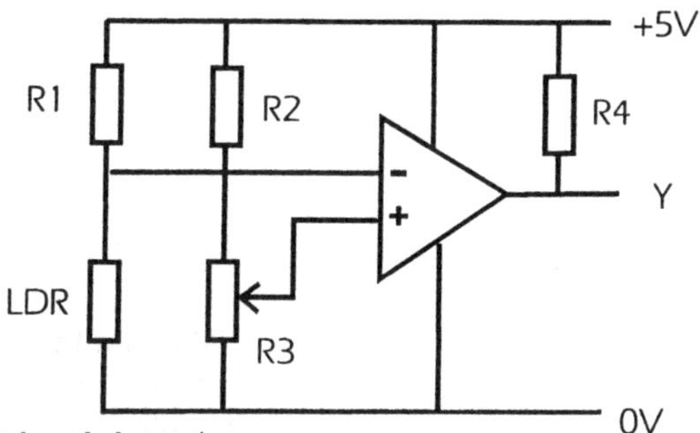

Fig. 11.3 Light-level detection.

interrupts, then the light-control task can be incorporated as one of the tasks within the background program.

11.4 Optocouplers

These devices are also referred to as optoisolators. They consist of an infrared-emitting diode and a light-sensitive switching device packaged in a single IC (a lightproof package). The circuit of Fig. 11.4(a) shows an optocoupler in a digital circuit. When a current I_f flows through the diode, it emits an infrared signal which falls on the base of the phototransistor, causing current to flow from the collector to the emitter. Digital signals are transferred from a part of the circuit which operates from a d.c. voltage supply V1 to another part operating from a different d.c. power supply V2. Energy is transferred in the form of infrared signals, eliminating the need for electrical connections. This feature makes optocouplers very useful in applications requiring the transfer of signals between different circuits while maintaining electrical isolation.

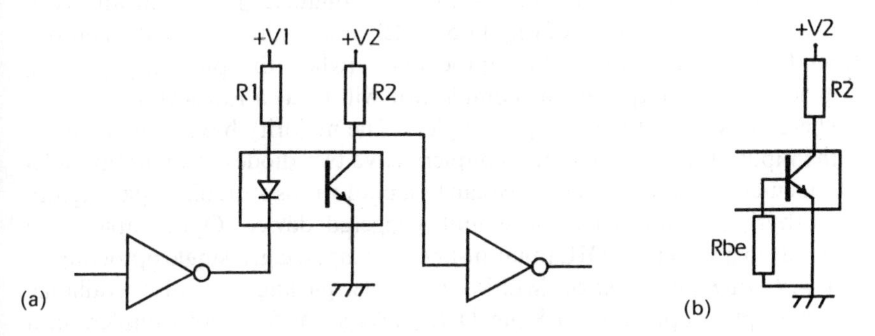

Fig. 11.4 Using optocouplers: (a) in a digital circuit; (b) using a base-to-emitter resistor.

They are, for example, incorporated to isolate the low-voltage circuits of a microcomputer from high-voltage power circuits, such as motor drives (Chapter 15). It is important to note that the isolation voltage is not the same for all optocouplers. The isolation can be increased by increasing the distance between the light-emitting device and the photodetector during the manufacturing process. The increase in distance will, however, reduce the current transfer ratio (CTR) of the optocoupler.

The voltage drop across the diode of Fig. 11.4(a) and the value of R1 limit the magnitude of I_f, which will, in turn, determine the current flowing through the phototransistor. The base current of the phototransistor, however, depends on the CTR of the device. A higher-gain device, such as a photodarlington, will provide a higher CTR and implies that a lower value of I_f will be sufficient to obtain the required value of current than if a phototransistor is used.

The rate of change of current through the phototransistor of Fig. 11.4(a) imposes a limitation on the frequency of the signal passed through the optocoupler. Figure 11.4(b) shows a method of speeding the operation by the use of an optocoupler which allows connection to its base. A base-to-emitter resistor R_{be} is employed to speed up the removal of the charge stored in the base of the phototransistor.

Figure 11.5 gives four circuits demonstrating the use of optocouplers with different devices. The circuit of Fig. 11.5(a) illustrates how an NPN transistor is employed to control the current conduction through the diode of an optocoupler. The circuit of Fig. 11.5(b) is slightly different, since it performs the function by the use of a PNP transistor. When comparing the two circuits, we can see that they differ in their requirement for the state of the input logic signal. A high-logic level at point X of Fig. 11.5(a) enables the conduction of current through the diode, while a signal of that level turns off the PNP transistor of Fig. 11.5(b) and stops the conduction. Figure 11.5(c) and Fig. 11.5(d) show two different ways of using the signal delivered by the secondary part of the optocoupler. The circuit of Fig. 11.5(c) incorporates the optocoupler to control the duration of conduction through an NPN transistor, while the circuit of Fig. 11.5(d) delivers a voltage level at its output depending on the state of the optocoupler (when the phototransistor is switched on, the output of the comparator will be at a low level).

There are several types of optocouplers. The majority have a single diode at the input although some optocouplers have two diodes in an antiparallel arrangement. The switching device can be a phototransistor, photodarlington, photothyristor, phototriac, or schmitt-triggered device. Optocouplers are available in a variety of DIL (dual-in-line) packages, e.g. a single optocoupler in a 4-, 6- or 8-pin package. Multi-optocoupler packages are also available, e.g. two optocouplers in an 8-pin DIL package, or four optocouplers in a 16-pin DIL package.

Optocouplers are used in a few examples within this book. The circuit of

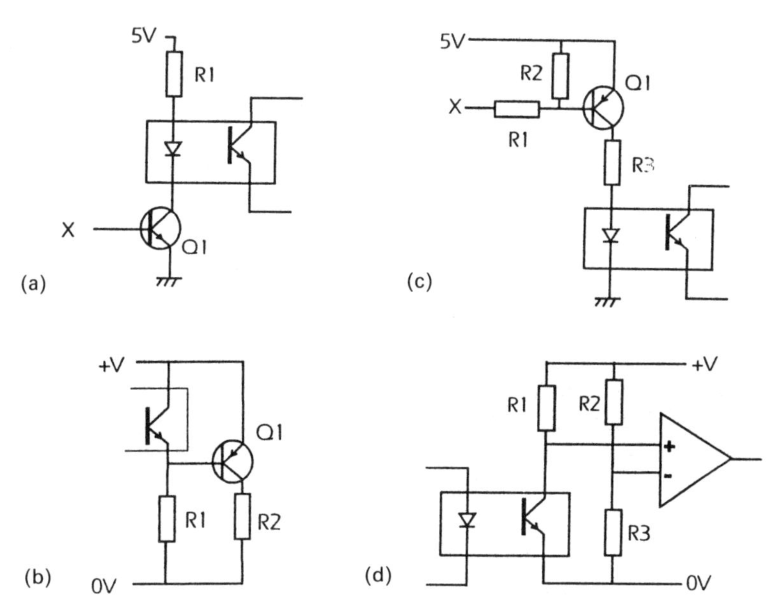

Fig. 11.5 Further optocoupler circuits: (a) primary switched by NPN transistor; (b) primary switched by PNP transistor; (c) secondary driving NPN transistor; (d) secondary driving comparator.

Fig. 14.8 is an example of utilizing an optocoupler in a transistor amplifier. The optocoupler is employed to transfer the signal from the low-voltage side to the power circuit while maintaining isolation for safer, less noisy operation. An optocoupler is also incorporated to transfer triggering commands from a microcomputer to an SCR in the circuit of Fig. 10.24(c).

11.5 Interfacing displays to microcomputers

A large number of microcomputers are designed for a specific operation and are not provided with displays. They may only have LEDs to inform an operator of the state of operation and to indicate faults. Displays can be added to such microcomputers and interfaced through output ports, or directly to the data bus through a suitable driver IC. An additional display may also be required in a multitasking application where a microcomputer has its own display reserved for one of the tasks. The add-on display may be needed to monitor the magnitude of various signals, or the state of events which the microcomputer detects or controls as a part of its multitasking activities.

Seven-segment LED displays are available in many sizes and in several colours (e.g. red, green or yellow). They are incorporated in a very wide range of applications, e.g. digital clocks, industrial controllers, televisions and hand-held instruments. They use a single LED per segment whose brightness is proportional to the current passing through it. The intensities of segments within a multisegment display unit are normally matched. In some displays, the voltage drop across each LED segment is not the same. This causes currents of different magnitudes to pass through the LEDs, leading to a mismatch in brightness. If necessary, a potentiometer can be placed in series with each LED and adjusted to obtain the best match.

A 7-segment display can be specified as a common anode or a common cathode. In the former, all the anodes of the LEDs are connected together internally, as shown in Fig. 11.6(a). The cathodes are left open for connections to resistors and driving devices. Common-cathode displays, on the other hand, have the cathodes of all the LEDs connected together internally, as shown in Fig. 11.6(b).

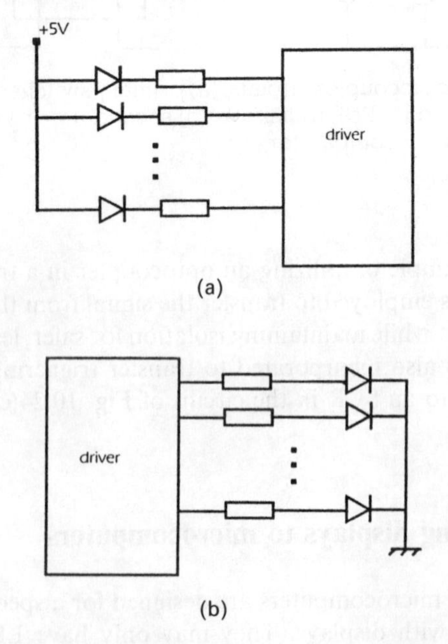

Fig. 11.6 Driving 7-segment displays: (a) common anode; (b) common cathode.

Example 11.2: **Describe how a 7-segment display can be interfaced to an 8-bit output port of a microcomputer. Assume that the output port does not possess sufficient drive capability to operate the 7-segment display.**

Solution 1

A method of interfacing the 7-segment display to an 8-bit output port is shown in Fig. 11.7(a). The state of each of the seven segments is controlled by a single digital-output line of a microcomputer. Two open-collector hex-inverter 74LS06 ICs are utilized to provide the required drive capability. The two ICs contain 12 open-collector inverting gates, seven of which are used here, leaving five gates available for use in other parts of the interface circuit.

Before supplying a value to the display, the microcomputer must convert that value into the right format in order to operate the corresponding segments of the display. The table presented in Fig. 11.7(c) gives the

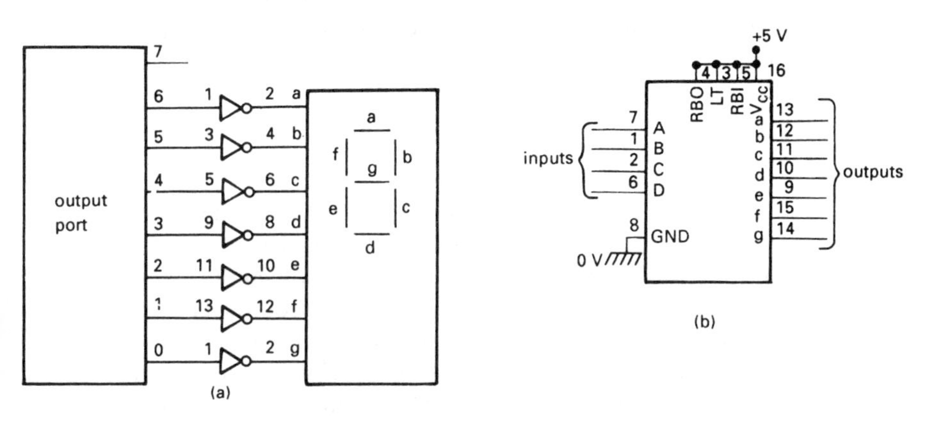

Number	a	b	c	d	e	f	g	output
0	1	1	1	1	1	1	0	7E
1	0	1	1	0	0	0	0	30
2	1	1	0	1	1	0	1	6D
3	1	1	1	1	0	0	1	79
4	0	1	1	0	0	1	1	33
5	1	0	1	1	0	1	1	5B
6	1	0	1	1	1	1	1	5F
7	1	1	1	0	0	0	0	70
8	1	1	1	1	1	1	1	7F
9	1	1	1	1	0	1	1	7B

(c)

Fig. 11.7 Interfacing a 7-segment LED display to a microcomputer: (a) using inverting gates; (b) using a decoder (7447 IC); (c) relation between number and port output.

relationship between the decimal value to be displayed and the corresponding segments in need of activation. The column marked 'output' represents the hexadecimal value which can be supplied to the port in order to produce each of the 10 decimal numbers. The table assumes that bit 7 of the output port is always at a low logic state. This bit can be used for other purposes, where it will be set or cleared by the use of a software OR or AND respectively (in a similar manner to the procedure presented within Solution 2 of this example). The conversion table can be stored in the memory of the microcomputer and a software pointer is introduced. The latter defaults to the base of the table. When the microcomputer needs to display a value, it adds that value to the pointer. The sum defines the location where the digital value is saved. The microcomputer will retrieve that value and transfer it to the output port.

Solution 2

Assume that the microcomputer needs to be utilized to implement other tasks besides sending values to the display. If the implementation of such tasks uses some of the output lines, then it will be desirable to reserve less lines to operate the display. This will be possible by employing a 4–7-line decoder, such as the 7447 IC (a BCD-to-7-segment decoder) shown in Fig. 11.7(b). It converts a binary number supplied to its four input lines to an equivalent 7-bit number. The number can be delivered to the 7-input lines of a 7-segment display through a 390-Ω current-limiting resistor per line. This decoder performs the function of the table presented in Solution 1. Its use will therefore eliminate the need for storing a table in memory. The result is simpler software, less work for the processor and a reduction in the number of interface lines at the expense of a small increase in cost and complexity of the hardware.

This solution only uses four out of the eight lines of the output port. Since the microcomputer writes to all the lines within an output port simultaneously, changing the displayed value should not alter the logic state of the other lines within the port. This can be achieved by reserving a byte ('y') in RAM in which the logic state of the eight output lines is stored. When a value is to be sent to the display, only the lower four bits of that byte need updating as follows, where '|' represents a software OR, and '&' represents a software AND.

```
1  READ y FROM RAM
2  y & 11110000
3  y|NEW VALUE
4  STORE BYTE IN RAM
5  SEND BYTE TO OUTPUT PORT
```

The procedure clears the contents of the lowest four bits in step 2 and replaces them with the new value in step 3. It assumes that the value is written in the four least significant bits of an 8-bit register (NEW VALUE) whose four most significant bits are set to 0.

11.6 Multiplexing displays

The previous example illustrates a method of interfacing a single-digit 7-segment display to a microcomputer. In the majority of applications, more than one digit is needed. Rather than introducing additional ports to a microcomputer, the digits can be multiplexed. In this way, the number of data lines stays the same as if a single digit is used. Additional selection lines need to be introduced, one for each digit. The circuit of Fig. 11.8 shows four common-anode 7-segment displays multiplexed to operate from a single 8-bit port of a microcomputer. The microcomputer selects one digit at a time through output lines Q0 to Q3 and sends the corresponding 4-bit data through lines Q4 to Q7. Rather than employing one decoder per digit, a single 7447 decoder/driver IC is time-shared between the digits. The circuit is therefore simpler and cheaper than interfacing to each display individually. The penalty is the additional processing time needed, since the microcomputer supplies information to each digit individually rather than writing to the whole display in one go.

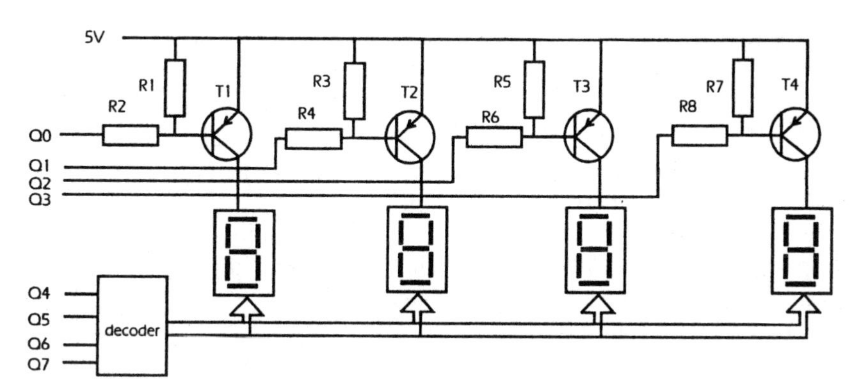

Fig. 11.8 Multiplexing displays.

Instead of connecting the common anode terminal of each digit directly to the power supply, the circuit of Fig. 11.8 uses PNP transistors as microcomputer-controlled digit-enable switches. Each transistor should be capable of conducting the total current needed by the seven segments of a single digit. The current that flows through each segment passes from the d.c. supply through a PNP transistor to the segment and then returns to the supply through an open-collector transistor within the 7447 IC. The magnitude of the current will therefore be affected by the value of the saturation voltage of both transistors.

Example 11.3: **What is the procedure for transferring information from the memory of the microcomputer to the circuit of Fig. 11.8 through the 8-bit output port?**

Let us assume that the number to be displayed in these four digits is stored in two successive bytes within the memory of the microcomputer. Let us refer to the byte containing the most significant two digits as 'A' and the one containing the least significant two digits as 'B'. A simple program needs to be written to separate the contents of each byte into two individual digits. The program can be represented by the following sequential steps, where 'X', 'Y0', 'Y1', 'Y2' and 'Y3' are bytes in RAM and the '≫' sign indicates a single-bit shift to the right.

1 X = A & 11110000
2 Y3 = X ≫ 4
3 Y2 = A & 00001111
4 X = B & 11110000
5 Y1 = X ≫ 4
6 Y0 = B & 00001111

Byte X is used for temporary storage and the result is stored in bytes 'Y3', 'Y2', 'Y1' and 'Y0'; bytes 'Y3' and 'Y0' contain the most significant and the least significant digit respectively. These values need to be transferred to the displays sequentially. The process will be repeated regularly, since the multiplexed digits need continuous refreshment even when there is no change in the value of the displayed information. The rate of transferring information to each digit need not be high, but the rate of sending data through the output port increases with the increase in the number of digits. If, for example, each of the four digits needs to be updated once every 10 ms, then the output port needs to send information once every 2.5 ms.

In many microcomputer-controlled applications, updating each digit of a multiplexed display can be combined with the occurrence of another time-dependent event, e.g. during a regular interrupt. If the rate of occurrence of the interrupt is high, then only one of the four digits can, for instance, be updated every time the interrupt is serviced. This is shown in the following procedure, where 'Z' contains the number of the digit that was last updated, and 'W' is a byte containing the information to be transferred to the output port. This consists of four bits of data plus four selection bits. The procedure takes place during the interrupt-service routine. It can be executed after completing the time-critical tasks of the interrupt and before returning to the main program.

```
1  W = 00001111
2  SEND W TO OUTPUT PORT
3  IF Z = 01, THEN Z = 02 AND LET W = W|Y2
       ELSE IF Z = 02, THEN Z = 04 AND LET W = W|Y1
       ELSE IF Z = 04, THEN Z = 08 AND LET W = W|Y0
       ELSE Z = 01 AND LET W = W|Y3
4  OUTPUT PORT = W|Z̄
```

The first two lines of this procedure disable all the digits and clear the last value sent to the 7447 IC. Step 3 advances the number of digits and refreshes the contents of the 'W' register. The value is sent to the output port in step 4.

The circuit of Fig. 11.8 reserves an 8-bit output port of the microcomputer for its operation. If the microcomputer is involved with other activities which require the use of some of its output lines, then the number of output lines reserved by the display circuit will need to be reduced further. One solution is to introduce a shift register to select the digits sequentially. Such a register will advance its content by 1 bit every time it is pulsed by the microcomputer through a digital-output line.

11.7 Bar graphs

A bar graph is a series of LEDs employed to form a simple interface between an electronic instrument and a user. It may consist of a series of individual LEDs, or a group of LEDs packaged in a single unit referred to as a 'bar graph array'. When compared to the former, the latter can be more attractive to use in many applications, since it reduces the number of components, eliminates alignment problems, needs a single rectangular hole when used in a front panel, and can be placed in a single socket or soldered directly into a board. Discrete LEDs, on the other hand, are preferred to satisfy certain needs, e.g. spacing or different-colour requirements.

Bar graphs are employed in a variety of applications, such as to indicate the level of an analogue signal. In such an application, bar graph elements are turned on or off sequentially as the level of the signal changes. A graphic equalizer in a music centre has a few bar graphs used in this manner. Some industrial units utilize one or more bar graphs to indicate various levels, e.g. the level of a liquid in a tank, the percentage load on the unit, or the deviation of temperature from a set point. Some types of electronic equipment save power by only lighting the highest element which corresponds to the measured level.

An alternative use of a bar graph is as a state indicator. An electronic unit may incorporate a bar graph to indicate such functions as power on, low battery, over-temperature, overload or an internal fault. Each element of the bar graph is turned on or off depending on the state of the flag it represents.

Clearly, there is a big difference between this application and the level indication explained above. An interface circuit designed to operate one of these applications may not be suitable to operate the other. The designer will therefore need to define the type of operation intended for the bar graph and design the circuit accordingly. If there is a possibility of changing from one type to the other, then it will be advantageous to design a microcomputer-controlled circuit to operate individual elements of the bar graph and then use software to light the required LEDs in the required manner.

Employing a bar graph as a state indicator needs a digital command to turn each of the elements on or off. Transistors and open-collector digital gates are examples of drivers for each element of the bar graph. Such drivers can be interfaced to digital-output lines of a microcomputer to enable a regular update of the displayed information. Some of the information may be obtained by an external circuit which detects when a level is exceeded, or when a condition is violated. The output of such a circuit can either be fed back to the microcomputer, or may be used directly to light the corresponding element of the bar graph. The latter approach saves processing power, but leaves the microcomputer without any knowledge of the measured value or the detected levels.

Example 11.4: **Show how a bar graph can be utilized in a microcomputer-based controller to display the magnitude of a d.c. voltage in discrete visual levels.**
One way of satisfying the requirement is to employ an A/D converter to digitize the d.c. voltage and then quantize the result according to the number of elements within the bar graph. This is followed by using digital-output lines to light the corresponding elements through suitable drivers (e.g. the 74LS06 IC used in Section 7.4).

Since the microcomputer does not need to use the digitized value, this solution may be considered tedious and can be replaced by using a group of comparators, as shown in Fig. 11.9. Resistors R1 to R5 are of equal value and are utilized to provide equal voltage steps. The input voltage signal (V_{in}) will be compared with all of these steps. If the value of the input voltage exceeds that of a step, then the corresponding comparator will supply a low-state output signal. Otherwise, the output of that comparator is at a high state. If the anode of an LED is connected to the d.c. supply and its cathode is connected to the output line of a comparator through a suitable current-limiting resistor, then that LED will light when the output of the comparator goes low. This implies that more of the bar graph LEDs will light when the input voltage increases. The voltage drop across resistor R1 sets the lowest limit of comparison with the input voltage. Similarly the value of resistor R5 sets the highest voltage limit. The values of R1 and R5 can therefore be selected according to the upper and lower limits of the displayed voltage.

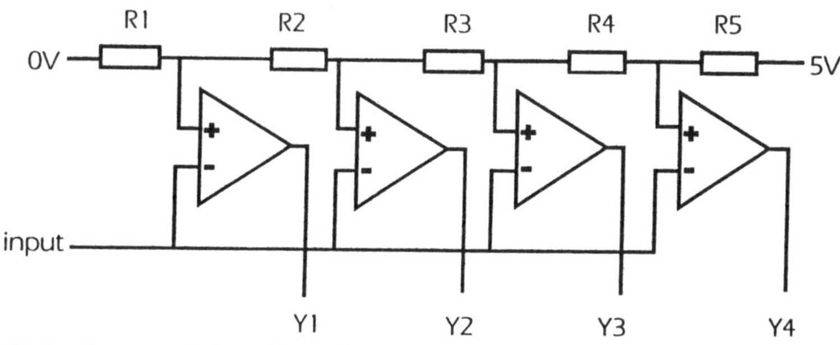

Fig. 11.9 Bar graph interface circuit.

Example 11.5: **Design a back-up circuit for a d.c. power supply incorporated as a part of a microcomputer-controlled system. The voltage level should be displayed through a bar graph and the back-up circuit operates only when the voltage drops below a fixed minimum limit. The required circuit employs two batteries; one can be used while the other is being charged. When the voltage of the battery in use drops below a minimum limit, the microcomputer uses the other battery as the source of power. The operation of the back-up circuit will stop when the d.c. power supply resumes operation. Assume that the controlling microcomputer monitors the state of the voltage of the d.c. supply line as a part of its operation.**

The two batteries are connected in parallel to transfer current to the d.c. line through thyristors, as shown in Fig. 11.10. The main part is the display circuit. It is represented by a block diagram and is the same as that of Fig. 11.9 (except that a 6-stage bar graph is used). It monitors the voltage of the d.c. line by reading the voltage drop across resistor R_y, which is connected in series with resistor R_x to form a voltage divider. The value of the two resistors is chosen to give a voltage drop on R_y of, say, 4 V when the d.c. power supply is operating normally. A 5.1-V zener diode is used to protect the display circuit. As long as the voltage of the d.c. line is higher than the minimum acceptable limit, no action will be taken other than operating six LEDs to display the magnitude of the voltage. The LEDs are arranged as a bar graph and will turn off sequentially as the voltage reduces (as in Example 11.4). The LED driven by output line Y6 is turned on when the d.c. line is at its highest level, and the LED driven by line Y1 is turned off when that voltage is dropped below the lowest acceptable level. This implies that the signal delivered by line Y1 can be employed to turn on one of the two thyristors in order to connect a battery to the d.c. line and use it as the source of power.

The voltage drop across resistor R1 of Fig. 11.9 defines the minimum voltage level. This level corresponds to the lowest acceptable operating voltage. By correctly choosing that level, the bar graph will only display voltages in the operating range. Figure 11.10 shows two digital-output lines

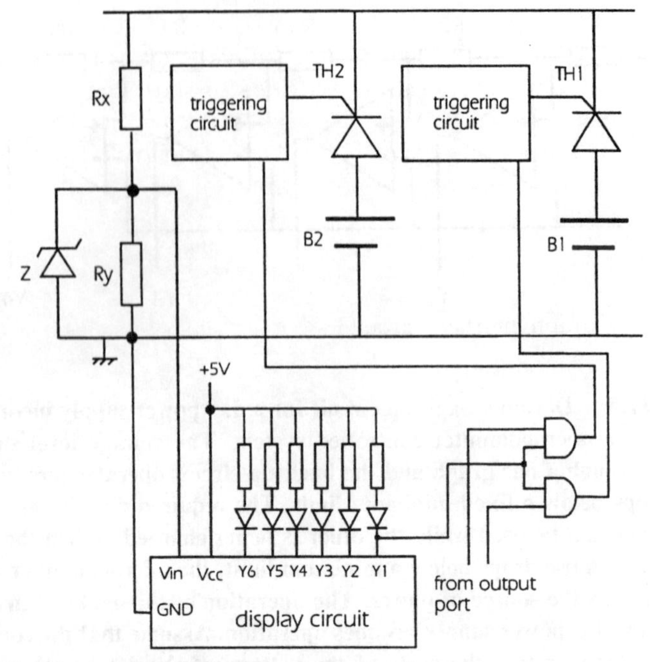

Fig. 11.10 Voltage-level display and control.

of the microcomputer connected to two AND gates. Only one of these lines is set to a logic-high state in order to trigger one of the two thyristors. The triggering signal is produced by the display circuit when the state of line Y1 changes to a logic high, i.e. when the minimum voltage limit is reached. Conduction through a thyristor increases the voltage of the d.c. line and lights the corresponding LEDs. Since the microcomputer monitors the d.c. line, it will detect the increase in the voltage level to that of the battery. Once the triggering of the thyristor is recognized, the microcomputer changes the state of both of the digital-output lines connected to the AND gates. This prepares the AND gates to trigger the other thyristor the next time line Y1 sends the command. With this arrangement, the microcomputer does not need to take fast action, thereby freeing its resources to execute other tasks.

Assume that battery B1 is used as the source of power after the failure of the d.c. power supply. It delivers current to the d.c. line through thyristor TH1 while battery B2 is fully charged and not in use. The microcomputer has already altered the state of its two output lines in order to select thyristor TH2 when LED 1 is turned off. The battery charge will reduce steadily at a rate defined by the requirements of the load. When the voltage drops below the minimum allowable value, a logic-high signal is delivered to the triggering circuit of TH2. This action turns on TH2, causing the voltage on the d.c. line to increase to that of the fully charged battery B2. The increase in the voltage level causes thyristor TH1 to be reverse biased and turn off. This separates battery B1 from the d.c. line. If the battery is of the rechargeable type, then a battery charger can be activated to start charging B1. If the charging takes too long, or the load demands are high, then the back-up capability can be increased by designing the circuit with more than two battery stages. This obviously requires additional AND gates and output lines from the microcomputer.

11.8 Optical encoders

Optical sensors are widely used to measure position and speed in various applications. This section considers a few types of encoder, highlighting their advantages and limitations and illustrating how they can be incorporated in microcomputer-controlled applications. The utilization of optical devices to detect motion is considered here. The use of other devices for the detection and measurement of motion is the subject of Section 13.7.

The increase in the use of optical encoders has led to the availability of ready-made encoders at relatively low prices and with different specifications. They are available in compact sizes and at different resolutions. They offer a choice of casings as well as the length and diameter of an encoder's shaft (encoders are also available with a hollow shaft for direct connection to a motor shaft). In some cases, however, it is not possible to find an encoder that satisfies the mechanical requirements. This leads to the design or purchase of an adapter to connect the encoder to a particular shaft.

11.8.1 Incremental encoders

This is one of the most widely used position- and speed-measuring devices. Its operation can be simply described as measuring displacement by counting pulses. The number of pulses within a fixed time interval gives a measurement of speed. The output of the encoder is a series of digital pulses that can be used by a microcomputer through a relatively simple interface circuit.

A 'linear encoder' is one type of incremental encoder. It is basically a strip with equally spaced slots. The slots are either located on the edge of the strip, or are in the form of holes along the strip. Figure 11.11 shows an example of such an encoder. A light source and a light-sensitive device can be placed on either side of the strip, a few millimetres apart. When the strip is moved, light passes from the light source to the light-sensitive device only through the slots or holes. Conduction through the light-sensitive device will be sensed by the interface circuit, which will count the number of times the photosensitive device is turned on and off, thus obtaining a count proportional to the displacement.

Fig. 11.11 Linear encoder.

Rather than using slots or holes, equally spaced light-reflective lines can be placed on one side of a strip. With this arrangement, the light source and sensor are placed on the same side (as in Fig. 11.2(c)). The sensor will be turned on and off as the reflective lines are moved.

Linear encoders are suitable for detecting motion along a straight line. A very large number of applications, however, use rotating shafts, e.g. robotics and machine control. They perform the angular measurement by the utilization of 'rotary encoders'. Such encoders are based on the same principle as the linear encoders, except that each encoder is in the form of a slotted disk rather than a slotted strip. The centre of the disk connects to the shaft and the slots are either on the outer edge of the disk or in the form of holes close to its outer edge. The method of interfacing and the process of counting the number of slots is the same as that of a linear encoder.

When compared to a rotary encoder, a linear encoder has a limitation in the displacement range it can measure. This range is defined by the length of the strip. A shaft encoder, on the other hand, has no end limit since it continues to provide pulses as long as the shaft is turning. Incremental encoders, however, only measure a change in position, but not the exact position. A temporary loss of power in the microcomputer or the interface circuit will, for example, erase the count, leaving the equipment without any knowledge of the actual position.

One of several procedures can follow a power loss. The moving object can, for instance, be moved to a default position before restarting the operation. This may not always be possible, since many processes do not allow repeating certain actions before completing others. The requirements of such systems may be better satisfied by a different type of encoder, such as the one explained in the next section.

Since the detection is based on the use of optical sources and sensors, the circuit needed to generate the digital pulses is similar to those described earlier in this chapter. The connection of the light source can be the same as that of Fig. 11.1(a). The interface of the light-detecting device can take the shape of the circuit of Fig. 11.1(c), where the output is taken at the point joining the transistor to resistor R2. Alternatively, the circuit of Fig. 11.5(d) can be adopted, where the optocoupler is replaced by a light-emitting device and a phototransistor. The output of either circuit is a digital signal which can be used to provide a digital representation of motion.

Example 11.6: **A microcomputer-controlled electric vehicle has a wheel circumference of 1 m. How can an optical encoder be used to allow the microcomputer to move the vehicle according to a set of programmed distances?** An incremental-rotary-encoder disk can be connected to the wheel of the vehicle. Both the light source and the photodetector can be mounted on the chassis of the vehicle and interfaced to the onboard microcomputer. Any movement of the vehicle will turn the wheel and the encoder's disk, thereby breaking the light beam at a rate proportional to the speed of the vehicle.

An encoder with 100 slots per revolution will provide 100 pulses per turn of the wheel, whose circumference is 1 m. This means that the encoder is capable of detecting a 1-cm movement of the vehicle. The microcomputer reads the required travelling distance and moves the vehicle while counting the number of pulses received from the encoder. If, for example, the vehicle should be moved by 4.2 m, then motion will continue until the count reaches 420. The microcomputer-controlled motion will follow an algorithm which may be set to accelerate the vehicle to maximum speed when the count is low, and then decelerate as the required count is approached.

Incremental encoders are not always used in their simplest form. Modifications have been introduced to increase their usefulness and provide the user with the ability to select the best encoder for the job. One such modification is the use of a marker, or an index. This is a single hole located at a different radius to that of the encoder's main slots or holes. It provides a single pulse per revolution and requires a dedicated light-emitting source and sensor. It is employed to count the number of revolutions, or to detect the angular displacement from a fixed point on the shaft. For these reasons it is also known as the zero signal, or the reference signal. The marker can therefore be employed to provide a synchronization pulse which may, for instance, be utilized as a start point in position-control applications. A controller can, for example, turn a motor shaft and examine a digital-input line until it detects the marker pulse. Operation can then start from that point, which will be considered as a reference. Pulses from the slots can be counted to measure the angular displacement, while marker pulses are used to count the number of turns, as shown in the following example.

Example 11.7: **Design a circuit to count the pulses delivered by an incremental shaft encoder which provides a marker signal.**

The angular displacement can be measured by employing an up-counter to count the number of pulses. The marker pulses can be used either to interrupt the microcomputer or be counted by a second counter. The former case has a few limitations, as there may not be an interrupt line available for the interface, or the rate of interruption may be high enough to delay other tasks handled by the microcomputer. The decision on adopting this approach will therefore depend on the rate of angular displacement and the type of tasks handled by the microcomputer.

If the use of interrupts causes limitations, then a circuit of two counters can be employed. Counter 1 counts the angular displacement, while counter 2 counts the number of revolutions. Let us assume that the measurement will start from a reference point, i.e. the marker. When the shaft is turned, pulses will be delivered to the interface circuit and counted by counter 1. This continues until a marker pulse is received. This pulse achieves two purposes. It clears the count of counter 1 and increments the count of counter 2. In this way, the output of the two counters gives the number of revolutions plus a fraction of a revolution represented by the sum of pulses from the last marker.

A microcomputer can read the two counters at any time to obtain a measurement of the angular displacement. The microcomputer can either reset the counters after each reading, or it can leave them counting and perform a subtraction to obtain the angular displacement since the last reading. Attention should be given to the instant of reading the counters. There is a possibility that the microcomputer reads counter 2 just before the arrival of a marker pulse and then reads counter 1 just after the marker is received. The effect is to lose the count of counter 1, which leads to an error equivalent to one revolution. This can be prevented if the microcomputer reads counter 2, and then reads counter 1 before repeating the reading of counter 2. If the successive readings of counter 2 are not the same, then the microcomputer repeats the reading of the two counters.

The optimum resolution of the counter depends on several factors, such as the number of slots within the encoder, the rate at which the microcomputer reads the displacement, the number of input lines available for the interface, and the cost and space limitations of the interface circuit. A 12-bit counter (such as the 74HCT4040 IC used in Example 9.3) can count 4096 pulses before overflowing to 0. Using such a counter removes from the microcomputer the need to read the counters at a high rate, since the circuit can measure the angular displacement of up to 4096 revolutions before resetting to 0.

11.8.2 Two- and three-channel encoders

One of the main problems with incremental encoders is their inability to detect a change in direction. This can lead to wrong measurements if the moving object reverses direction or oscillates. A very attractive solution is to incorporate a 2-channel encoder. It contains two optical sensors positioned to provide two outputs with 90° phase shift. The detection of the direction of rotation is based on recognizing which of the two signals is leading the other.

Example 11.8: **Design a circuit to detect the direction of rotation and speed of a motor shaft where the speed-measuring device is a 2-channel shaft encoder.** Measuring the speed is performed by counting the number of pulses from one of the two channels of the encoder, in a similar way to that of the previous example. The resolution of the counter and the rate at which the microcomputer needs to read the count depend on the range of speeds that the motor shaft will be rotating at.

Figure 11.12 is an example of the output signals from such an encoder, which can be interfaced to two digital-input lines of a microcomputer. The latter can read the lines regularly to examine the change in state of the two channels and deduce the direction of rotation. In the waveforms of Fig. 11.12(a), channel A leads channel B by 90°. Let us assume that the microcomputer reads the two channels at the instant when channel A is at a high state and channel B is at a low state. When motion occurs, the state of the two channels will change in four sequential steps:

10 11 01 00

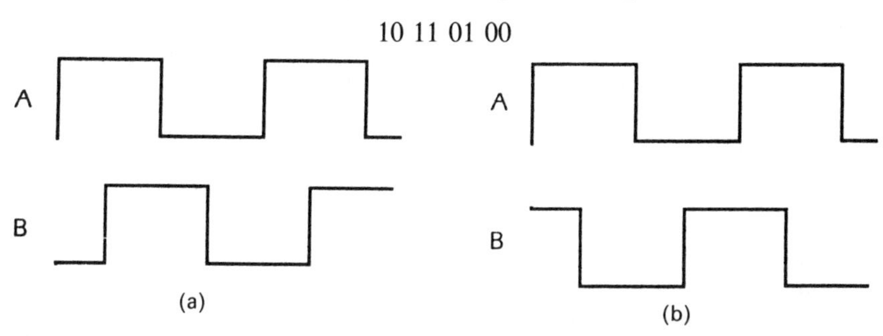

Fig. 11.12 Two-channel encoder: (a) forward; (b) reverse.

The steps will repeat in the same sequence as long as motion continues in the same direction. If the direction of motion is reversed, then the output of the two channels will be as that of Fig. 11.12(b). The signals received by the input terminals of the microcomputer will change in the following sequence (assuming that the microcomputer starts to read the two channels at the point where channel A is at a high state and channel B is at a low state):

10 00 01 11

The direction can be realized by comparing the sequence of the received signals with either of the two codes.

Example 11.9: **An interface circuit uses the computational power of a microcomputer to examine the output signals delivered by a 2-channel rotary encoder. How can the interface circuit be provided with a hardware means to detect the phase shift between the two signals? The modified circuit should also provide the microcomputer with an indication representing the direction of rotation.**
The detection of the direction of rotation is based on recognizing which of the two encoder signals leads the other. One solution is based on the use of a D-type latch clocked by one of the two encoder's signals (through the CLK line). The D input line of the latch is connected to the other signal from the encoder, as illustrated in Fig. 11.13. The state of the output line of the latch (line Q) will be set according to the state of signal A when signal B occurs. In Fig. 11.12(a), signal A leads B. Supplying these signals to the latch will send output line Q to a logic-high state. Reversing the direction of rotation will make signal A lag signal B (as shown in Fig. 11.12(b)). This makes the latch supply a logic-low signal when clocked by signal B. A microcomputer can therefore test the output of the latch to determine the direction of rotation. A measurement of speed can be obtained by reading a counter which is counting the number of pulses of both of the two channels.

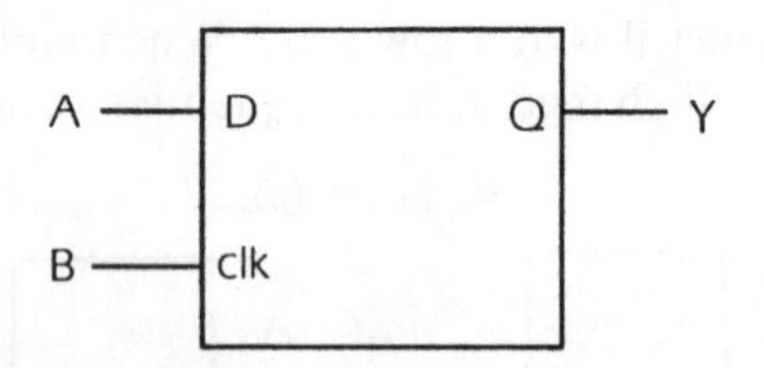

Fig. 11.13 Detecting direction of motion.

Using a counter to measure the speed with this approach will, however, lead to a false reading when the rotating shaft changes its direction of motion. Although the output of the latch will inform the microcomputer about the change, the counter will continue to count pulses regardless of the change in direction. This will obviously lead to a wrong reading whenever a change in direction takes place. This can be prevented if the circuit is modified to use an up/down counter where the direction of changing the count is set by the state of the output of the latch. Such a counter is employed in Example 12.3 and is shown in Fig. 12.1. Clearly, the resolution can be improved by cascading a few counters.

The 2-channel encoder provides a measurement of displacement and an indication of the direction of rotation. A third channel can be added to generate the marker indication. The result is an encoder capable of providing displacement, direction and a reference point. This encoder can also be used to measure the number of revolutions.

Using an optical encoder in a noisy environment may lead to a false representation, due to the imposition of spikes on the train of pulses delivered to the interfacing circuit. Twisted pair cables are used to protect the signal from noise. The length of these cables is kept as short as possible and they can be screened for further protection. A further measure can be introduced by utilizing an encoder with complementary output signals. Such an encoder gives two output signals 180 electrical degrees apart. The interface circuit needs to be modified to invert one of the signals before ORing it with the other. This action is taken in order to make the noise glitches cancel each other, leaving a noise-free train of pulses. The effect of noise will be considerably reduced if an absolute encoder (Section 11.8.4) is used, because the interface circuit reads the actual position rather than counts pulses.

11.8.3 Improving resolution

The resolution of the measured displacement depends on the type of encoder used. In an incremental encoder, the resolution is defined by the number of slots or holes within a single turn of the encoder. If, for instance, a shaft encoder with 360 slots is selected, then the resolution is 1° per pulse. The resolution will increase to 0.18° per pulse if the encoder is replaced by another of 2000 pulses per revolution.

Figure 11.14 shows three signals. The one at the top is a square wave representing encoder pulses delivered to a counter. For a fixed rate of change in position, the resolution of the encoder defines the rate of pulses supplied to the counter. The error in the measured angular displacement is therefore a function of the distance or duration between two successive pulses that can be delivered by the encoder at a given speed. Reducing this error will obviously improve the measurement, since the microcomputer will be able to detect smaller changes in position. An easy way of reducing the error is to employ a higher-resolution encoder, but this is more costly. An alternative solution is to modify thc interface circuit to detect both edges of the received pulses. In this way, the microcomputer will be able to detect smaller changes in position, since the resolution is effectively doubled. Adopting the same approach with a 2-channel encoder implies the detection of both edges of each of the two signals. This will quadruple the resolution without the need to change the encoder.

There are several ways of detecting both edges of a square wave. One way is to delay the pulses, and then Ex-OR them with the original pulses. The result is pulses at twice the frequency, as shown in Fig. 11.14. An alternative solution is to supply the received signal to two monostables, one set to be triggered by the rising edge, while the other is triggered by the falling edge, in a similar way to that of Fig. 9.18.

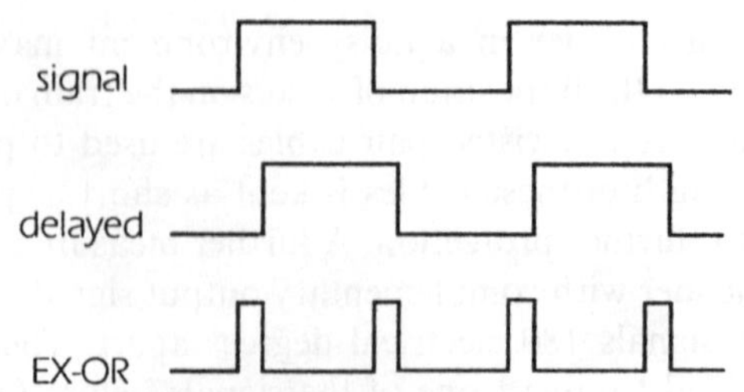

Fig. 11.14 Increasing resolution.

11.8.4 Absolute encoders

These encoders are coded to represent every position by a unique code. In this way, a controlling device, e.g. a microcomputer, can read the exact position as soon as power is supplied to the unit. An example of a linear absolute encoder is shown in Fig. 11.15, where the holes are made on three horizontal lines. This gives eight different positions; each is read as a binary number, where a hole permits conduction of light, thereby causing a photosensitive device to turn on.

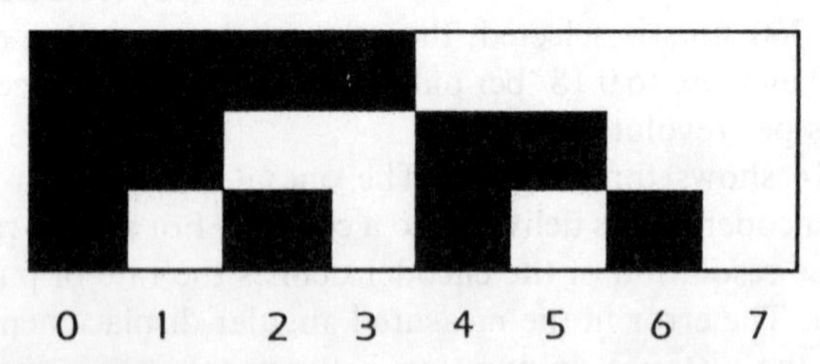

Fig. 11.15 Absolute encoder.

This particular encoder needs three channels, each of which detects the light conduction through one line. The number of channels defines the resolution and is referred to as the number of bits of the encoder. An 8-bit encoder, for instance, has eight channels and provides 256 different readings per revolution. A 12-bit encoder, on the other hand, has 12 channels and can read 4096 positions. Increasing the resolution will therefore increase the cost of the interface circuit, but the interface circuit does not need to adopt the pulse-counting approach used in incremental encoders. This is because, with an absolute encoder, a microcomputer reads the position directly from the encoder. The two main advantages of an absolute encoder are its abilities to give the exact position of an object in one reading and to continue to provide that reading even after turning the power off and then on again.

The states of the channels can be transmitted to the microcomputer through parallel lines, the number of which depends on the resolution. A microcomputer can read the coded position in one step and calculate the positional displacement accordingly. The undesirable increase in the number of interface lines has led some manufacturers to produce high-resolution absolute encoders which change the coded position to a serial form and transmit it to the microcomputer serially. This reduces the number of interface lines considerably, but creates the need to transform the signal back to the parallel form either by the microcomputer's interface circuit or by software. The former approach increases the number of components, while the latter reserves some of the processing power of the microcomputer for the conversion.

In many applications, such as a multi-degree-of-freedom robot arm (Chapter 13), the microcomputer is required to know the actual positions of two or more objects simultaneously. An absolute encoder can provide a reading of the actual position of each degree of freedom in one step. The number of software steps needed to transfer the coded-position information from both encoders to the microcomputer depends on the resolution of both the encoder and the input ports of the microcomputer. Two 8-bit encoders interfaced to a microcomputer through two 8-bit input ports will enable the microcomputer to read each encoder in one step. This could, however, give an error if the objects were moving while the microcomputer obtained the two readings. Latches can be incorporated to store both readings simultaneously before being read by the microcomputer. The number of required latches will obviously increase with the resolution.

Since an absolute encoder is capable of detecting the actual position within a revolution, a microcomputer will have to read the encoder before the encoder completes one turn. This can overload the microcomputer if the rate of change of position is high and/or the microcomputer is involved with the execution of other time-consuming tasks. An attractive solution can be based on the use of a 'multiturn absolute encoder'. Such an encoder contains a second disk employed in counting the number of turns. It is incremented by 1 every time a revolution is completed. The utilization of such an encoder will remove from the microcomputer the need to read the encoder regularly. The second disk will also provide a direct reading which is not affected by resuming operation after a power failure. The utilization of the second disk will considerably increase the resolution of the encoder. For example, an encoder with a 12-bit disk for measuring position and a 9-bit disk for measuring the number of revolutions will provide a 21-bit resolution, which corresponds to the ability to detect 2 097 152 positions.

Chapter 12

Generating waveforms

12.1 Introduction

Numerous microcomputer-controlled applications impose the requirement of generating certain time-varying signals by the microcomputer. These signals may, for instance, be used as demand signals for analogue equipment. The required shape, amplitude, frequency and offset vary from one application to another. For example, a microcomputer may be utilized to supply angular-demand signals to a robotic mechanism possessing its own analogue control circuitry (Chapter 13). The need to command the mechanism to move from one position to another in the shortest possible time may impose the requirement of generating a step demand signal from the microcomputer. The angular displacement is specified by the amplitude of the step, and the direction of motion depends on whether it is an upward or a downward step. Supplying the mechanism with a square wave will therefore command it to move to a destination and to return to the original position at the highest possible travelling speed. In this case, the time which the mechanism spends in either direction depends on the frequency of the demand signal.

If the requirement is to move the mechanism at a specified rate of displacement, then the demand will be in the form of a ramp signal which increases at a constant rate. In this case, the need for commanding the mechanism to move between two locations repeatedly can be satisfied by supplying the mechanism with a triangular-wave demand signal.

The generation of the required wave-shapes can be achieved in several ways. One way is to produce the waveform by software with the aid of an appropriate output device, e.g. a D/A converter. Alternatively, a waveform-generating circuit can be purchased, or designed, to produce the required waveform under the control of the microcomputer. This chapter considers some methods of generating microcomputer-controlled waveforms. It gives examples and explains how to provide the user with the means of choosing any of several waveforms, introducing offsets, and setting the amplitude and frequency.

12.2 Methods of generating waveforms

Let us start with a description of methods of generating waveforms under the control of a microcomputer.

12.2.1 On-line calculations

One way of generating a waveform is to utilize the processing power of a microcomputer to calculate instantaneous values of the waveform and to supply each value to an output device at the end of a predefined time interval. The output device is normally a D/A converter which generates an analogue value directly related to the digital number supplied by the microcomputer. The overall processing time that the microcomputer spends on the generation of a waveform is proportional to both the rate of generating samples and the time needed for calculating each sample. The on-line calculation method is attractive to use in many applications, particularly when the calculations are simple to implement and the sampling rate is not high. This, however, is not the case in a large number of applications. Consequently, one needs to estimate the overall time needed for the implementation of this method and its effect on the performance of the system.

12.2.2 Generating and using a look-up table

A look-up table is a group of sequential locations in memory at which information is stored and can be retrieved by the microcomputer at any time. This approach eliminates the need for on-line calculations, thus saving computation time. However, it occupies an area in memory whose length is defined by the number of data values within the table.

One method of generating a look-up table is based on using the microcomputer to calculate each of the values off-line. Alternatively, the values can be supplied by the user or from an external device. Numerous systems use look-up tables for storing valuable information needed for operation or for the generation of waveforms. Such tables are normally saved on a disk or permanently stored in EPROMs. When a look-up table is used, the microcomputer reads the values from that table and supplies them to an output device at the required rate. When compared to the on-line calculation method, this method reserves more memory, but consumes less execution time. The execution time of both methods is proportional to the rate at which the microcomputer updates the output.

12.2.3 The hardware approach

The implementation of either of the two methods described above can occupy a large portion of the processing time when the sampling rate is very high. If this is the case, then the system throughput can be greatly increased when hardware is utilized to generate the required waveform or to assist the microcomputer in performing the waveform-generation task. The complexity of waveform-generating circuitry depends on the shape of the required waveform as well as on the number and type of additional features imposed by the design specifications. For instance, the circuitry may need to provide the user with the ability to specify the amplitude and the frequency of the waveform.

12.3 Sampling and quantization error

It is described in Section 1.10 how the quantization error in a microcomputer-generated analogue signal depends on the resolution of the device performing the conversion. It is also shown in Chapters 6, 7 and 8 that the resolution of a data converter can be limited by the number of lines in the data bus of the microcomputer. Consequently, using a high-resolution D/A converter reduces the number of steps in a microcomputer-generated analogue waveform. However, it increases the price of the interface circuit. The number of steps may also depend on the sampling rate of the output, which in turn depends on the time needed by the microcomputer to perform the calculations and/or to attend to other tasks besides generating the waveform.

In some systems, this imposes a restriction which can be removed by the use of waveform-generating hardware. Although the hardware approach increases the component cost and the board area, it forms the optimum solution for applications where the microcomputer cannot spare the time to generate the waveforms at the required rate. Example 12.3 illustrates the introduction of additional hardware to ease the task of using a look-up table and assist in saving computation time.

12.4 Examples of generating waveforms

12.4.1 Generating a square wave

Methods are given in earlier chapters illustrating the generation of a square wave by the use of a microcomputer. In Examples 3.1 and 3.3, a square wave is produced by software means and is delivered to the output through a digital-output line. Example 6.4, on the other hand, employs a D/A converter to deliver a software-generated waveform to the outside world. Using such a converter provides the microcomputer with the ability to adjust the amplitude and the offset of the waveform without the need for additional hardware. An alternative method of generating a square wave is presented in Example 3.2, where the circuit of Fig. 3.1 is employed to generate a square wave. The example also explains how a microcomputer can be given the ability to control the generation of such a waveform.

Example 12.1: **A microcomputer is utilized to generate on/off command signals in an industrial application. The process needs to be repeated at a rate of 50 Hz with the on/off period set according to a request from an operator through a potentiometer. How can the requirement be satisfied assuming that the command signal needs to be in the form of a square wave whose mark-to-space (M/S) ratio represents the on/off periods?**

- *Reading the user's command*:
  The potentiometer can be interfaced to the microcomputer by the use of an A/D converter whose conversion speed does not need to be high. If an 8-bit converter is employed, then the conversion produces values in the range 00H to FFH depending on the setting of the potentiometer. In order to allow the change of the M/S ratio from 0 to 100% we need to take two steps. Firstly, let the duration of a full cycle of the waveform be represented by parameter V which is equal to the maximum digital number produced by the A/D converter (FFH in this case). Secondly, let the digitization result represent the 'MARK' of the waveform (M). This implies that the 'SPACE' of the waveform (S) is represented by:

$$S = V - M$$

  Therefore, a digitization result of FFH (equivalent to 255 decimal) demands $M = 100\%$ and $S = 0$, whereas a result of 7FH demands $M = S = 50\%$.

- *Scaling factor*
  In order to enable a microcomputer to implement this task, a count needs to be introduced to represent the time required before switching between the two output states. This count is inversely proportional to the frequency of the waveform and its magnitude depends on the frequency of the microcomputer's clock. If, for instance, the microcomputer decrements the count once every microsecond, then the duration of the 50-Hz waveform is 20 ms and can be represented by a count of 20 000. This implies that when the conversion result is 255, then $M = 20\,000$ and $S = 0$. One can therefore say that

  M = conversion result × scaling factor
  where the scaling factor equals 20 000/255 = 78.43

  In some situations, it is preferable to save computation time by approximating the multiplication factor to the nearest integer number, thus permitting the use of integer multiplication.

- *Implementation*
  The procedure of generating the required M/S ratio can therefore be based on performing an A/D conversion, calculating M and S according to the result, and generating the waveform. This can be represented by the following procedure:

  1 START AN A/D CONVERSION
  2 WAIT TILL EOC
  3 M = (CONVERSION RESULT) × (FACTOR)
  4 S = 20 000 − M
  5 LOAD A SOFTWARE COUNTER WITH M
  6 SUPPLY THE AMPLITUDE OF THE 'MARK' TO A D/A CONVERTER
  7 DECREMENT COUNT UNTIL = 0
  8 LOAD A SOFTWARE COUNTER WITH S
  9 SUPPLY THE AMPLITUDE OF THE 'SPACE' TO A D/A CONVERTER
  10 DECREMENT COUNT UNTIL = 0
  11 GO TO STEP 5

With this procedure, the microcomputer only reads the potentiometer at the start to recognize the request of the user. After that, it continues to generate the same waveform. It can be seen from steps 7 and 10 that the program-controlled approach is used in generating the waveform. Example 12.3 uses the interrupt-controlled approach. As an exercise, rewrite the above procedure to allow the microcomputer to generate the waveform by using the interrupt-controlled approach. Between attending to the interrupt-service routine, let the microcomputer monitor the setting of the potentiometer to permit on-line variation of the M/S ratio.

12.4.2 Generating a triangular wave

Examples of generating ramp signal have been presented in previous chapters. The generation process is based on driving the digital inputs of a D/A converter from a binary counter. It is shown that the counter can be an IC, or its function can be performed by software. Section 8.2.1 presents an example of a commercial data converter containing a counter and a D/A converter in the same IC. The use of such an IC reduces the number of components and eases the ramp-generating task. The number of components can be further reduced when the software approach is adopted. However, this monopolizes the microcomputer while the waveform is generated. This is acceptable in applications having no requirement for using the microcomputer during the wave-generation interval. Alterations can be made to the presented ramp-generating methods in order to produce triangular-wave generators.

We start with the software approach. A typical procedure for generating a full cycle of a triangular wave by a microcomputer consists of three parts. The first is the initialization, which gives the starting point, taking into consideration the required level of offset. Let us assume that no offset is required and the initialization value is equal to 0. The second part involves incrementing the amplitude regularly until it reaches the required level. This is followed by the third part, where the microcomputer decrements the value until it reaches the lowest limit. The continuous generation of a triangular wave will therefore consist of executing the second and the third parts repeatedly.

Example 12.2: **Describe how a microcomputer can generate a 1-Hz triangular wave by using a unipolar 5-V 10-bit D/A converter. The required amplitude is 0 to 3 V.**
In the 5-V 10-bit D/A converter, an amplitude of 5 V is represented by 1024 levels, i.e. each level represents 4.88 mV. Consequently, the required 3-V amplitude is represented by 614 levels. Each cycle needs to be generated in 1 s to achieve the 1-Hz frequency. This means that the requirement can be satisfied when the output is incremented in steps for the first 500 ms and then decremented in similar steps in the next 500 ms. Therefore, if the value is to be changed by one level each time, then there will be a total of 1228 steps per cycle. In this way, a value needs to be supplied to the output once every 814 μs. Alternatively, the value can be incremented by two levels every time, therefore permitting updating the output once every 1628 μs. It is obviously desirable to increment the output in as small steps as possible. In the case under consideration, incrementing the value once every 814 μs does not consume much execution time when the interrupt-controlled approach is adopted. It therefore represents the optimum choice.

As in any interrupt-controlled method, the procedure consists of two parts. These are the initialization program and the interrupt-service routine. It uses two values 'x' and 'y'. The first represents the number of steps, whereas the second contains the digital value calculated by the microcomputer and supplied to the D/A converter.

- *Initialization*
  1 CALCULATE A COUNT CORRESPONDING TO 814 MICROSECONDS
  2 SET THE MICROCOMPUTER TO ACCEPT INTERRUPT FROM TIMER

```
3  LOAD TIMER WITH COUNT
4  START TIMER
5  LET x = 0, AND y = 0
6  CONTINUE
```

- *Interrupt routine*

```
1  RELOAD AND RESTART TIMER
2  IF x < 614
         THEN INCREMENT x AND y
   ELSE
         IF x < 1228
               THEN INCREMENT x, AND DECREMENT y
         ELSE
               LET x = 0, AND y = 0
3  OUTPUT y TO D/A CONVERTER
4  RETURN FROM INTERRUPT
```

The procedure results in generating a triangular wave of a fixed amplitude and frequency. Many applications introduce the requirement to provide the user with the ability to alter these functions. As an exercise, modify the procedure by replacing the constant number of steps with a value obtained from a user-controlled potentiometer.

Example 12.3: **How can the ramp generator described in Section 7.3.1 be modified to generate a triangular wave with minimum intervention from the microcomputer?**

The solution given in Section 7.3.1 is based on employing an 8-bit binary counter whose output lines drive the digital-input lines of a D/A converter. If the binary counter is replaced by a microcomputer-controlled up/down counter, then a triangular wave can be generated easily. The operation is based on the microcomputer instructing the counter to count up for a specified time interval and then commanding the counter to count down for an equal time period. The rate at which the counter counts is specified by the frequency of the clock signal supplied to its input. It can be derived from the clock of the microcomputer by the use of a divide-by-N counter.

Figure 12.1 presents an example of an 8-bit counter formed by cascading two 4-bit up/down counter ICs (74LS191) with their output lines driving the inputs of an 8-bit D/A converter. The cascading of such counters is made simple by the provision of the RC (ripple clock) signal at pin 13 of the IC. This pin is normally at a logic 1 state. It produces a low pulse (a logic 0) when the counter overflows or underflows. The figure shows that the clock input line of the second counter is supplied from the RC line of the first. The direction of change of count is specified by the microcomputer through controlling the down/up input line of the counter from a digital-output line. A logic 1 state commands the counter to count down, whereas a logic 0

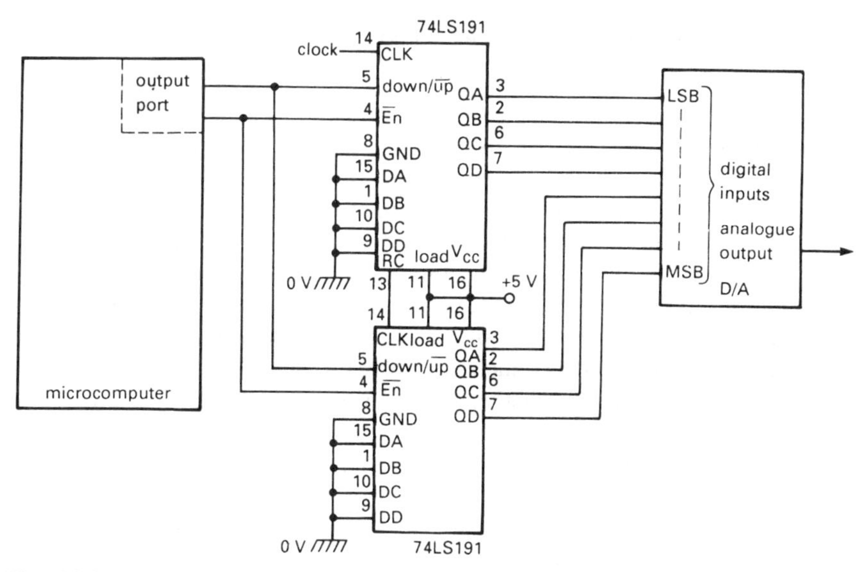

Fig. 12.1 Using up/down counters.

instructs it to count up. The microcomputer also controls the active-low enable line of the counters. The counter will count as long as the state of that line is a logic 0. (More information on the use of this counter is given in Example 12.7.)

By using a hardware counter, execution time will be saved as the microcomputer only needs to enable the counter and to specify the direction of the count. This implies that the microcomputer needs to complement the state of the down/up counter twice every cycle. Although this creates the need for the use of an internal interrupt within the microcomputer, it does give the microcomputer the ability to specify the operating frequency of the resulting waveform. The frequency of the clock should be correctly chosen in order to prevent exceeding the maximum count. In an 8-bit counter, the maximum number of steps is 256 in each direction, i.e. 512 steps per cycle. If, for example, the clock frequency is 1 kHz, then the duration of a full cycle is 512 ms, which corresponds to a frequency of 1.95 Hz.

12.4.3 Generating other waveforms

The selection of a method of generating a waveform depends on many factors, such as the shape of the waveform, cost, development time, capability of a microcomputer, and whether or not the microcomputer can spare the processing time needed for the execution of the wave-generating program.

(1) On-line calculations or look-up tables

In many situations, calculating the instantaneous amplitude of each point of a waveform is a simple task, e.g. the generation of the waveforms given in Examples 12.1 and 12.2. This is not the case for other types of waveforms, such as sine wave. The time needed for the on-line calculations is relatively long and can be considered unacceptable in many applications, e.g. real-time control systems. In such applications, the optimum solution is likely to be in the form of employing a look-up table containing a group of data values representing the shape of the waveform. Alternatively, a mathematical coprocessor may be used to simplify the on-line generation of the required function. The high cost of such a processor limits its use in low-cost products. The decision on whether to perform the on-line calculations or to use a look-up table also depends on the capability of the microcomputer and the available memory area. If, for instance, the system program reserves the whole of the memory area, then the designer may be forced to implement the on-line calculations with a sacrifice in performance, or to introduce hardware to generate, or assist in generating, the waveform.

(2) Using a look-up table

The data values forming the look-up table can reside either in a memory device (e.g. one or more EPROM ICs) or on a disk. In the former case, the memory device is either placed in the microcomputer or forms part of the interface circuit. In the latter case, the information is loaded from the disk to a fixed area in RAM.

The interface of an EPROM to a microcomputer can be achieved by connecting the EPROM as a memory device. Reading a value from the look-up table is based on executing a memory read instruction from a specified address (see Example 13.3). This commands the EPROM to deliver the corresponding data to the processor through the data bus. The processor then transfers the data to the digital-input lines of a D/A converter to produce the waveform. Computation time can be saved if the generated data values are supplied from the EPROM directly to the D/A converter as shown in Fig. 12.2. Thirteen digital-output lines are used to address each of the 8192 bytes within the EPROM. In this way, an analogue value is generated as soon as the microcomputer supplies an address to the EPROM. As digital-output lines are dedicated to address the EPROM, the latter will continue to supply the D/A converter with the corresponding value until the EPROM receives the next address from the microcomputer. With this approach, the microcomputer only supplies the sequential addresses. The frequency of the generated waveform is therefore proportional to the rate of supplying the addresses.

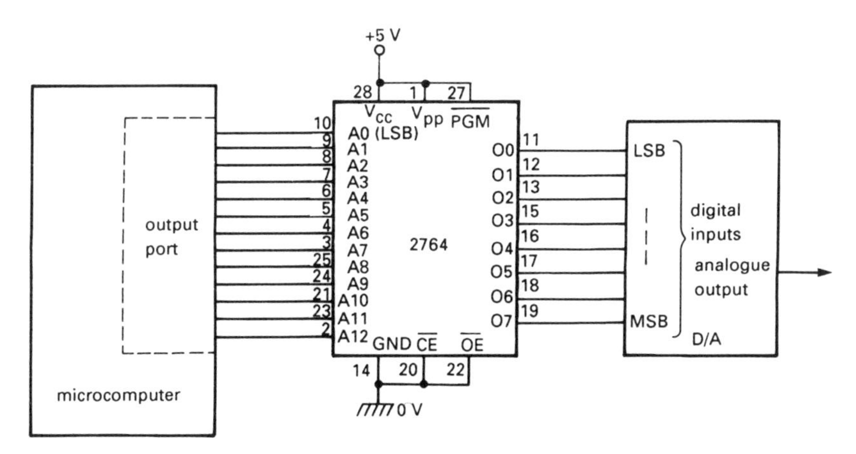

Fig. 12.2 Using an EPROM in a waveform generator.

(3) Introducing additional hardware

The use of the EPROM removes from the microcomputer the need to calculate the instantaneous values of the waveform. However, it uses 13 digital-output lines from the microcomputer. This restriction can be removed by introducing a counter to supply sequential addresses to the EPROM. It starts with a count of 0 and is incremented each time a clock pulse is supplied to its input. The count increases sequentially, so the addresses are supplied to the EPROM in sequence. This makes the EPROM step through the look-up table and supply each value to the digital inputs of a D/A converter. The latter generates the waveform as its output. The counter can be formed by cascading 74LS93 ICs, as in Fig. 7.1. This counter, however, counts in one direction. More flexibility can be given if that counter is replaced by the up/down counter used in Example 12.3. In this case, the microcomputer has the ability to define the direction of stepping through the table.

(4) The sampling interval

The rate at which the EPROM steps through the look-up table depends on the frequency of the clock supplied to the counter. The rate at which the microcomputer supplies the counter with pulses depends on the frequency of the required waveform and on the number of data values in one cycle of the waveform. If, for instance, the waveform is represented by $n = 1000$ values and the required frequency is $f = 10\,\text{Hz}$, then the sampling interval $S = T/n = 1(fn) = 100\,\mu\text{s}$, where T represents the duration of one cycle of the waveform.

The use of software or hardware means of controlling the amplitude and the frequency of a waveform generated from a look-up table is explained in

Sections 12.5.2 and 12.5.3 respectively.

Example 12.4: **How can a microcomputer generate a fixed-amplitude sine wave by the use of a look-up table stored in RAM?**

The requirement of this example can be satisfied by reading the table at regular intervals. Assuming that the sampling interval is not very short, the interrupt-controlled output approach can be used to interrupt the microcomputer at the end of each interval. The procedure of implementing this approach consists of the following two parts, where 'y' represents the offset from the beginning of the look-up table:

- *Initialization*
  1 LOAD TIMER WITH A COUNT CORRESPONDING TO SAMPLING INTERVAL
  2 SET THE MICROCOMPUTER TO ACCEPT INTERRUPT FROM TIMER
  3 START TIMER
  4 LET y = 0
  5 CONTINUE

- *Interrupt routine*
  1 RELOAD AND RESTART TIMER
  2 GET VALUE OF ITEM y IN TABLE
  3 OUTPUT VALUE TO D/A CONVERTER
  4 INCREMENT y
  5 IF y > LENGTH OF TABLE
     THEN LET y = 0
  6 RETURN FROM INTERRUPT

The procedure assumes that the table contains values corresponding to a complete cycle of a sine wave. Since such a wave consists of two identical, but inverted, half-cycles, the generation of the waveform can be achieved by the use of a table of half the length. In such a case, the sign of the sequential values is inverted each time the end of the table is reached. This approach saves memory area at the expense of a small increase in the execution time. As an exercise, modify the above procedure to allow use of this approach.

12.5 User interface

A microcomputer having the ability to generate several waveforms normally provides the user with manual means of selecting a waveform, introducing offsets, and setting the amplitude and frequency.

12.5.1 Choosing a waveform

A microcomputer can be programmed to monitor an input device through which a user can provide a command for the generation of a waveshape.

Example 12.5: **An operator needs to be provided with the ability to command the generation of any one of five waveforms, and to switch off the output when the generation is not required. How can this be achieved by a microcomputer?**
One way of implementing this task is to utilize a 6-way rotary switch, as shown in Fig. 12.3(a). Each of the first five positions results in selecting a specific waveform, whereas the sixth position commands the microcomputer to stop the waveform-generation process. In such a configuration, the microcomputer can, for instance, be programmed to interrogate the state of the input lines through an input port once every 100 ms. One of these lines must be at a logic 0 state; it represents the user's choice. If that input is not line 6, then the microcomputer generates the waveform corresponding to the selected position. If, on the other hand, position 6 is selected, then the microcomputer will stop generating waveforms, but will continue to monitor the inputs until a change in the setting of the switch is detected.

Example 12.6: **The solution presented in the previous example uses six digital-input lines. Suppose that all the digital inputs of the microcomputer are used for other applications and there is only one analogue-input line available. How can that be utilized to satisfy the requirement?**
In this situation, the rotary switch can be connected as shown in Fig. 12.3(b). In this configuration, resistors of equal values are connected in series through the terminals of the switch. The terminals at the end of the series are connected to +5 V and 0 V. A user can position the switch to connect any of the six points to the input of the A/D converter. Since the resistors are connected in series, each position of the switch produces a d.c. voltage of a magnitude different from the others. The converter can therefore digitize the d.c. value and compare it to the corresponding values stored in memory to determine which of the five waveforms is selected, if any.

As an exercise, write the procedure needed to achieve the detection of the position of the rotary switch in both of the above examples.

12.5.2 Setting amplitude

The amplitude of a computer-generated waveform can be altered by software or by hardware means. The latter can be applied by introducing an amplifying stage at the output of the D/A converter. As described in Chapters 6 and 7,

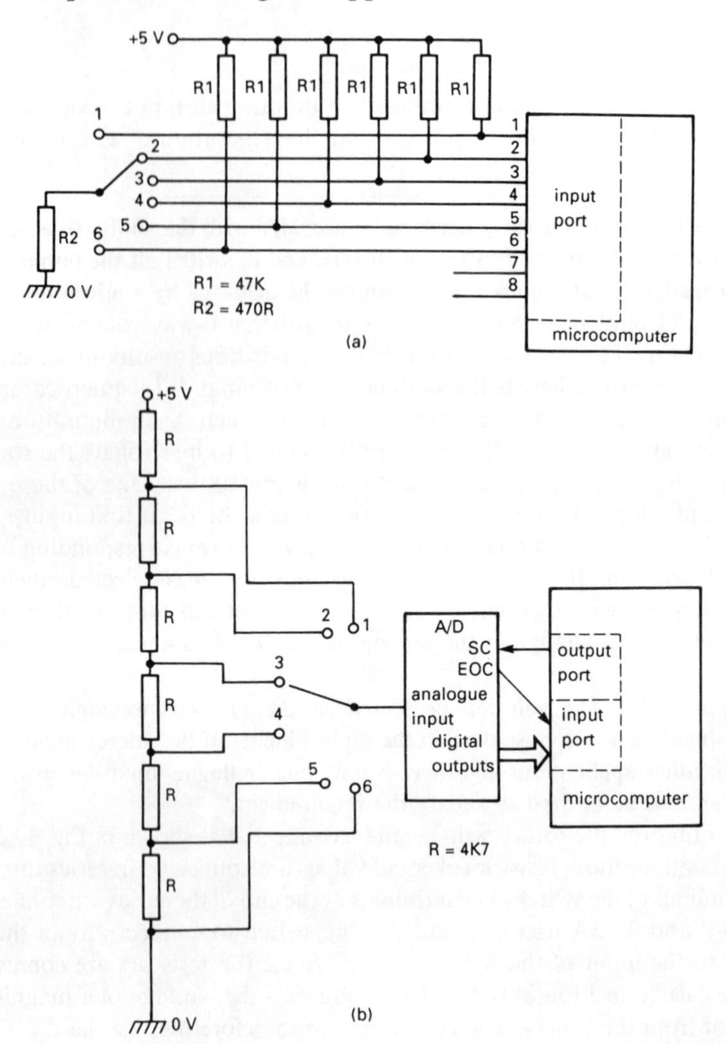

Fig. 12.3 Using a rotary switch: (a) digital interface; (b) analogue interface.

the gain of the amplifier can be fixed, manually adjustable, or software adjustable. Potentiometers, jumpers or switches are normally employed to enable an operator to manually define the amplitude of a waveform. For example, a potentiometer can be used to alter the gain of an amplifier, thus changing the amplitude of the output linearly without the intervention of the microcomputer.

Let us now consider using the software approach to alter the amplitude of a waveform represented by sequential data stored in a look-up table. The magnitude of the amplification is specified by an amplification factor. The value given to the factor can be defined within a program executed by the microcomputer, calculated on-line, or supplied by an operator through an input device such as the keyboard or a potentiometer. Software amplification is achieved by performing a mathematical operation (multiplication and possibly a division). Such operations extend the execution time and, if not correctly arranged, can reduce the accuracy of the resulting value. For instance, if integer numbers are used and the calculation is in the form of $a \times b/c$, then achieving the division before the multiplication results in a larger error than if the multiplication takes place first. Verify this for $a = 2$, $b = 3$ and $c = 2$.

The time taken to perform the multiplication and/or the division depends on the type of processor used and the number of bits in each of the values. For example, if both values are represented by 16-bit numbers, then carrying the multiplication by an 8-bit processor takes longer than if it is achieved by a 16-bit processor, particularly if the former does not include a multiplication instruction and the latter does. The time needed to perform a mathematical operation becomes significant when the output needs to be updated at a high rate. It could lead to reducing the overall performance of the microcomputer.

In a poorly designed system, attempting to generate a high-amplitude waveform may result in demanding an amplitude which exceeds the capability of the D/A converter. Such a requirement can be met without distorting the shape of a microcomputer-generated waveform when an amplifier is introduced to magnify the output of the converter to the required level. On the other hand, attempting to produce a waveform of a very small amplitude directly from a D/A converter may result in distoring the waveshape due to the quantization error. This problem can be solved by generating a waveform of a large magnitude and then scaling it down by the use of an operational amplifier.

12.5.3 Setting frequency

Several examples have been presented in previous chapters illustrating the setting of the frequency of waveforms. It has been shown that manual setting is achieved by the use of potentiometers or in-circuit links. They can either be applied to inform a microcomputer of the magnitude of the required frequency, or they can be placed to alter the frequency of a hardware-generated waveform without the intervention of the microcomputer. It has also been shown that changing the frequency of a software-generated waveform is based on changing the contents of a memory location representing the time interval between delivering successive samples to an output device.

A user can specify the required frequency through an input device such as a potentiometer interfaced to an A/D converter. The program needed to implement the frequency change should be arranged such that the sampling interval decreases when the digitized value increases. If, for example, an 8-bit A/D converter is utilized, then one can say that the sampling interval is given by (V – conversion result) where V is a constant specifying the boundary of the frequency-variation range. If, for instance, the requirement is to generate a waveform defined by a data block within a look-up table, then V can be set to a value, say 102H. This implies that when a request for maximum frequency is received from the potentiometer, the sampling interval is equal to 3 μs (102H – FFH). This gives the microcomputer sufficient time to read a precalculated value from memory and deliver it to an output device. If one cycle of the waveform is represented by 1000 samples in the look-up table, then one cycle of the waveform is generated in 3 ms. This corresponds to an output frequency of 333 Hz. On the other hand, the minimum possible frequency generated with the arrangement is achieved when the digitized value is 0, i.e. the sampling interval is equal to V, which in this case is 102H. This implies that a full cycle of the waveform is produced in 258 ms, which corresponds to a frequency of 3.87 Hz.

A higher value of V can be used when the frequency range needs to be set at a lower level.

12.5.4 Introducing an offset

Numerous applications have the need for providing a user with the ability to offset waveforms. This can be accomplished either by software or by hardware. The method of supplying the offset depends on whether the waveform is generated by software or by hardware means.

(1) Offsetting a microcomputer-generated waveform by software

In this case, the offset is in the form of a digital number within the microcomputer. The latter adds the offset value to, or subtracts it from, each digital sample before its delivery to an output device. A user can specify the offset value through an input device, e.g. keyboard, external switches, or a potentiometer as shown in previous examples.

When offsetting a microcomputer-generated waveform by software, a software clamp needs to be introduced to prevent misrepresentation. For example, if an 8-bit D/A converter is employed to generate a value of A0H with an offset of 70H, then the microcomputer adds the two values producing 110H, which is a 9-bit number. The program therefore needs to detect the 9th bit and clamp the value to FFH. Otherwise, only the lowest eight bits will be supplied to the converter, i.e. 10H.

(2) Offsetting a hardware-generated waveform by a microcomputer

One way of offsetting a waveform is to program a microcomputer to generate a d.c. value through a bipolar D/A converter. This allows the microcomputer to produce a positive or a negative value which is converted to an equivalent analogue level and then added to a hardware-generated waveform as shown in Fig. 12.4. In this configuration, the waveform-generating hardware may or may not be interfaced to the microcomputer. It may, for example, be used to allow the microcomputer to start and stop the waveform-generation hardware, or to select the frequency of the generated waveform. The magnitude of the offset can be specified by the user through an input device.

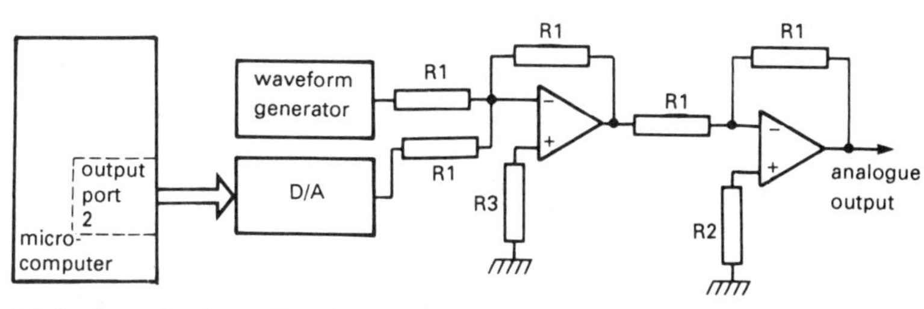

Fig. 12.4 Introducing offset by a microcomputer.

(3) Offsetting a waveform by hardware

In this case, a d.c. level is added to or subtracted from the waveform by analogue means in the same manner as that presented in Examples 6.5 and 7.7. The hardware-introduced offset is the same whether the waveform is generated by software or by hardware means. This method allows the user to set the magnitude of the offset by a potentiometer which determines the voltage to be added to, or subtracted from, the waveform.

Example 12.7: **How can the circuit of Fig. 12.1 be modified to provide the microcomputer with the ability to offset the generated waveform?**

Figure 12.1 shows pins 1, 9, 10 and 15 of each counter connected to the 0-V line. These pins are the digital-input lines of the counter. They are provided to permit a controlling device to introduce an initial binary count. The counter accepts the inputs when the load line (pin 11) is taken to a logic 0 state (this line is shown connected to the +5-V line in Fig. 12.1, as no offset is introduced in Example 12.3). After loading the value and returning the load line to a logic 1 state, the count will resume from the new value.

Let us assume that we want to apply this method by the use of the microcomputer of Fig. 6.1(a). The microcomputer provides 16 output lines divided among two ports. Besides the two output lines shown in Fig. 12.1, the modification needs nine output lines. Eight of these supply data to the two counters (four to the input lines of each counter, and one to drive the load line of both counters). Port 2 of the microcomputer can be utilized to supply the data lines of the counters, whereas port 1 can be employed to provide the commands to the load line, enable line, and down/up line.

12.6 A waveform generator

The method of providing a microcomputer with the ability to generate several waveforms depends on many factors, such as whether the waveforms are to be generated simultaneously or not, and on the required range of amplitude and frequency of each waveform.

Suppose that we want to produce a signal generator in which only one analogue waveform needs to be generated at any instant of time. Assume that 8-bit resolution is sufficient to satisfy the design specifications. In this case, the look-up table approach forms an attractive solution where a data block is needed for each waveform. Let us choose to represent each waveform by 256 steps. If we want to generate 32 different waveforms, then we need a storing area of $256 \times 32 = 8192$ bytes, which can be obtained by the use of a 2764 EPROM IC connected as shown in Fig. 12.5. The data blocks can be stored in that EPROM in sequence.

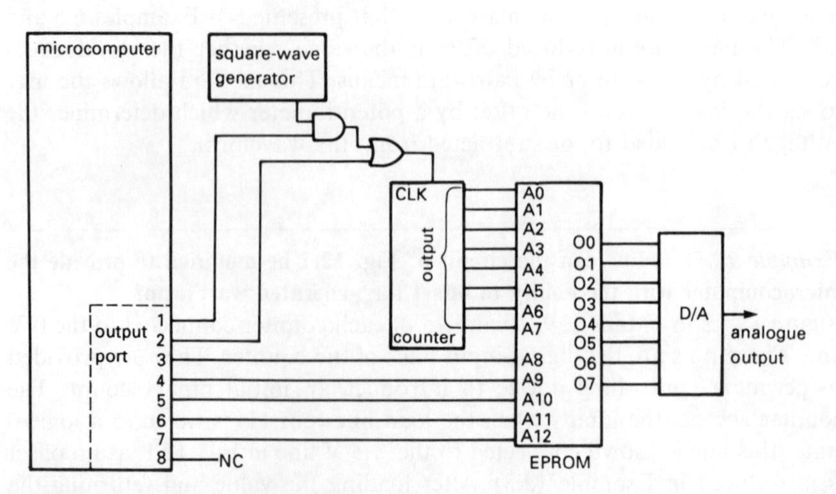

Fig. 12.5 A waveform generator.

The EPROM IC has 13 address lines (A0 to A12) and eight data lines (O0 to O7). The 256 sequential locations of each table can be addressed through the least significant eight address lines. In the configuration of Fig. 12.5, sequential addressing is provided by driving these lines from an 8-bit counter. The selection of any of the 32 waveshapes is achieved by using five digital-output lines from the microcomputer where the binary combination of these lines identifies the required data block.

Besides using the five block-addressing lines, the configuration of Fig. 12.5 uses two output lines of the microcomputer and two gates. These are employed to provide the microcomputer with the ability to choose between supplying the counter with a hardware-generated clock or with sofware-generated pulses. If pins 1 and 2 of the microcomputer are set to a logic 1 and a logic 0 respectively, then the hardware-generated pulses will pass through the gates to the clock input line of the counter. If, on the other hand, line 1 is set low then the microcomputer drives the counter through pulsing line 2. Operating from the hardware-generated clock consumes a very small portion of the execution time of the microcomputer, as the microcomputer only needs to take an action when a change of waveform is requested by the user.

Using the information given in this chapter, a signal generator can therefore be designed to provide a user with the ability to select a waveform, offset it, and set its amplitude and frequency. If the up/down counter of Fig. 12.1 is used in generating the sequential addressing, then the microcomputer will have the ability to reverse the addressing sequence by changing the logic state of the down/up line of the counter. If the counter is connected as described in Example 12.7, then the microcomputer can also specify a point on the waveform from which the generation starts.

The above example considers the generation of one waveform at a time. The use of look-up tables to generate several waveforms simultaneously is similar to the above except that a value needs to be read from each table and supplied to a corresponding output. This requires the use of a multichannel D/A converter and implies that the EPROM is interfaced to the microcomputer as a memory device from which the microcomputer reads the data values and supplies them to each of several output devices. This method is implemented in Example 13.3, where the analogue output is used to move a robotic mechanism under the control of a microcomputer.

Chapter 13

Microcomputer-controlled robotic mechanisms

13.1 Introduction

A large number of industrial processes and much equipment are controlled by microcomputers. Robotics and machine tools are examples of using microcomputers as intelligent position controllers in the automotive manufacturing industry. The use of robotic mechanisms in industrial environments may specify that a mechanism and its controller (which includes a microcomputer) are designed as a self-contained unit. This implies that a single-board microcomputer can be utilized which includes neither a monitor nor a full keyboard. Instead, operator instructions may be supplied through switches. A simple display may be included. The use of microcomputers as controllers for real-time applications is considered in this chapter, where position control in robotic mechanisms is taken as an example. It is assumed that such mechanisms have built-in control circuitry and power drivers. They are designed to accept digital commands (e.g. start and stop) as well as digital or analogue data (e.g. demand signals) from an external controller such as a microcomputer.

13.2 Designing control equipment

The type and complexity of a controller depends on the nature of the operation it needs to perform and the required degree of controllability. For instance, an industrial multi-axis robotic mechanism may contain motors controlled by in-robot servo-mechanisms. Each motor performs the movement of a single degree of freedom and is driven by a dedicated digital or analogue control circuit. The internal operation of the robot may be synchronized by an internal microprocessor which acts according to information supplied from a supervisory microcomputer. The microprocessor can command each of the control circuits to perform the required operation. If a local microprocessor is not used, then the supervisory microcomputer needs to supply information (commands and data) to each of the individual control circuits. The latter may be designed to send information back to the microcomputer for monitoring, testing or control purposes.

When employing this type of a robotic mechanism, users need to take two courses of action. Firstly, they should provide a circuit for interfacing the robot to the microcomputer. Secondly, they need to produce a program which generates suitable information for directing the robot to perform the required task. The interface can be similar to the examples presented in the previous chapters. The required hardware and software depend on the capability of the utilized mechanism, the type and form of the signals it accepts or transmits, and the degree of sophistication of the microcomputer.

Before describing the microcomputer-controlled operation, the rest of this section is devoted to taking a general look at the control of equipment. Feedback loops are widely used for the implementation of such control. The operation of each loop is based on comparing a demand with a feedback signal. The latter is obtained from a transducer to provide information on a signal such as position or speed. Many systems operate without using feedback loops (open-loop systems). They are briefly described in the following section.

13.2.1 Open-loop control

An open-loop controller has no requirement for testing a feedback signal, and thus eliminates the need for employing feedback transducers. When open-loop control is performed by a microcomputer, the latter supplies a demand signal to a circuit to fulfil a certain requirement. Such a system does not have the ability to respond to changes in the load or to verify the achievement of the required action. The response and performance of such a controller are obviously not as good as that of a closed-loop controller. However, much equipment operates effectively in an open-loop configuration. For instance, an open-loop controller can be used to move a projector lamp in an application where absolute positioning is not an issue. On the other hand, a sophisticated closed-loop controller is required for the automatic insertion of electronic components on PCBs.

13.2.2 Closed-loop control

Closed-loop control is based on continuously comparing demand and feedback signals and generating a drive signal according to the result of the comparison. The drive signal aims to remove any difference between the demand and the feedback quickly and efficiently. Fast response and high performance can be obtained by measuring both the demand and the feedback signals correctly, sensing the changes as they occur, and responding within the shortest possible time. The performance, development time and construction cost of a closed-loop controller depend on the type of feedback loop used and on the control algorithm. The latter depends on the type and capability of the device being controlled as well as the task it needs to perform. Feedback loops can be classified into two main types, digital and analogue.

13.2.3 Digital closed-loop

Let us consider a feedback loop incorporated in a single-degree-of-freedom robot arm. Assume that the in-robot drive circuit contains a microprocessor which accepts demand signals as well as control commands from a microcomputer. It achieves the required positioning by employing a digital feedback loop. Such a loop deals with samples of the signals and is represented in block diagram form in Fig. 13.1(a). The in-robot microprocessor uses the information it receives from the supervisory microcomputer and compares it with digital samples of the feedback signal. The result of the comparison is used as a part of a calculation process (the control algorithm). The output of the process is supplied to the system through a D/A converter. In this case, the microprocessor forms an important part of the control loop, where it achieves the required control by software through working on digital samples of both the demand and feedback signals.

The performance of the loop depends on the resolution of the two data converters (an A/D and a D/A), their conversion times, and the rate of sampling the signals and executing the control algorithm. Besides implementing the control, the microprocessor may need to attend to other tasks. These are normally of a lower priority and may be defined as background tasks for execution by the microprocessor in between its implementation of the control algorithm. The latter may be executed regularly by the use of interrupts. The time needed for the efficient execution of the additional tasks may affect the performance of the feedback loop. It can limit the maximum attainable rate of obtaining samples and implementing the control algorithm (see Section 13.3).

13.2.4 Improving performance

The capabilities of such a system are limited by several factors and may not be suitable for some applications. It is therefore important to carry out an initial study to determine the requirements of implementing a digital loop

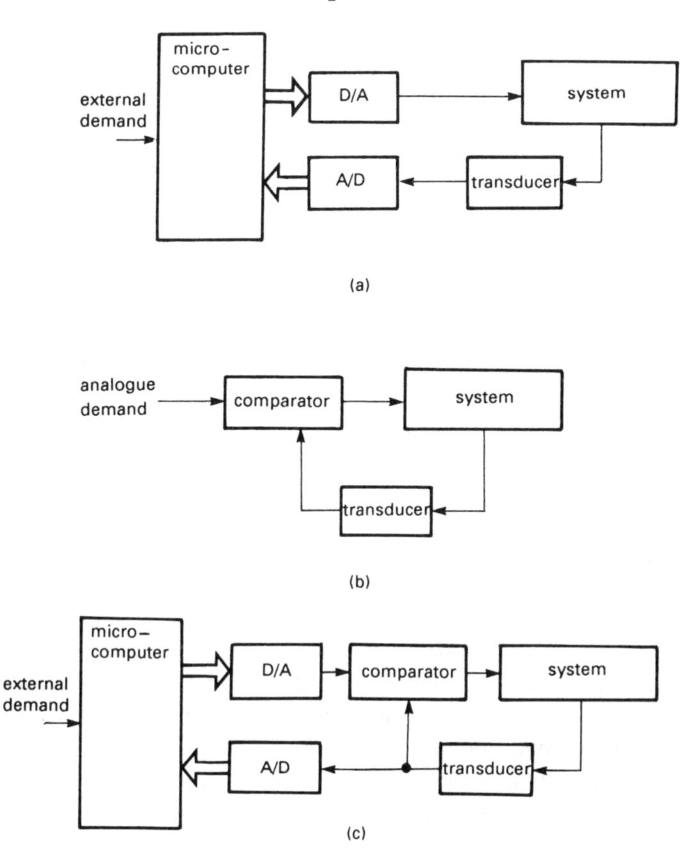

Fig. 13.1 Feedback loops: (a) digital; (b) analogue; (c) analogue with digital interface.

efficiently. If it proves that the microprocessor is not capable of meeting the design specifications efficiently, then (without changing the microprocessor) the designer has several options. The designer may, for instance, decide to replace some of the time-consuming software tasks by electronic circuits, or to include a maths coprocessor thereby increasing the processing power of the circuit. Alternatively, the designer can implement an analogue feedback loop (Section 13.2.5), or may be able to purchase a ready-made local controller which can handle a portion of the processing load and contribute to satisfying the performance requirements.

13.2.5 Analogue closed-loop

An analogue loop is represented by the block diagram of Fig. 13.1(b). It deals with continuous analogue signals rather than sampled data. The comparison of signals and the implementation of the control algorithm is done by hardware. A robotic mechanism based on the use of analogue control loops can be designed to operate without a local microprocessor. It accepts demand signals from a supervisor (such as a microcomputer) and performs the comparison and the control by hardware. If the supervisor only supplies demand signals, then it does not receive an indication of the magnitude of the controlled function, nor can it influence the performance of the feedback loop (unless additional hardware is included).

13.2.6 Comparison

Utilizing an analogue loop removes from the microcomputer the need for carrying out tedious calculations and forms an attractive solution to numerous control problems. Such loops are therefore applied in many controllers. However, digital loops also provide attractive features. For example, they permit the implementation of sophisticated control algorithms more easily than analogue loops. They offer good repeatability and their characteristics do not change drastically with changes in temperature. Furthermore, design modifications of a digital loop can normally be achieved by changing a program without altering the hardware. This is particularly attractive in industrial controllers, as the change can take the form of replacing EPROMS rather than having to cut tracks, replace devices, or solder additional components at the customer's site. In the last few years, many analogue controllers have been replaced by microcomputer-controlled equipment employing digital feedback loops. Such loops use software techniques to implement algorithms aimed at achieving better performance and higher productivity.

13.2.7 The monitoring feature

The configuration shown in Fig. 13.1(b) can be modified to provide the microcomputer with the ability to examine the controlled function. This can, for instance, be used to permit the microcomputer to verify that the output of the system is following the demand signal. This helps the microcomputer to detect faulty operation and act to rectify the fault or raise the alarm. The ability to examine the output signal can be provided by employing an A/D converter as shown in Fig. 13.1(c).

The rate at which the microcomputer should sample the feedback signal depends on the type of application. Let us consider two examples for which the configuration of Fig. 13.1(c) is used. In the first example, the aim is to allow the microcomputer to perform occasional checking of the output to ensure correct operation. In this case, the microcomputer reads the feedback signal in between its execution of tasks, because fast and regular monitoring is not necessary. In the second example, the reason for using the A/D converter is to introduce a special mode of operation to a microcomputer-controlled robot arm. In this mode, the microcomputer needs to collect detailed information on the path of motion. It commands the arm to travel along a given path and, at the same time, it monitors the travelling sequence along the path. The required monitoring can be performed by obtaining fast successive samples of the feedback signal during the travel of the arm. The collected data can be stored in RAM (as shown in Section 17.2) and used to prove the successful implementation of the control algorithm as well as to verify the accuracy of the designer's calculations for position estimation and collision avoidance.

13.3 Microcomputer-based control systems

Industrial control is one of many application areas which have benefited from the recent developments in microcomputers. A designer of a microcomputer-based control system has a choice between a large number of microcomputers offering a wide range of features and capabilities. In such systems, the degree of the microcomputer's involvement in the implementation of the control algorithm varies from one control application to another. For instance, the operation of numerous controllers is based on using the microcomputer to read external (input) signals, process them, and produce an output signal. The source of the input signals can be an operator or external equipment. In some cases, the processing is simple and may take the form of adding two input signals and supplying the result to an external control circuit. In other cases, the calculation may be in the form of implementing tedious mathematical operations and supplying the result to a D/A converter to drive an external device.

Computation time can be saved if the tedious calculations can be limited to taking place only when the magnitude of the input signals has changed rather than after each reading of the inputs. Such a change can be recognized by comparing the new received values with their corresponding old values. If no difference exists, or if the difference is smaller than a predetermined value, then the calculations are bypassed. Otherwise, the new values are stored and the calculations are performed. However, in a large number of control applications, the input signals are continuously changing. This means that implementing the digitizations and the calculations can be costly in terms of execution time and may take a large share of the processing power of the microcomputer. This could result in the overall performance of the system deteriorating. Consequently, the design of much computerized control equipment is based on employing a microcomputer as a supervisor which commands and synchronizes the operation of several local controllers. Tedious calculations may be replaced by introducing add-ons or look-up tables (Section 13.6). A wide range of local controllers is available ready-made for direct connection to certain microcomputers. Their capability and price vary over a wide range and they normally require minimal control from the microcomputer. The transfer of information (data and/or commands) between such controllers and the microcomputers can be based on any of the methods presented in Section 16.3.

Sharing the workload between the microcomputer and additional devices or equipment allows the implementation of multiple functions in parallel. This can improve the system performance and permit directing the computational capability of the microcomputer towards improving the system throughput as well as performing various tasks efficiently, e.g. user interface, error detection and data transfer. Clearly, the cost-effectiveness of the system can be improved if the microcomputer can function as a stand-alone controller. This is feasible in a large number of systems, particularly those in which the microcomputer is devoted to the implementation of a single task.

13.4 Implementation of tasks

When employing a microcomputer as a controller, or as a supervisor in a control application, a good understanding of the application and careful planning are essential to the realization of the optimum performance of the system. A balance is needed between sampling signals, executing algorithms and attending to the background tasks efficiently. A typical application is a microcomputer-controlled multi-axis robot arm applied to pick individual objects from a defined location and deposit them in any of several user-selectable destinations. Before moving the arm to pick an object, the microcomputer needs to know the actual position of the end-point and to determine the path which it needs to follow in order to reach the required destination. During the movement of the arm, the microcomputer may need to maintain up-to-date information on the location of the end-point. It therefore reads feedback signals from transducers (one from each joint within the arm) and uses the information to determine the actual location of the end-point at a given time.

In addition to performing the controlled movements, the microcomputer may need to handle other, perhaps less important, tasks. It is therefore important to organize the rate and sequence of executing tasks in a way that will satisfy the requirements efficiently. When developing software for a system requiring high performance, the minimization of the execution time of the system tasks is normally given a higher priority than the compactness of the software.

Most microcomputer-based control systems (especially those involving multitasking real-time control) are interrupt driven. They react to external asynchronous events, or in response to a timing device which synchronizes the operation of the system. Normally, the use of interrupts enables a control system to respond quickly and efficiently to events which may happen at any time during operation. Many microcomputers include several interrupt-request input lines available for the user. The interrupt processing can be tailored to achieve the best performance by the use of priorities (Section 1.9.3). The operation of a real-time control system can be based on generating a regular interrupt to inform the microcomputer of the need for sampling the control signals and executing related tasks. This interrupt is normally given the highest priority to ensure that its execution will not be disturbed by other activities.

Obviously, this does not mean that the program-controlled approach is not suitable for real-time control applications. It forms a feasible answer to many application problems, especially those where the tasks need to be executed in sequence and the servicing of external events need not take place as soon as an event occurs. For instance, the microcomputer-controlled pick-and-place operation mentioned at the beginning of this section can be applied by the use of the program-controlled approach. The microcomputer tests a status flag to determine whether or not an object is available for picking. When an active flag is detected, the microcomputer commences the execution of a program which moves the end-point of the robot to pick the object before transferring it to the required destination. If the operation is arranged so that a new object arrives at the picking area before the robot completes a pick-and-place operation, then there will be no waiting intervals. This arrangement makes the program-controlled approach very attractive, since adopting the interrupt-controlled approach will result in generating an interrupt when a new object is available and the microcomputer is not ready to begin a new pick-and-place operation. In this case, the interrupt-service routine sets a flag to indicate the availability of an object to pick. After its current task, the microcomputer tests that flag to determine the presence of a new object.

The operation of many computerized controllers is based on the adoption of both the interrupt-controlled and the program-controlled approaches. The former handles the time-dependent tasks, whereas the latter performs the rest of the operations.

13.5 Examples of position control

This section presents examples of using a supervisory microcomputer to achieve positional control of robotic mechanisms. One or more demand signals are supplied from the microcomputer and the required part of a robotic mechanism moves accordingly. In many applications, the microcomputer reads operator-set data in engineering units, e.g. degrees, metres, etc. A conversion to a suitable digital format is required, as shown in Example 13.1.

13.5.1 Single degree of freedom

Let us consider an example where the microcomputer supplies a single-degree-of-freedom robotic mechanism with an analogue demand signal through a D/A converter.

Example 13.1: **A microcomputer is employed to control the horizontal movements of a robot arm between two angles. The arm has its own drive circuit which locates the arm at the centre position when supplied with 0 V. The angular displacement varies linearly from +60° to −60° when the analogue voltage supplied to the device is varied from +2.5 to −2.5 V. How can the microcomputer control the movement of the arm linearly between +15 and −15°?**
An 8-bit bipolar D/A converter is considered suitable for positioning the robot arm. The interface of such a converter to a microcomputer is shown in Fig. 6.2. The digital value supplied by the microcomputer needs to be related to the actual angular displacement of the arm. Since an 8-bit converter is utilized, the 120° range (+60° to −60°) is represented by 256 levels. Therefore, each level represents 0.469° (i.e. 120/256). Changing the angle from −60° to +60° can be accomplished by supplying the D/A converter with consecutive values starting with 00H and ending with FFH. Locating the arm at the centre position is therefore achieved by supplying the D/A converter with 80H (half of full-scale).

If the projector is to be located at +15° then the microcomputer needs to supply the converter with the hexadecimal value A0H. This is the sum of 20H and 80H, where the 20H represents the equivalent value of 15° and the 80H is the equivalent value of 0 V. Similarly, a movement to an angle of −15° can be performed by supplying the converter with the hexadecimal value 60H (which is 80H − 20H). The microcomputer supplies each value to the D/A converter in the same manner as that of Example 6.2.

Having defined the start and end values, our next task is to gradually move the arm between −15° and +15°. One way of performing the required movement is to supply the digital demand from the microcomputer as a sequence of values. The rate at which the values are supplied depends on the mechanism and the required displacement rate. The procedure can be represented by the following steps. It is assumed that, at the start, the arm is located at the centre position. The procedure uses a predetermined variable z containing a value inversely proportional to the rate of displacement. The procedure also uses a byte y to include a digital value representing the actual displacement.

```
 1  w = z
 2  INCREMENT y
 3  IF y > A0H
         THEN GO TO STEP 7
 4  SEND y to D/A
 5  DECREMENT w UNTIL = 0
 6  GO TO STEP 1
 7  w = z
 8  DECREMENT y
 9  IF y < 60H
         THEN GO TO STEP 1
10  SEND y TO D/A
11  DECREMENT w UNTIL = 0
12  GO TO STEP 7
```

By altering the value of z, a user can set the rate of displacement of the arm.

13.5.2 Multi-degree of freedom

Example 13.1 has presented a method of moving a mechanism possessing a single degree of freedom between two points. Moving a multi-degree-of-freedom mechanism is, however, more difficult. In many situations, the associated calculations impose a restriction on the on-line performance. Such a restriction is often overcome by the use of look-up tables (Section 13.6). Let us consider an example of manipulating a two-degree-of-freedom mechanism.

Example 13.2: **Write a procedure for operating a robotic mechanism comprising two horizontal axes with a gripping mechanism forming the third axis perpendicular to the first two. A microcomputer controls the position of the end-point by supplying three analogue signals (one per axis). Positioning along each axis is proportional to the magnitude of the applied analogue signal. The mechanism is employed in a pick-and-place operation where the availability of an object is indicated by an active-high digital-input line.**

The generation of the analogue signals is accomplished by interfacing three 8-bit D/A converters to the microcomputer. This can be implemented by using the information given in Chapters 6 and 8. Let us define four bytes for use in controlling the movements along the two horizontal axes. The first two bytes ('xk' and 'yk') are related to the 'picking' location. They represent the digital values which need to be supplied to the two corresponding D/A converters in order to move the mechanism to that location. The other two bytes ('xp' and 'yp') correspond to the digital values needed to locate the mechanism at the 'placing' location. We can define two more bytes ('wu' and 'wd') to represent the values needed to move the gripper up and down respectively.

The procedure can be represented by the following, where step 1 prevents the microcomputer from taking an action when there is no indication of the availability of an object to pick. For clarity and simplicity of explanation, the procedure assumes that the microcomputer only needs to supply demand signals. Each axis-control circuit responds to a demand supplied to its input by producing a drive signal to achieve the required displacement. The procedure also assumes that the drive circuit generates a status flag 'k' whose function is to indicate the arrival at the requested destination. The microcomputer therefore monitors that flag, seeking a change in its state.

```
1  IF DIGITAL INPUT = 0
       THEN GO TO STEP 8
2  SUPPLY xk AND yk TO THE TWO D/A CONVERTERS
3  WAIT TILL FLAG k IS SET
4  PICK OBJECT (SUPPLY 'wd', WAIT, GRAB, SUPPLY 'wu')
5  SUPPLY xp AND yp TO THE TWO D/A CONVERTERS
6  WAIT TILL FLAG k IS SET
7  PLACE OBJECT (SUPPLY 'wd', WAIT, RELEASE, SUPPLY 'wu')
8  CONTINUE
```

The calculation of the relation between the demand signal and the actual displacement is similar to that of the previous example.

13.6 Defining the path of motion

There are several ways of defining the path of motion of a robotic mechanism. On many occasions, the application imposes constraints such as the need to avoid obstacles, or the requirement to follow the shortest possible path. The calculation of the path may therefore involve tedious mathematical operations. The complexity of these operations can increase with the increase in the number of moving sections of the mechanism. One way of defining the path of motion is to calculate the path on-line. Generally, this reserves a big portion of the processing power of the microcomputer. An alternative method is to carry the calculation off-line and to store the result of the calculation (the path) in memory as a look-up table (see Section 12.4.3). This considerably reduces the on-line processing and can contribute to improving the performance of the system.

Rather than calculating the values, a teaching method can be used. In this method, the robot arm is manually moved along the path and the microcomputer is instructed to read and store the coordinates in a table in memory. After collecting the data, the microcomputer can, at any time, use the data to execute the recorded sequences, thereby directing the arm to duplicate the manual operation. Since the microcomputer controls the supply of data to the arm, it is easy to move the arm at a different speed to the teaching speed. Besides eliminating the need for the calculations, the teaching method removes from the operator the requirement of physically measuring the coordinates of certain points along the path of motion, e.g. start-point, end-point and the location of obstacles. The teaching process is, however, time-consuming and is not attractive in situations where the path of motion changes frequently.

The use of look-up tables is not always attractive, as the memory space they reserve may be needed for applying other tasks. Attention should therefore be paid to the storage capability of the microcomputer (see Section 17.3). In situations where the memory area is limited, the location of certain points along the path can be 'taught' and stored in memory. The intermediate points can then be calculated by the microcomputer. The selection of the 'taught' points can be made such that the required calculation is simple, e.g. calculating points along a straight line.

The decision on whether to use a look-up table or to perform the position calculation on-line depends on many factors, such as the number of degrees of freedom, the processing power of the microcomputer, and the size of memory area available for storage. Look-up tables can either be temporarily loaded into the microcomputer or be permanently stored. Temporary storage means that the contents of the table are supplied to the RAM of the microcomputer from a disk, through a serial link, or from the keyboard. Permanent storage, on the other hand, means that such information resides in an EPROM to allow the microcomputer to read it at any time without the need for reloading. This is ideal for routine tasks such as paint spraying or arc welding.

Since the use of look-up tables does not need tedious calculations, and since the majority of robotic applications involve repetitive movements, one can deduce that look-up tables form an attractive solution to numerous applications.

Example 13.3: **Consider an application where the end-point travel of a three-degree-of-freedom robot arm is defined by a table consisting of 1024 points. The operator must be provided with the ability to define the direction of motion through two PTM switches. How can this be achieved by a microcomputer? Consider both 8-bit and 16-bit microcomputers.**

The solution can be divided into five parts, as follows:

(1) Length of table

What length of table do we need to reserve for a 1024-point path? The answer to this question depends on how many microcomputer-controlled joints need to be moved to achieve the required motion. Since the robot under consideration possesses three joints, the table consists of three parts, each containing the values corresponding to the required displacement of one joint of the arm.

During each step of motion, the microcomputer reads an individual value from each of the three parts of the table and supplies each value to its corresponding control circuit. This means that we have $1024 \times 3 = 3072$ values. If each value is stored as a byte, then the length of the table is 3072 bytes. If, on the other hand, higher resolution is used, then each value is stored as 2 bytes. In this case, the length of the table is 6144 bytes.

(2) Type of interfacing

Assuming that the latter case is adopted, the 6144 locations of the table can be stored in one or more EPROMs (see Section 12.4.3). Eight-bit interfacing means that the microcomputer reads an EPROM as an 8-bit device, i.e. 16-bit values are read as two successive bytes. In this case, only one EPROM can be used, as shown in Fig. 13.2(a). Alternatively, microcomputers with 16-bit data buses can use 16-bit interfacing, i.e. reading 16-bit values simultaneously. This saves processing time, but it introduces the requirement of using two 8-bit EPROMs, as shown in Fig. 13.2(b). In this case, one EPROM contains the lower eight-bits of the value, whereas the other EPROM contains the rest of the bits (up to 8 bits). In both of these figures, the EPROM is interfaced directly to the data bus. Compare this connection with that of Fig. 12.2, where the EPROM is connected to supply information directly to an external device.

(3) Reading from tables

In order to describe how values can be read from a look-up table, let us assume that the table is stored at a memory area starting at address 'y' in the form of three consecutive data blocks. Since each value is stored as a 16-bit number in 2 bytes, then each table consists of 2048 locations. This means that the first block of the table covers area 'y' to 'y + 7FFH' (where 7FFH represents 2048 bytes). Similarly, the second block of the table covers area 'y + 800H' to 'y + OFFFH', and the third block covers area 'y + 1000H' to 'y + 17FFH'. Consequently, to obtain a reading corresponding to point 'x' in the path, the microcomputer executes memory-read instructions to obtain the three values. Each value consists of two bytes. The first value is located at address 'y + 2x' and 'y + 2x + 1', the second value is at 'y + 800H + 2x' and 'y + 800H + 2x + 1'), and the third value is at 'y + 1000H + 2x' and 'y + 1000H + 2x + 1'.

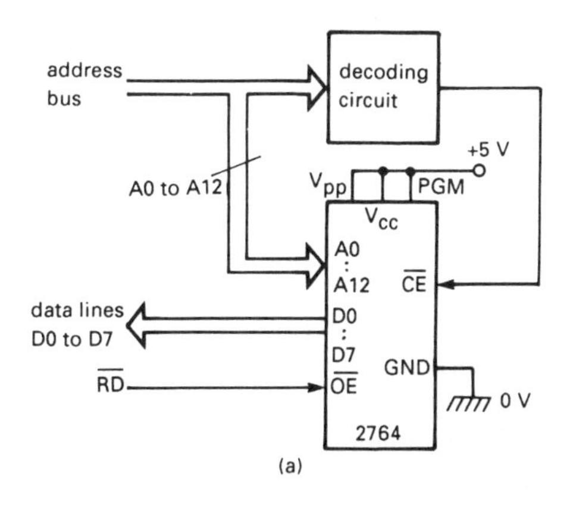

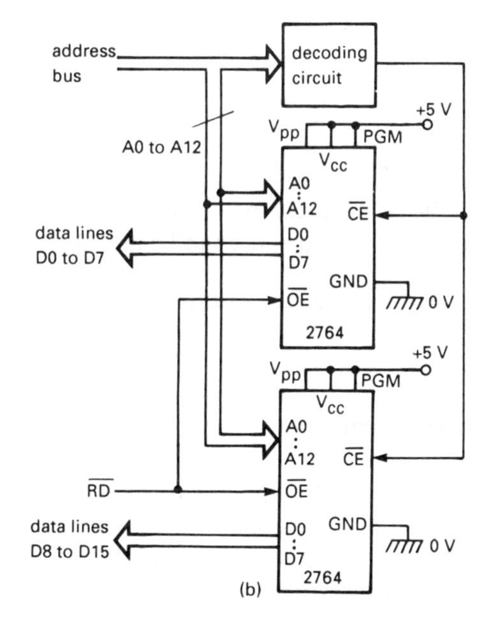

Fig. 13.2 Storing look-up tables in additional EPROMs: (a) 8-bit interfacing; (b) 16-bit interfacing.

(4) Reading the switches

The procedure of performing the monitoring by a microcomputer has been presented in Chapters 4 and 6, where it is explained that, since the switches are set by an operator, they do not need to be monitored at a high rate. During its operation, the microcomputer has the ability to read the table and supply the values to the corresponding control circuits at a much higher rate than the mechanism is capable of moving at. Consequently, software-wait loops need to be introduced, or the microcomputer employed to carry out useful tasks during these intervals. One such a task is to monitor the state of the switches.

(5) Operation

Assuming that the program-controlled approach is used, then during operation the microcomputer reads the state of the PTM switches and takes a decision accordingly. It makes use of a software pointer which contains an address in the table corresponding to the position of the end-point of the arm. If the first switch (S1) is pressed, then the microcomputer increments the counter. On the other hand, after detecting a closure of the second switch (S2) the microcomputer decrements the count.

Following this, the microcomputer reads the contents of locations 'y + count' and 'y + count + 1' and forms a 16-bit number. It then supplies that number to the first control circuit. After completing this task, the microcomputer reads the contents of location 'y + 800H + count' and 'y + 800H + count + 1', forms a 16-bit number and supplies it to the second control circuit. Finally, it forms a 16-bit number after reading the contents of locations 'y + 1000H + count' and 'y + 1000H + count + 1' and supplies it to the third control circuit. If the microcomputer does not need to perform other tasks, then a software-wait loop may need to be introduced before repeating the procedure.

13.7 Position transducers

Various types of transducer are employed to provide feedback signals in position-control applications. They can be grouped according to the type of signal they produce as digital and analogue transducers. An example of the former is the optical encoder described in Section 11.8. The latter is considered in this section.

A simple and a widely used analogue position transducer is the potentiometer. During its operation, such a transducer is supplied from a d.c. voltage source and connected so that any movement will result in varying the position of the potentiometer's wiper, thus altering the resistance. This produces a voltage proportional to the displacement. Analogue systems can directly use the analogue signal produced by a potentiometer. In a digital system, however, this signal needs to be converted to a digital form by using an A/D converter. The resolution of the result depends on the resolution of the A/D converter.

Potentiometers form the most economic position-sensing solution to many problems. There are, however, some limitations associated with their use, e.g. linearity and magnitude variation with temperature. Another limitation is that most potentiometers are not continuous. High-linearity continuous potentiometers are generally expensive. The non-continuity restricts the allowable angular displacement of the corresponding mechanical joint to a certain number of degrees, e.g. 270°. However, it does not restrict the operation of numerous robotic mechanisms. For instance, a robotic mechanism can be employed for collecting individual objects and depositing them in a predefined destination by performing an axial rotation of 180°. This type of mechanism has no requirement for a continuous potentiometer, as its angular displacement is restricted to 180°.

In some applications, the non-continuity constraint can be removed by the use of a multiturn potentiometer. This is obviously of a higher price, but allows a joint to rotate several turns, e.g. 10 turns. One should note the effect of using such a potentiometer on the resolution of the result. Let us assume that an 8-bit A/D converter is used to provide a microcomputer with a digital value corresponding to the location of the wiper of a potentiometer. If the potentiometer allows movement for 270°, then these readings will be represented by 256 digital levels, so each level corresponds to 1.055°. If, on the other hand, a 10-turn potentiometer is used, then the 256 digital levels correspond to 3600°, i.e. each level corresponds to 14.06°. This means that many applications will not allow the use of an 8-bit A/D converter with this type of potentiometer. Instead, a higher resolution converter can be used; for example, a 12-bit converter provides 4096 digital levels to correspond to the 3600°, i.e. each level corresponds to 0.879°.

One of the main advantages of using a potentiometer as a position transducer is that at any instant of time its reading corresponds to the position of the shaft. This is not the case for many other types of position transducers (such as a slotted-disk encoder), as they produce signals corresponding to the change in position from the last reading rather than the actual position. The advantage of being able to recognize the absolute position is clear when a digital system is switched on and the actual position is not stored in memory. With a potentiometer, the actual position can be determined by reading the output of an A/D converter. In a system using a slotted encoder, however, a command can be issued to move the object to a default position.

A decision on whether to present the displacement to the microcomputer as a digital or an analogue value depends on the type of interface circuitry used in the system, the nature of the application, and the capability of the microcomputer.

13.8 Sequential processing

In a microcomputer-based control system, the microcomputer is programmed to generate data and commands to add-on circuits and/or external controllers to command and synchronize their operation. Microcomputer-generated signals are produced by employing internal or external information. The former uses data stored in memory or generated by the microcomputer, whereas the latter utilizes commands and data from an operator or an external device. Consider, for example, a microcomputer-controlled robotic mechanism moving in a straight line at a speed set by means of a potentiometer. The microcomputer responds to one of two commands from an operator. The first moves the mechanism in the forward direction, whereas the second command moves it in the reverse direction. While in motion, the operation of that mechanism will terminate when the microcomputer receives a signal from a sensor indicating the arrival at the required destination. The procedure of operating such a mechanism is based on scanning digital-input status lines from the user and the sensors by the microcomputer. The latter responds to changes in the state of these inputs. A request from the user causes the program to read the setting of the speed-request potentiometer, and generate an analogue signal. The sign of this signal depends on which of the digital inputs is active and its magnitude depends on the requested speed. On the other hand, an indication from the sensors causes the microcomputer to generate a command to bring the mechanism to a halt.

Many industrial processes are based on performing sequential operations. The duration of each operation is likely to be different from that of the others. By programming a microcomputer to command the start and the end of each operation, the control circuit is simplified since the sequence-generating circuit is replaced by software. In addition to simplifying the circuit, the software solution eases the introduction of future alterations to the sequence of operations or to the duration of each operation. The choice between implementing the interrupt-controlled or the program-controlled methods is greatly dependent on the duration of each operation. The generation of long sequential intervals is best achieved by using the interrupt-controlled method, whereas the requirements of short sequential durations are best satisfied by adopting the program-controlled method.

Let us consider an example of the generation of sequential operations. In order to simplify the solution and clarify the implementation, operations of equal duration are assumed.

Example 13.4: **Consider a control process consisting of eight sequential stages. The duration of each stage is 20 μs and the process needs to be executed repeatedly. The operation of each stage is initiated when a logic 0 signal is supplied, and is terminated when the state of that signal changes to a logic 1. How can this process be controlled by a microcomputer?**

Solution 1 The software approach

The microcomputer can generate the stages by supplying eight individual commands from eight individual digital-output lines, e.g. all the lines of an

8-bit output port. Since the microcomputer needs to generate an output once every 20 μs, the program-controlled output method is more suitable for this application than the interrupt-controlled method.

The procedure can be represented by the following where n is a number corresponding to the output line which needs to be activated.

```
1  LET n = 0
2  n = n + 1
3  IF n > 8,
       THEN LET n = 1
   ELSE ONE MICROSECOND DELAY
4  LET LINE n OF OUTPUT PORT = LOGIC 0
   AND LET THE OTHER 7 LINES = LOGIC 1
5  WAIT FOR (20 – x) MICROSECONDS
6  GO TO STEP 2
```

Note that in step 5 the delay is $(20 - x)$ μs, where x represents the time needed to execute steps 2 to 6. In step 3, the program can take either of two paths depending on whether $n > 8$ or not. By introducing a 1-μs delay, the execution time of both paths become the same. In this way, the duration between supplying one signal and the next is 20 μs even between stages 8 and 1 as the process is repeated.

Solution 2 The microcomputer-controlled hardware approach

The requirements are satisfied in Solution 1 by employing eight output lines from the microcomputer. If, in a given application, these lines are required for use as general-purpose output lines, then the requirement must be satisfied by using an alternative solution. It can be based on introducing additional hardware in the form of a 3-to-8 decoder, such as the 74LS138 IC shown in Fig. 2.10. Enable line G1 of the decoder is connected to the +5-V supply, whereas lines G2 and G3 are connected to the 0-V rail. In this way, the decoder is always enabled. The microcomputer employs three of its digital-output lines to supply the decoder with the address of the output line it wants to select (as explained in Section 2.2.5). The state of the selected output line will change to a logic 0, leaving all the other output lines at a logic 1 state. The application requirements can therefore be satisfied by programming the microcomputer to use a software counter and change the address sequentially. As in Solution 1, the alteration of the sequence or the duration of each stage can be performed by software. The procedure needed to implement this solution is simpler than that of Solution 1. As an exercise write the procedure.

Obviously, the second solution increases the cost of the hardware, but it satisfies the requirements with a smaller number of output lines from the microcomputer. It therefore forms the optimum method for many applications. If, however, the designer is faced with the difficulty of having to satisfy the requirements with only two output lines instead of three, then additional hardware is needed. If it is not possible to introduce additional ports, then a hardware counter can be dedicted to producing the sequential addressing in the same manner as that of Example 6.9.

Example 13.5: **A three-degree-of-freedom robot arm is to be used in an exhibition. It is required to move a single joint at any instant of time. The 'shoulder' should move first for 1 s, then the 'elbow' for 2 s, followed by turning the 'wrist' for 3 s. Design a circuit to supply logic commands to initiate and terminate the three tasks repeatedly. Assume that a logic-high signal commands a joint to move.**

Solution 1 The software approach

In this application, actions need to take place every second or every integer multiple of a second. Consequently, the timing requirement can be satisfied by interrupting a microcomputer once every second. The microcomputer needs to be provided with the ability to decide which joint to enable and when. Once a decision is made, output commands can be sent through three digital-output lines, where only one line should be at a logic-high state at any instant of time. The decision-making process is based on the use of a software counter within the interrupt-service routine. It is incremented and tested every time the microcomputer is interrupted as represented by the following steps:

```
1  INCREMENT COUNTER
2  IF COUNT = 6
        THEN LET COUNT = 0,
        STOP WRIST AND START MOVING SHOULDER
   ELSE IF COUNT = 1
        THEN STOP SHOULDER AND START MOVING ELBOW
   ELSE IF COUNT = 3
        THEN STOP ELBOW AND START MOVING WRIST
```

The procedure is simple, but requires three digital-output lines and a 1-Hz interruption signal. The latter can be derived from a higher-frequency signal, e.g. the clock of the microcomputer (by using frequency dividers, as shown in Chapter 9). This approach can be adopted only if there is access to an interrupt line of the microcomputer as well as three digital-output lines. Such access is not always available. Note that, in this application, the use of an interrupt-based software solution does not load the microcomputer, since it needs a low rate of update.

Solution 2 The hardware approach

A more practical solution may be based on the use of the simple circuit of Fig. 13.3. The 1-Hz signal is employed to drive the 74HCT4017 IC (the

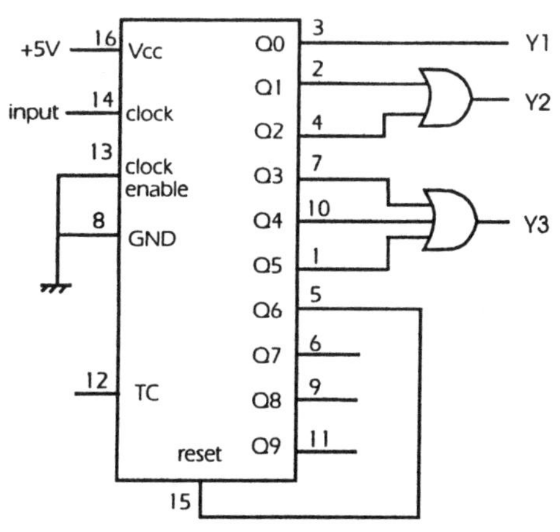

Fig. 13.3 Sequential control of a three-degree-of-freedom robot arm.

operation of this IC is explained in Example 9.4). Six output lines of the IC are required. The seventh output line Q6 is connected to the reset line in order to limit the effective outputs of the IC to six lines generating sequential pulses. Each output will be at a logic-high state for 1 s. Output Q0 is utilized to provide the command to move the shoulder. Two OR gates are also included. One possesses two input lines employed to combine the signals delivered by output lines Q1 and Q2. It generates the 2-s command required to drive the elbow. The second OR gate possesses three input lines to combine outputs Q3, Q4 and Q5 producing the 3-s signal needed to rotate the wrist.

Exercise

Assume that you have access to another digital-output line of the microcomputer. Modify the circuit of Fig. 13.3 to allow the microcomputer to start and stop the robot arm when requested to do so. (Hint: The STOP command can only be effective if the three output lines are forced to a logic-low state.)

The circuit of Fig. 13.3 provides a fixed duration for each of the events. In many microcomputer-based sequential operations, programmable intervals are required. They can be introduced in several ways, depending on the type of application and the availability of interface lines within the microcomputer.

Example 13.6: **Modify the solution of Example 13.5 to provide the microcomputer with the ability to alter the rate of moving the joints of the robot arm. Assume that there is access to an interrupt line plus a single digital-output line.**
Altering the rate can be achieved by controlling the frequency of the clock pulses supplied to the 74HCT4017 IC. The availability of access to an

interrupt line allows the use of the software approach. The function of the microcomputer, however, differs from that of Solution 1 of the previous example. It is employed as a frequency changer rather than as a decision maker. The frequency of the interruption signal can be set to, say, 10 Hz to activate an interrupt every 100 ms. The requirement is fulfilled in a few steps. First, the microcomputer reads the required rate (e.g. from memory, or through the keyboard of the microcomputer). It then quantizes the received number into a multiple of 100 ms and stores it as the required count. It will be decremented every time the interrupt is activated. The corresponding digital-output line is activated only when the count is decremented to 0. This line is connected to the input line of the circuit of Fig. 13.3. In this way, the rate of advancing the outputs of the 74HCT4017 IC is set according to the requirement and can be altered in steps of 100 ms.

Example 13.7: **A microcomputer-based controller is employed to start the second stage of a sequential process. The required delay between the beginning of the first stage and start of the second depends on the type of process. Design a circuit to allow an operator to program any of 200 equally spaced delays through the keyboard of the microcomputer. Assume that there is access to an 8-bit output port, but not to interrupt lines.**
Since the delays are equally spaced, the design can be based on the use of a counter where each delay unit is represented by a clock pulse. The need for 200 different stages can be satisfied by the utilization of an 8-bit counter. It provides up to 256 stages. The first 200 are of interest here. The number entered by the operator should define the count at which the counter will be reset. This number will be latched so that the operation will continue to have that delay until the operator enters the next number. Before reading the next paragraph, can you think of a solution?

The microcomputer reads the value of the delay from the keyboard and sends it to an output port where it will be latched. The circuit of Fig. 13.4 incorporates an 8-bit counter whose eight output lines (Q0 to Q7) are connected to an 8-channel digital comparator (the 74HCT688 IC, also used in Examples 9.11 and 9.12). The IC compares the count with the value latched by the output port of the microcomputer. When the two numbers are equal, the comparator delivers a logic-low pulse through its output line Y (pin 19).

The line indicating the occurrence of event 1 is connected to the CLK line of a D-type latch. When the latch senses the event, the state of its output line Q will change to a logic high. This change of state permits conduction of clock pulses through the AND gate and into the counter, which starts counting from 0. The counting continues until the digital comparator recognizes that the count is the same as the requested delay. The state of its output line (Y) will change to a logic low, thereby resetting the latch (forcing the state of its output line to a logic low). This terminates the conduction of pulses through the AND gate. The change of state of line Y also resets

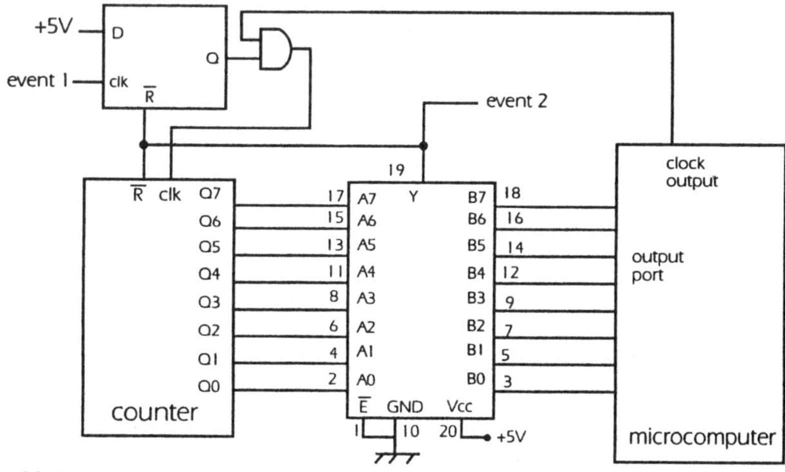

Fig. 13.4 Programmable delay circuit.

the counter, whose output will no longer equal the number delivered by the microcomputer. The result is a change of state of line Y to a logic high again. The circuit is then ready for the next occurrence of the first event in order to repeat the action. The circuit provides a pulse through line Y to initiate event 2.

Example 13.8: **Light-emitting devices and optical sensors are incorporated in a simple in-factory transportation system. They count the number of items passing between assembly stations and loaded into a carriage. The number of items that should be passed is set by an operator through a thumbwheel switch. Starting with an empty carriage, the microcomputer starts the transportation line with a logic-high command. When the required number is reached, the microcomputer stops the line temporarily and issues a logic-high command to replace the carriage with another before repeating the process. Show the required interface.**

Solution 1

The circuit of Fig. 13.5 illustrates the use of the microcomputer with minimum hardware. For simplicity, let us consider the utilization of a single-decimal-digit thumbwheel switch. Such a switch provides four lines carrying a BCD (binary coded decimal) number representing the setting of the switch. The microcomputer reads the required number through four digital-input lines and starts the operation by changing the state of a digital-output line (Y1) from a logic low to high. An interrupt is received through line X every time an item is detected by the light sensor. The rate of interruption depends on the type of items being assembled and transported. It is likely to be low, say once per second. This rate will not overload the microcomputer, which increments

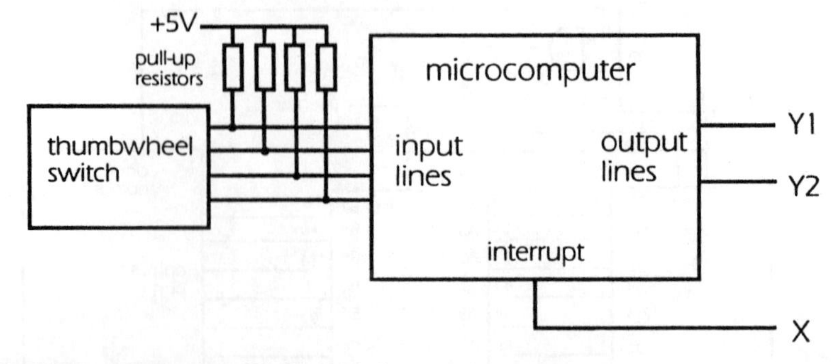

Fig. 13.5 User-controlled operation.

a software counter within its interrupt-service routine before comparing the resulting count with the number set by the operator. When the two are equal, the microcomputer will stop the line by changing the state of line Y1 to a logic low. It then sends a logic-high signal through line Y2 to command replacement of the carriage. The process can then be repeated.

Solution 2

Let us assume that no digital-input lines are available in the microcomputer. This gives the microcomputer no means of reading the setting of the thumbwheel switch. Consequently, counting the pulses and comparing the result to the setting of the switch should take place outside the microcomputer. The latter only needs to be interrupted when the count is reached. It will command the stoppage of the line and the start of the carriage-replacement process before commanding the restart of the line to repeat the operation. As in Solution 1, each of the commands is sent through active-high digital-output lines (Y1 and Y2).

The circuit is shown in Fig. 13.6, where Ex-NOR gates are employed (four such gates are available in a single 74LS66 IC; see Appendix). The output of an Ex-NOR gate is set to a logic-high state when its input lines are at the same logical state. A 4-bit counter is employed to count the number of pulses received from the sensors. It will be cleared by the microcomputer before it commands the line to start. The output of the counter is incremented every time a pulse is received from the sensors. This will continue until the count becomes the same as the setting of the thumbwheel switch. This condition changes the state of the output lines of the four Ex-NOR gates to a logic high. This sets the outputs of the three AND gates to a logic high which will interrupt the microcomputer. The latter will respond by stopping the line, commanding the replacement of the carriage, and resetting the counter before repeating the process.

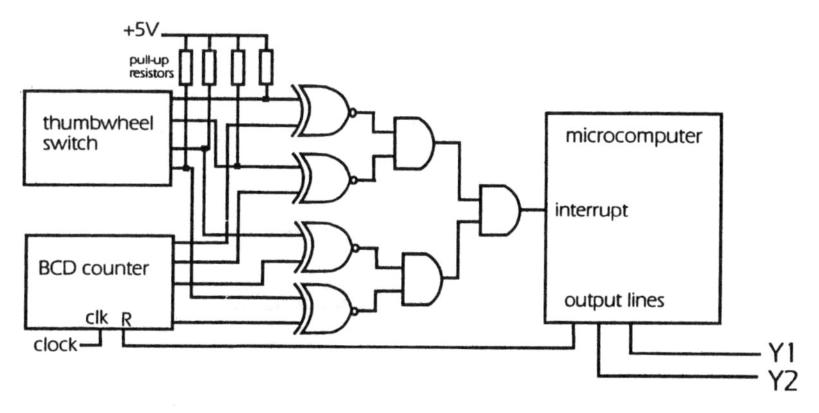

Fig. 13.6 Using a digital detection circuit.

Chapter 14

Temperature measurement and control

14.1 Introduction

Storage, programmability and on-line decision-making capabilities make microcomputers very attractive to use in temperature-measurement and temperature-control applications. The required temperature levels can, for instance, be programmed to change in the required manner, or according to the time of day. Microcomputer-based temperature controllers are commercially available with the ability to control the temperature of a number of objects or a whole building. They can be programmed for a year in advance and can be easily altered to meet changes in requirements.

This chapter starts with a general look at microcomputer-based temperature controllers. It proceeds to consider popular temperature sensors and their use in microcomputer-controlled applications. It then explains methods of controlling temperature from a microcomputer and presents a few examples. The degree of control given to the microcomputer is not the same for all the presented examples, since the optimum processing power needed to control the temperature depends on the requirements of other tasks handled by the microcomputer. The last section of this chapter considers the generation and dissipation of heat in electronic equipment. The heat needs to be measured or compared to a fixed level in order to allow a microcomputer to detect an over-temperature and start a cooling device, sound an alarm or shut the unit safely.

14.2 Microcomputer control of temperature

A typical microcomputer designed specifically to control temperature is based around a microcontroller and is likely to have most or all of the interface and decision-making circuitry on a single board. Alternatively, the temperature interface circuit can be designed as an add-on card for inclusion within the casing of a general-purpose microcomputer. The card provides screw terminals for connecting wires from sensors or external circuits. The card can include jumpers to allow the user to set the range of temperature, the type of control or the sampling rate. In a distributed measurement or control system, readings need to be obtained from several locations and commands need to be sent to a few circuits. In such a system, the in-computer interface card may be in the form of a microcomputer-controlled input and output multiplexer. It can be designed to read from and write to an external interface unit with a backplane capable of accepting, say, 8 or 16 plug-in modules. Each module has an interface circuit, an A/D converter, terminals to connect sensors and, perhaps, a digital line to indicate a fault or an unacceptable temperature level. The use of a microcomputer-based controller allows the modules to be made field-configurable by software. This approach permits the manufacturer to produce a standard unit suitable for a wide range of applications. This approach is also advantageous to the user, since the same unit can still be used if the measurement requirement is altered or the control range is extended.

The method of operation of microcomputer-based temperature controllers can be based on the generation of a time-proportional signal where the microcomputer controls the percentage-ON time of a fixed time cycle. When a temperature controller is first turned on, the controlling microcomputer can be programmed to set the percentage-ON period to its highest possible level (highest mark-to-space ratio). This leads to changing the temperature at the highest possible rate. In order to permit reaching the set point without an overshoot, the rate is reduced as the temperature approaches the set point. Once the set point is reached, the microcomputer needs to maintain operation as close to that point as possible by switching the temperature changer on and off. Switching at a high rate is not favourable and can be avoided if a 'hysteresis band' is allowed. This is represented by an upper and a lower limit where the temperature is permitted to change between them. Introducing such a band, allows the temperature to deviate from the set point but removes unnecessary switching. Clearly, the allowable band should vary according to the type of application. Consequently, the use of microcomputer control permits the introduction of a software-selectable hysteresis band and cycle time.

14.3 Temperature sensors

There are several types of temperature sensors available at relatively low cost. They provide a wide choice to suit a variety of applications, such as temperature measurement, temperature control, over-temperature protection, over-current protection or liquid-level detection. With the aid of a simple circuit, a sensor can be interfaced to a microcomputer to perform a simple monitoring task or accurate control of temperature.

The positioning of such sensors depends on the application. For example, they are placed on heat sinks or chassis, on high-current windings in transformers or motors, or along the air-flow path within various equipment. A temperature sensor employed for liquid-level indication is placed in the container to indicate whether a level is reached or not. A more accurate measurement of the level will need several sensors placed at different heights. A microcomputer can examine the state of each of the sensors and decide the level of the liquid accordingly. Another application of temperature sensors is in the detection of air flow. The method is based on sensing the reduction (or rise) of temperature as air flows and the increase (or decrease) in temperature when the air stops or when its flow rate is reduced.

Temperature sensors can also be employed as time delay devices. The idea is based on the time needed for the temperature to rise in an electric unit after it is powered on or after an increase in its load. A temperature sensor can be connected to an analogue comparator to detect when the temperature exceeds a certain level. The time taken to reach this level is the time delay. Selecting the level will therefore define the time needed before the output of the comparator changes state.

14.3.1 Thermistors

This sensor is a sensitive low-cost temperature-dependent resistor which is available in a broad range of resistance values. It covers a limited temperature range, but it is quick to respond to temperature alterations and gives a large change in value with temperature (e.g. 4% per 1°C). It can therefore be successfully employed to detect small changes in temperature. The change it exhibits is, however, non-linear. Since a thermistor is a resistor, internal heat is generated when current passes through it. The designer should therefore pay attention to this effect in the design, since a failure to dissipate this heat will affect the temperature measurement.

Most thermistors have a negative-temperature coefficient. This means that their resistance decreases with temperature. They are employed in a wide range of applications, e.g. sensing, measuring, control, protection and temperature compensation. An 'inrush current limiter' is an example of a negative-temperature-coefficient device. It is utilized to limit the power-on current in capacitor-based circuits, e.g. power supplies. When power is applied to such a circuit, the capacitors form a very low impedance which draws high currents (Section 4.6.2). Negative-temperature-coefficient devices can be connected along the current flow path to limit the current when the power is turned on. During operation, the increase in temperature will decrease the resistance of these devices to a low level which has a minimum effect on the operation of the circuit. Such devices are available in different ratings and resistive values. If one device is not sufficient to limit the inrush current, then additional devices can be connected in series along the current path. Connecting such devices in parallel is not recommended, since one device may end up conducting all the current.

Thermistors with a positive-temperature coefficient also find uses in a variety of applications. The increase of their resistivity with the increase in temperature can be utilized to protect against a short-circuit. It can be connected along the path of current flow, leading to a small voltage drop across it during normal operating conditions. In case of an over-current or a short-circuit, the sharp increase in current causes a sharp rise in the resistivity of this type of thermistor. This leads to a reduction in the current to a safe level. When the short-circuit is removed, the current level reduces to its normal operating value. The reduction of the current leads to a decrease in the heat of the thermistor, which will in turn reduce its resistivity to its normal value. Normal operation can then resume. This is therefore a simple method of protecting the circuit which does not need intervention from a microcomputer to limit the current. The microcomputer can, however, detect the increase of current to inform an operator or to stop the cause of the short-circuit.

Example 14.1: **During a microcomputer-controlled industrial process, the temperature of a device may increase sharply due to blockage of an air-flow path, or due to failure in a cooling device, e.g. a fan. Design a circuit to detect the increase in temperature and inform the microcomputer, which will take the necessary action. Assume that the microcomputer is dealing with time-sensitive events and its operation should not be interrupted unless necessary.**

A negative-temperature-coefficient thermistor 'Th' can be employed, as shown in Fig. 14.1. The sharp increase in temperature will quickly decrease the resistance of the thermistor, thereby reducing the voltage drop across it. This increases the voltage across resistor R1 until it exceeds the level set by the voltage divider of R3 and R4. This causes the comparator to alter the state of its output line from a low to a high level.

Since the cause of the temperature rise may only be a temporary blockage of the air flow, the response from the microcomputer need not be instantaneous. Depending on the system and the nature of the tasks the microcomputer is dealing with, it can be possible for the microcomputer to regularly check the output line of the comparator through a digital-input line to detect the increase in temperature. In some applications, however, the microcomputer may handle high-priority time-consuming tasks which will delay it from reading the temperature indication. Allowance has been made in the circuit of Fig. 14.1 by including a counter which counts only when the output of

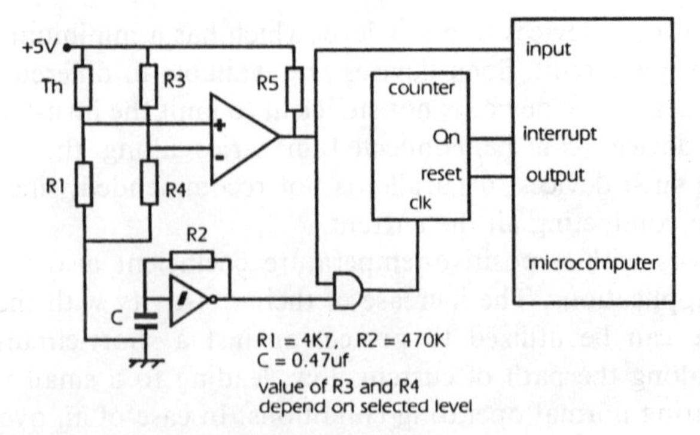

Fig. 14.1 Detection of over-temperature.

the comparator is high. The role of the counter is to introduce a delay after the occurrence of the temperature rise. If the microcomputer reads the input line and detects the rise in temperature within the set delay, then it will take the necessary action and reset the counter through a digital-output line. If, on the other hand, the microcomputer does not respond within the set time, then the counter will interrupt it and force it to respond.

This implies that the output of the comparator gives a warning signal to the microcomputer, but the output from the counter is a high-priority alarm which requires a very fast response. The resolution of the counter and the frequency of the clock signal supplied to it depend on the delay that can be allowed before interrupting the microcomputer. The delay depends on the type of equipment to be protected and the allowable temperature rise.

Example 14.2: **Design a test circuit to measure the time at which the temperature exceeds a manually set limit. The measured time needs to be read by a microcomputer and stored for future analysis.**

One way of meeting the requirement is to adopt an approach similar to that of the previous example, where the required temperature level is set manually by a potentiometer and is compared to a temperature-related level generated by the sensor. The output of the comparison is employed to control the flow of clock pulses to a counter with the aid of an AND gate (as in Fig. 14.1).

An alternative approach is to control the operation of the clock which supplies the counter. This will remove the need for the AND gate. The method can be implemented by the use of the temperature-controlled switch of Fig. 14.2, which utilizes a negative-temperature-coefficient thermistor 'Th' as the temperature transducer. With operation from a fixed d.c. voltage supply, the voltage across the potentiometer depends on its setting as well as the resistance of the thermistor. In other words, the detection level depends on both the manual setting and the measured temperature. An increase in temperature will decrease the resistance of Th, thereby increasing the voltage

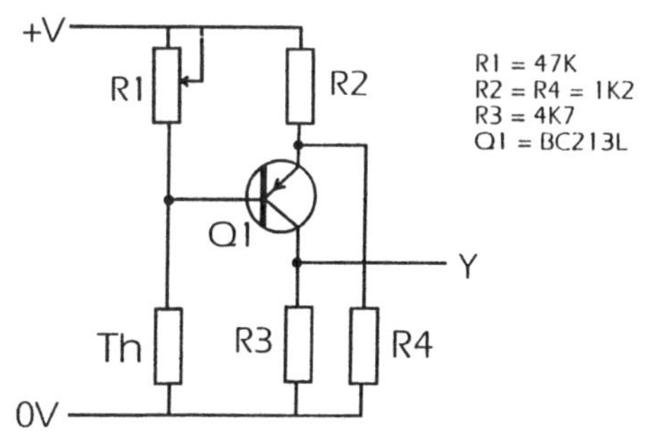

Fig. 14.2 Temperature-controlled switch.

drop across the potentiometer. This switches on the PNP transistor to pass current from the d.c. supply to the circuit driven by the signal at point Y.

This signal can be used to control the operation of a 555 timer. Such a timer is shown in Fig. 3.1 where pin 4 is connected to V_{cc} to allow continuous operation. A controlled operation can be achieved by connecting pin 4 to point Y of Fig. 14.2 rather than to V_{cc}. This alteration makes the 555 timer run only when transistor Q1 is on (in a similar way to that explained in Section 9.3.2 with reference to Fig. 9.8). In other words, the 555 timer will operate only when the temperature exceeds the manually set level. The output of the timer (pin 3) will be connected to the clock input line of a free-running counter which will count the supplied pulses. The microcomputer will read the counter regularly and calculate the time in two steps. It first finds the difference between the present and the previous readings and then multiplies that difference by the duration of each clock pulse (1/frequency of the clock). The effect of an overflow in the counter must be taken into consideration. It can be sensed when reading a count lower than the previous one. The program can detect the overflow and adjust the calculated value. Clearly, the required time interval between successive readings of the counter depends on the frequency of the clock and the resolution of the counter.

14.3.2 Thermocouples

This is a simple, inexpensive sensor which measures temperature over a wide range. It produces a very low voltage which increases non-linearly with temperature (the change is in millivolts when the temperature changes in hundreds of degrees). The idea of a thermocouple can be explained with the aid of Fig. 14.3(a), where two wires of dissimilar metals are welded together. If the joint is heated, a voltage difference will appear between the two terminals. The magnitude of the voltage depends on the material of each of the two wires. Thermocouples are identified by a letter according to the type

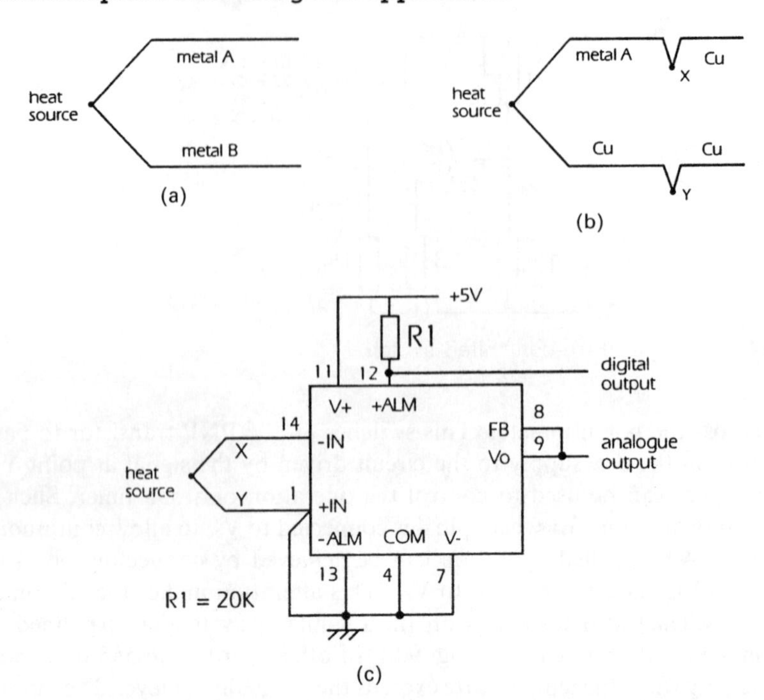

Fig. 14.3 Thermocouples: (a) two wires welded together; (b) with a reference junction; (c) interface circuit (the Analog Devices AD594/AD595).

of the two wires. For example, type E is formed by combining chromel and constanan wires, type J from iron and constanan, type K from chromel and alumel, and type T from copper and constantan. Different types differ in their volts per temperature characteristics. In terms of temperature range (highest temperature):

$$\text{type K} > \text{type E} > \text{type J} > \text{type T}$$

At a given temperature, the voltage magnitude is given by:

$$\text{type E} > \text{type J} > \text{type T} > \text{type K}$$

There are calibration tables relating the output voltage to the measured temperature according to the thermocouple combination. A table equivalent to the junction in use can be stored in the memory of a microcomputer and used to aid in converting the measured voltage to temperature.

Thermocouples can be classified according to the type of junction. An 'exposed-junction' thermocouple is fast to respond, but it is subject to corrosion. A 'grounded-junction' thermocouple has a sheath touching the thermocouple junction. It provides protection, but has a reduced response. The response is further reduced if an 'ungrounded-junction' thermocouple is used. In this type of thermocouple, the junction is insulated from the protective sheath in order to prevent noise. Thermocouples with faster response can be produced from thinner thermocouple wires, but such wires limit the temperature range.

In measurement and interfacing applications, the effect of the measuring leads can be eliminated by employing a 'reference junction', as shown in Fig. 14.3(b). Since junction Y is between two wires of similar material, it will not show a voltage difference. On the other hand, junction X is between two wires of dissimilar materials and will therefore generate a voltage. It is important to remember that junction X joins copper to metal A, while the measuring point joins metal A to copper. This implies that the junctions will form voltages opposing each other. The measured temperature is the difference between the two voltages. Subtraction can be avoided if the reference junction (also known as the cold junction) is kept at 0°C. Since this may be difficult to achieve in practice, the temperature of the reference junction can either be kept fixed at a more convenient temperature, or measured by the testing device, e.g. a microcomputer. In this way, the effect of the junction can be compensated for in order to generate an accurate reading of temperature. Measuring the temperature of the reference point can be performed by employing an additional temperature sensor. It may not seem right to use a temperature sensor to aid another temperature sensor to measure temperature! Thermocouples are, however, suitable for measuring temperature over a very wide range. They are simple to construct and can be easily connected to various positions without special arrangements. The reference voltage and its associated sensor can be fitted in a unit and shared between several thermocouples sequentially. Such thermocouples are of the same type and of the same lengths. They can be utilized to measure the temperature at various points within an environment, e.g. within the casing of a power controller. The readings can be supplied to a microcomputer, which samples and reads the temperature regularly. The results can be averaged, or stored for future inspection or analysis.

ICs have been developed to ease the task of interfacing thermocouples to various circuits. Figure 14.3(c) uses the Analog Devices AD594 or AD595 monolithic thermocouple amplifiers with cold junction compensation (the AD594 is for type J thermocouples, while the AD595 is for type K thermocouples). The IC amplifies the reading of temperature, producing an analogue signal of 10 mV/°C which can be read by the microcomputer. The IC can be operated from +5-V to ±15-V power supplies. In order to ease the task of interfacing the IC to a microcomputer, the circuit of Fig. 14.3(c) uses a +5-V power supply. The use of positive and negative supplies, however, allows the measurement of both positive and negative temperatures. In order to increase its usefulness, the IC delivers a fault indication in the form of a voltage level. The circuit of Fig. 14.3(c) delivers a logic level (from pin 12) which can be read by a microcomputer or by a digital circuit. This line is normally at a logic-high state. If an open circuit in the thermocouple is detected, then the state of the line will change to low. The line can either be connected to a digital-input line of the microcomputer for regular monitoring, or it can be connected to an interrupt line of the microcomputer to enable the latter to detect the fault as soon as it occurs. Note that Fig. 14.3(c) marks the thermocouple lines as X and Y. They are constantan and iron respectively in a type J thermocouple, and alumel and chromel respectively in type K thermocouples.

14.3.3 Resistive temperature sensors

This type of sensor covers a wide range of temperature. It is of low cost and is also referred to as RTD (resistance temperature detector). It is a resistance whose value changes with temperature. Current is required to flow through the sensor, which will in turn produce a stable reading whose change with temperature is more linear than that of thermocouples. The change is, however, small and relatively slow. When current passes through this sensor, heat is generated and, if not successfully dissipated, will lead to self-heating. The generated heat is proportional to the square of the current and can be reduced by avoiding unnecessarily high levels of current.

Since the change in magnitude is small, temperature alterations can be detected by the use of a Wheatstone bridge, as shown in Fig. 14.4. The sensor is connected away from the other resistors in order to protect them from the effect of the temperature source. The circuit is energized by a d.c. voltage V_{in} and produces a d.c. output voltage V_x whose magnitude depends on the value of the resistors. It is represented by:

$$V_x = V_{in}\left(\frac{R_T}{R3 + R_T} - \frac{R2}{R1 + R2}\right) \tag{14.1}$$

If the resistance of the leads is ignored, then the bridge will be balanced (i.e. $V_x = 0$) when $R1 = R2$ and $R3 = R_T$. Changes in temperature alter the value

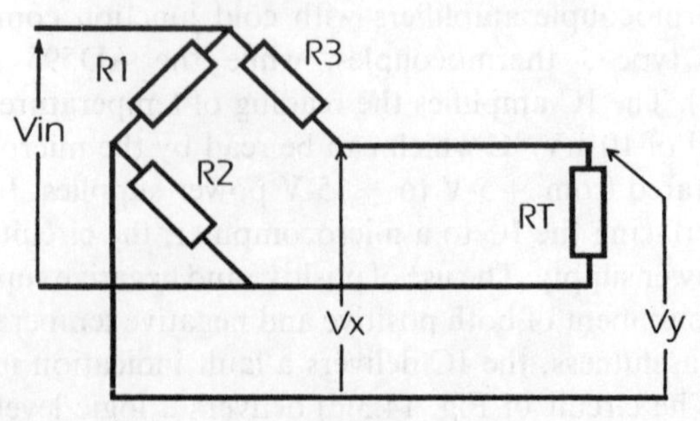

Fig. 14.4 Using a resistive temperature sensor.

of R_T and take the bridge out of balance. The value of V_x reflects the change in R_T.

The problem with this configuration is the effect of the resistance of the leads connecting R_T to the other resistors. It adds a resistive value which will also change with temperature. Its effect can be reduced by taking the output from point V_y rather than V_x in the circuit of Fig. 14.4. In this way, the resistance of one lead is added to R3 and the resistance of the other lead is added to R_T, thereby reducing the error. It is also useful to measure the resistance of the leads and include it in the calculations. The output of the circuit can be applied to an amplifier to assist in the detection of small changes in value. The output of the amplifier can be supplied to the microcomputer through an A/D converter.

Rather than using a Wheatstone bridge, a current source can be incorporated to energize the sensor. A microcomputer can then read the voltage across it. This voltage will be proportional to the value of the resistance (which changes with temperature).

14.3.4 Linear temperature sensors

This inexpensive semiconductor sensor produces a voltage that is linearly proportional to temperature. It has a limited temperature range and, like most temperature sensors, suffers from self-heating. An example of such a sensor is the National Semiconductor LM335 Precision Temperature Sensor. It can be operated continuously over a temperature range of −40 to +100°C. There are other members of the same series of Precision Temperature Sensors. The National Semiconductor LM135 and the LM235 cover a temperature range of −55 to 150°C and −40 to +150°C respectively. When compared to the lower cost LM335, both the LM135 and the LM235 offer a lower non-linearity than the LM335 (1 to 1.5°C) and a lower temperature error with calibration (1.5 compared to 2°C). The time constant of the sensor is measured in seconds and depends on the surrounding medium (e.g. less if forced air than still air). Each of these sensors produces an output voltage which changes by 10 mV per 1 K change in temperature and has a current range of 0.4 to 5 mA.

The circuit of Fig. 14.5(a) incorporates a 4.7-kΩ resistor R1 to connect the sensor to the +5-V supply and pass a current of around 1 mA (at 1 mA, the dynamic impedance is 0.5 Ω for the LM135 and LM235, and 0.6 Ω for the LM335). An operational amplifier is connected as a non-inverting buffer in order to supply an analogue voltage proportional to temperature. This voltage can be delivered to a microcomputer through an A/D converter. The circuit utilizes a potentiometer R2 (10 kΩ) to calibrate the sensor (T); time should be allowed for the sensor to settle between turning the power on and performing the calibration. With knowledge of the surrounding temperature,

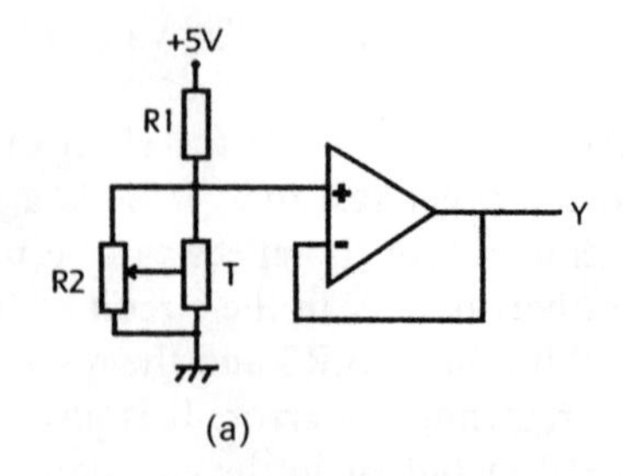

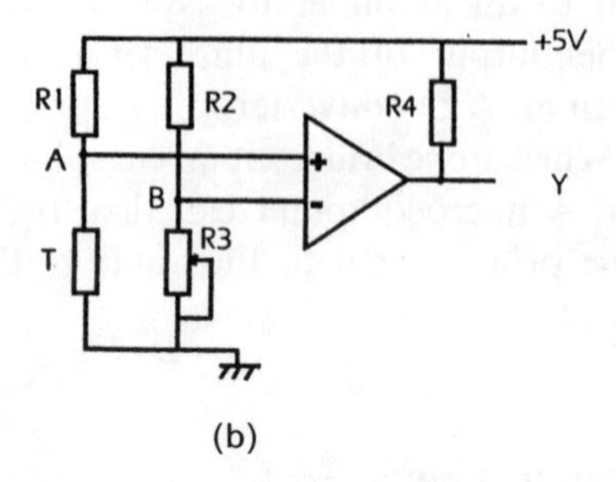

Fig. 14.5 Using the National Semiconductor LM335 linear temperature sensor: (a) interfacing and calibration; (b) level detection.

one can read the voltage generated by the sensor and adjust the potentiometer. If the calibration is carried out at temperature T1 to produce a voltage V1, then at temperature T2:

$$V2/T2 = V1/T1 = 10\,\text{mV/K} \tag{14.2}$$

where K represents temperature in kelvins.

Consequently, when the microcomputer measures a voltage V2, the measured temperature is:

$$T2 = V2/(10\,\text{mV/K})$$

i.e.

$$T2 = 100 \times (V2)\text{K}$$

If, for example, one reads a voltage of 3 V, then for a correctly calibrated sensor this temperature corresponds to T2 = 300 K. When interfaced to a microcomputer, the latter can perform a simple calculation to convert the measured voltage to temperature which can be in kelvins or converted to degrees centigrade (C) using:

$$C = K - 273 \tag{14.3}$$

The measured temperature is therefore 27°C.

The circuit of Fig. 14.5(b) illustrates the use of this sensor to inform a microcomputer when a temperature level is exceeded. The required level is set manually by potentiometer R3, which is connected in series with resistor R2 to form a voltage divider. Like the circuit of Fig. 14.5(a), the circuit under consideration connects the temperature sensor *T* to the +5-V supply through a 4.7-kΩ resistor (R1). This limits the current passing through the sensor and forms a voltage divider whose centre point A is at a voltage which increases linearly with temperature. Since point A is connected to the positive terminal of the analogue comparator, the output of the comparator is low as long as the temperature is less than the manually set level. The increase in temperature will increase the voltage at point A until the level exceeds the voltage of point B. When this happens the output of the comparator will change state from low to high. A digital-input line of a microcomputer can be utilized to test the state of this line.

Example 14.3: **Consider an application similar to that of Example 14.1 with a slightly different requirement. The microcomputer should be warned when the temperature increases above a predetermined level, and interrupted when the temperature exceeds an alarming level rather than after a predetermined time interval. How can this be achieved?**

The analogue signal delivered by a linear temperature sensor is supplied to the input terminal of the circuit of Fig. 14.6. It is connected to the positive terminal of two analogue comparators. The first comparator (whose output line is marked Y1) gives a warning when the temperature increases above a fixed level. This level is set by the voltage divider formed by resistors R3 and R4. This active-high warning is connected to a digital-input line of the microcomputer in a similar way to that of Example 14.1. The second comparator generates an active-high signal (Y2) to interrupt the microcomputer when the temperature exceeds a second higher level (formed by the values of resistors R1 and R2). The values of the resistors are chosen according to the required levels (resistors R5 and R6 are pull-up resistors).

Examples 14.4: **How can a microcomputer measure the difference in temperature between two points?**

One solution is to incorporate two linear temperature sensors, one at each of the points of interest. This, however, requires the microcomputer to read two analogue signals, thereby reserving two analogue-input channels. An alternative solution is given in Fig. 14.7, where the two sensors are connected to a differential amplifier of unity gain (R3 = R4 = R5 = R6). The amplifier will produce an analogue signal representing the difference between the two temperatures. In this way, only one A/D conversion channel is required. Note that the amplifier in Fig. 14.7 has a unity gain. If the difference is small, then one way of amplifying the difference is to increase the value of resistors

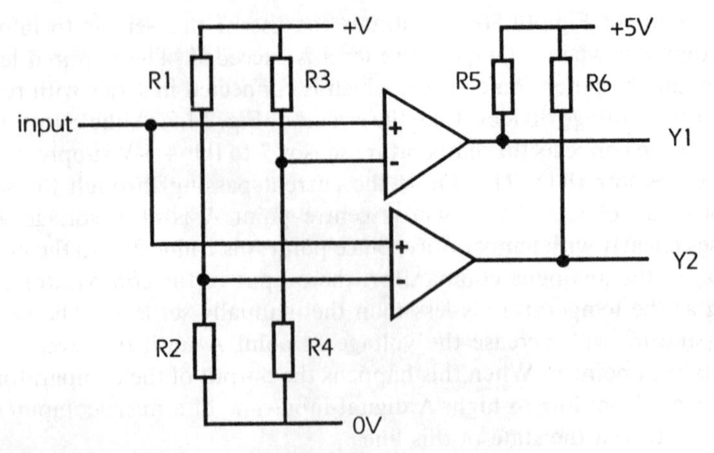

Fig. 14.6 A two-level over-temperature detector.

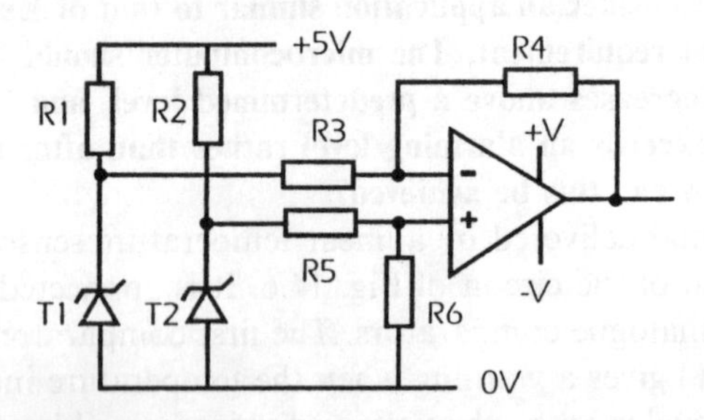

Fig. 14.7 Measuring difference in temperature.

R4 and R6. Clearly, one should ensure that the allowable voltage range of the A/D converter is not exceeded. If there is a possibility of producing higher than acceptable levels, then zener diodes can be utilized to clamp excessive values to safe levels.

A manual offset adjustment can be included by replacing one of the resistors of the differential amplifier (e.g. R6) by another of a lower value in series with a potentiometer.

14.4 Methods of changing temperature

There are several methods of controlling temperature with a microcomputer. The response time and the required degree of intervention of the microcomputer in the operation of the controller can differ from one system to another. In a large number of applications, the microcomputer is programmed to turn a heating (or a cooling) device on and off during specified time intervals, e.g. during working hours in an office. During that interval, the microcomputer controls the heat-changing device in order to maintain the temperature at a constant level, or between an upper and a lower limit. It achieves this by examining the temperature and comparing it with the required level. It then performs the temperature-control task by turning the heat-changing device on or off. The required temperature level may be programmed, or set manually through a potentiometer or a keypad.

The rest of this section presents a few methods of controlling temperature. The control can be performed either open-loop or closed-loop.

14.4.1 Pulse-width method

This is an example of controlling temperature when operating from a d.c. supply. A temperature-changing device, e.g. a heater, is supplied with pulses of power to turn it fully on or fully off. The heat generated by the heater is proportional to the mark-to-space ratio of the pulses. The circuit of Fig. 14.8 is an example of interfacing a heater to a microcomputer through an optically isolated transistor amplifier. The microcomputer generates, or controls the generation of, pulses which are delivered to this circuit. During the mark of the pulse, current passes through the optical diode, which will transmit light energy to turn the phototransistor on. The phototransistor forms the first stage of an amplifier operating from an isolated d.c. power supply (+V). When the phototransistor is turned on, current starts to flow through resistor R2 causing the PNP transistor Q1 to turn on. Current passes through resistor R3 and into the base of the NPN transistor Q2, which turns on, allowing current to pass through the heater. Turning Q2 off is accomplished by supplying a logic-low signal to the optical diode (i.e. during the space of the pulse). This stops energy from being transferred to the phototransistor, which will turn off. Current stops flowing through R2 and Q1, causing Q2 to turn off and stopping the current conduction through the heater.

The circuit will therefore allow a microcomputer to control the flow of an amplified current through the heater. This method of control has the advantage of a fast response, since a change of temperature can be implemented within the duration of a single pulse. This can be a very short period if high-frequency pulses are used.

14.4.2 Phase-angle method

This method is based on changing the triggering angle of two back-to-back thyristors, or a single triac for operation from an a.c. supply. The circuit

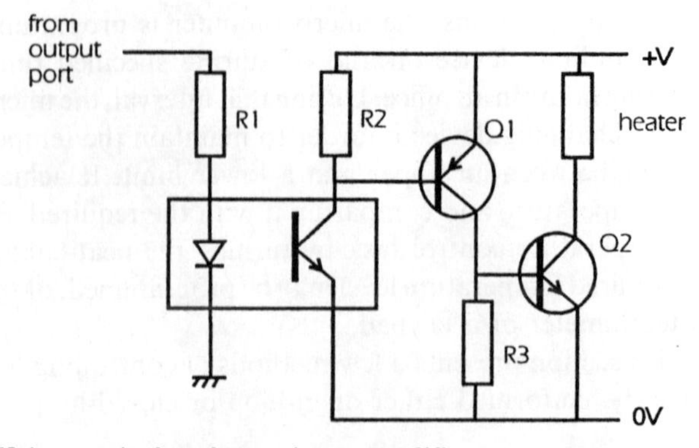

Fig. 14.8 Using an isolated transistor amplifier.

and the waveforms are the same as those of Fig. 10.23; in this case, the load is a temperature-changing device, e.g. a heater. Open-loop control is performed by changing the triggering angle according to a user's setting, or in a preprogrammed manner without examining the resulting temperature. Closed-loop control, on the other hand, examines one or more feedback signals and adjusts the triggering angle accordingly.

This method is simple to apply and only needs one triggering signal per half-cycle of the a.c. supply. This corresponds to one signal per 10 ms for a 50-Hz supply (or 8.33 ms for a 60-Hz supply). Consequently, the adoption of this approach does not consume much of the microcomputer's processing power and can be applied to achieve a relatively accurate control of temperature. The minimum step change is limited by the resolution of the circuit generating the phase angle (the delay). In a digitally controlled system, this circuit is likely to be based around a programmable timer or a presettable counter. Unless connected directly to the data bus, a higher-resolution device requires more output lines from the microcomputer, and this may not be possible nor economical.

Triggering a thyristor or a triac away from the instant where the a.c. voltage signal crosses the 0-V line can produce noise as well as a high rate of change of current, especially with a resistive load, e.g. a heating element. This can introduce a limitation into the adoption of this method. It is, however, easy to implement and permits a controlling device to alter the temperature once every half-cycle of the a.c. supply. This is considered a fast response for the majority of temperature-controlled operations.

14.4.3 Zero-switching method

Like the phase-angle method, this method is also applied to operation from an a.c. supply. It ensures that the switching devices (thyristors, triacs or relays) are switched at, or very close to, the point where the a.c. voltage crosses the 0-V line. In this way, only complete sine waves are delivered to the load. The variation of the temperature is achieved by the alteration of the percentage number of complete a.c. cycles where the devices are conducting. In other words, the devices will be switched on for X cycles and then switched off for Y cycles where the X/Y ratio depends on the required temperature. If, for example, the operation is from a 50-Hz mains, then there are 50 cycles per second. One can, for instance, let $X + Y = 50$ and vary X from 0 to 50 according to the requirements. This particular 50-cycle controller can vary the temperature in steps of 2%. Clearly, the size of the steps is inversely proportional to the sum of X and Y. Increasing the sum, however, means requiring more cycles to apply a change, which leads to a slower response.

Zero-switching implies a maximum conduction of power to the load for a number of cycles. If the current requirement is high, then every time the heater is switched on, a relatively large current will be drawn from the supply which can cause a dip in the supply voltage. Large current changes can be avoided if several zero-switching circuits are used. Each circuit supplies a part of the load and is switched on and off at different times from the rest, but at the same on/off ratio.

14.4.4 On/off method

This method is based on switching a temperature-changing device on and off at a very low rate without examining the zero-crossing points of the a.c. supply. It is similar to the zero-switching method, but has no need for a zero-crossover detector or a means of synchronization. It is applied for operation from an a.c. supply in applications where the interference generated from the asynchronous switching is not a problem. It requires a simple control circuit and can be used in lower-cost systems where frequent switching is not a requirement (e.g. where the temperature change is very slow).

14.5 Examples of temperature control

This section gives examples illustrating the change of temperature according to the requirements. The implementation of control is based on activating and deactivating a heat-changing device under the control of a microcomputer. This is performed with the aid of power-switching devices interfaced to the microcomputer in a similar manner to that explained in Chapter 10.

Example 14.5: **In an open-loop temperature-control application, d.c. power is applied directly to a heater. This needs to be replaced by microcomputer**

control where the percentage-ON period should be controlled in discrete steps according to a manual setting by the operator. How can this be achieved? Assume that a 1-MHz clock signal is available from the microcomputer.

Solution 1 The software approach

Since the power is supplied in a d.c. form, the pulse-width method can be selected. A counter can be incorporated as a frequency changer (Section 9.4) and used to derive a regular signal from the 1-MHz clock. This signal can be utilized to interrupt the microcomputer. The pulse-width method provides flexibility, since it allows one to set the frequency according to the requirement of the application. Assuming that the heat does not change at a very high rate, the frequency of the interrupting signal can be set low (e.g. once per 10 ms) in order to avoid unnecessary usage of the microcomputer's processing power. Having defined the rate of interruption, one can now set both the cycle time and the required on/off time of the heater as multiples of the period of the interrupting signal. A software counter is utilized to count the number of interrupts and is reset when the microcomputer is turned on and at the end of each heating cycle. The procedure can be represented by the following steps where 'x' and 'y' are RAM locations containing the required percentage-ON time and the cycle time respectively. A decision on whether to use 'x' and 'y' as one or more bytes depends on the rate of interruption and the selected cycle time.

```
1  TURN ON HEATER
2  INCREMENT COUNTER
3  IF COUNT > x, THEN
        TURN OFF THE HEATER
4  IF COUNT = y, THEN
        RESET COUNTER
```

The heater will be on for 'x' successive interrupts and off for 'y − x' successive interrupts.

Having decided on the method of changing the temperature, the next step is to decide on how to allow a user to set the discrete levels of temperature. This can be accomplished by interfacing a 10-way switch through an A/D converter in a similar manner to that of Fig. 16.4. The power can then be switched on and off according to the demand and under the control of the microcomputer by a relay.

Although the required processing power is low, the interface needs a dedicated interrupt line to implement the control, and an analogue-input line to read the user's command. If these are not available, or are reserved for other activities, then the following solution can be adopted.

Solution 2 The hardware approach

A simple circuit can be constructed to reduce the number of interface lines and the involvement of the microcomputer. Figure 14.9 shows such a circuit where the input to the circuit is a 10-kHz clock derived from the 1-MHz clock signal by using the 74HCT390 IC (described in Example 9.8). The output of this IC is supplied to a 74HCT4017 IC through pin 14 (marked 'input' in Fig. 14.9).

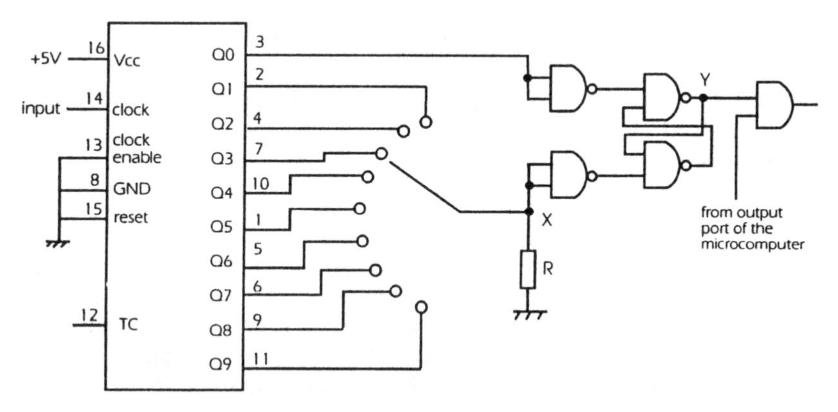

Fig. 14.9 Discrete-level temperature controller.

The output of the 74HCT4017 changes sequentially with each clock pulse. Let us start at a point where a clock pulse causes line Q0 to be at a high level. Point X is connected to one of the other nine outputs of the 74HCT4017. A logic-high state at line Q0 triggers the flip-flop circuit, sending point Y to a logic-high level. The next clock pulse sets line Q1 high and all the other outputs of the 74HCT4017 to low. Point Y will continue to be high until point X is set to a logic-high level. The instant at which this happens depends on the setting of the rotary switch. Connecting point X to line Q1 will therefore produce the shortest pulsewidth at point Y. Turning the switch away from Q1 increases the width of the pulse delivered at point Y, thereby increasing the percentage-ON period of the heater.

Note that the setting of the switch will only change the percentage-ON period, but has no effect on the frequency (which is the rate of setting line Q0 high).

This approach implies that line Q0 will always be set to a logic-high state once every 10 pulses of the 10-kHz clock, i.e. once every millisecond. The circuit performs the required actions and eliminates the need to interrupt the microcomputer. The role that the microcomputer plays is reduced to deciding whether or not to allow the signals to operate the heater. The decision depends on other measurements and calculations that the microcomputer is handling and is applied through an AND gate whose output line is interfaced to a relay through a driver circuit such as that of Fig. 10.26.

Example 14.6: **Repeat the previous example by designing another circuit where the user can alter the temperature linearly rather than in a discrete number of steps.**

Solution 1

The linear setting can be applied through a potentiometer. Instead of reading the setting of the potentiometer through an A/D converter, a circuit such as that of Fig. 14.10 can be used. It utilizes the LM311 analogue comparator (also employed in Example 4.3) to compare two signals. The first is a d.c. level representing the setting of potentiometer R1, while the second signal is a slowly rising ramp supplied through line X. Line W is introduced to allow the microcomputer to control the operation through a single digital-output line regardless of the state of the input lines of the comparator (the LM311 allows this type of control by providing a strobe line at pin 6). If the microcomputer sets line W to a logic-high level, the PNP transistor starts conduction through resistor R3 (1.2 kΩ). This brings the voltage at the strobe line to a low level, thereby turning off the output stage of the comparator. This stops the flow of current through the coil of the relay, which will open its contacts. Operation of the heating system will start when the microcomputer sets the state of line W to a logic low.

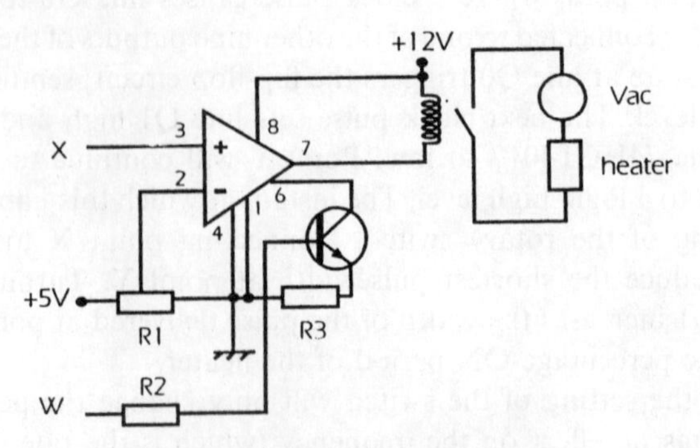

Fig. 14.10 Linear control of temperature.

To explain the operation of the circuit, let us start at the instant where the ramp signal is rising from a lower level than the voltage setting across the potentiometer. The output of the comparator will be at a low level, allowing current to flow through the coil of the relay. This energizes the relay, which will close its contacts and permit power to reach the heater. When the ramp exceeds the d.c. level set by the user, the comparator changes the state of its output line to high. This results in de-energizing the relay which will stop the power conduction through the heater. The circuit of Fig. 4.10 satisfies the requirements by allowing an operator to set the percentage time of operating the heater with very little intervention from the microcomputer.

Solution 2

An alternative method is to employ a timer, such as the LM122H described in Section 9.3. The circuit is shown in Fig. 14.11, where the output of the timer is used to operate the relay directly. The microcomputer sends a regular triggering signal to the trigger input line (pin 2) of the timer. The timing is achieved by choosing the values of potentiometer R1, resistor R2 and capacitor C. These values depend on the rate at which the signal is supplied from the microcomputer to pin 2 (see Section 9.3).

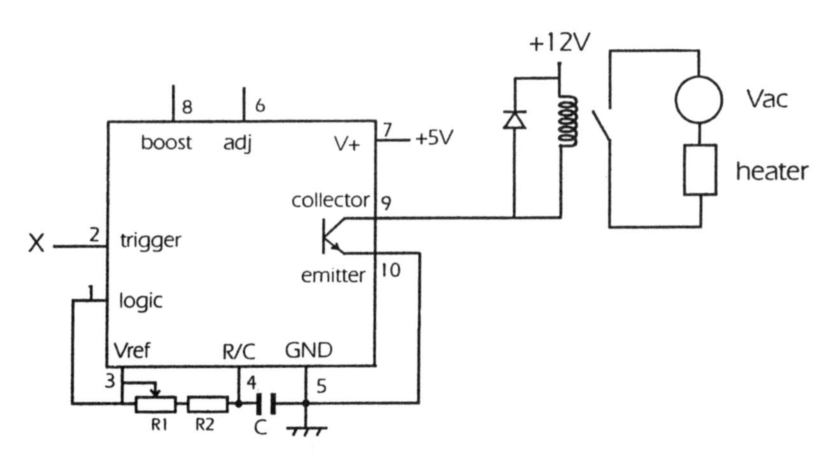

Fig. 14.11 Timer-based temperature controller.

The logic input line of the timer is connected to line V_{ref}. This will let the collector line (pin 9) be at a high level during the charging of capacitor C. The relay is connected between this line and the d.c. supply, while the emitter of the output transistor (pin 10) is connected to the 0-V line. The on/off period can be altered manually by adjusting potentiometer R1.

Example 14.7: **Design a triac-based programmable temperature controller.** The design of the controller is based on interfacing a triac to a microcomputer and controlling the temperature by the phase-controlled method. The control algorithm is based on changing the phase angle of triggering the triac by the use of a microcomputer-controlled delay circuit. One such circuit is presented in Fig. 13.4 and explained in Example 13.7. It incorporates an 8-channel digital comparator which compares the output of a counter to a digital number delivered by the microcomputer. When the two numbers are equal, it delivers a logic-low pulse through output line Y (pin 19 of Fig. 13.4).

The requirement of this example differs from that of Example 13.7 in the need to start the delay close to the point where the supplied a.c. voltage crosses the 0-V line. This creates the need for a zero-cross detection circuit to generate narrow pulses at the zero-crossing instants. These pulses can be supplied to the CLK input line of the D-type latch in circuit of Fig. 13.4 to start the counting process. When the required count is reached, the state of line Y changes to a logic low. It resets both the latch and the counter as well as requesting the triggering of the triac. Its duration, however, is too short to be used as the triggering pulse. A monostable (such as the one used in Fig. 9.17) can be incorporated within the circuit and triggered by that pulse. The required width can be set by the value of the resistor and capacitor of the monostable, as described in Section 9.5.1.

The next step in the design process is to relate the number delivered by the microcomputer to the frequency of the clock, taking into consideration the duration of a single cycle of the a.c. supply. Since an 8-stage counter is employed, the maximum count is 255. This corresponds to 180 electrical degrees of the a.c. cycle (i.e. half a cycle) which is 10 ms for a 50-Hz supply (8.33 ms for a 60-Hz supply). The frequency of the mains, however, can suffer from a small variation from time to time. Setting a lower count, e.g. 240, to correspond to the 10-ms period allows for a safety margin. This implies that a count of 240 corresponds to 180 electrical degrees, which means that a count of 1 corresponds to 0.75°. Having 240 pulses per 10 ms corresponds to having 24 000 pulses per second. This implies that the required clock frequency is 24 kHz. If the microcomputer is not capable of delivering this, then it may be derived from the clock signal supplied by the microcomputer (as shown in Chapter 9). If it is not possible to derive the exact value, then a lower value can be used which will reduce the resolution. Knowing the required angle, the microcomputer can perform simple calculations to convert it to the equivalent digital number.

This method of programmable delay allows a microcomputer to control the temperature accurately without frequent intervention, since all the microcomputer needs to do is supply the value to the output port. The value will be latched by the output port and will only need to be altered by the microcomputer if the temperature setting needs to change.

Example 14.8: **Redesign the circuit presented in the previous example in order to reduce the number of components.**

The microcomputer-controlled delay circuit can be based around an external timer or counter loaded with a digital number through an output port of the microcomputer. This number represents the required phase angle and is calculated in the same way as in the previous example. The circuit of Fig. 14.12 uses an 8-stage synchronous down-counter (the 74HCT40103 IC). The microcomputer delivers an 8-bit digital number from its output port to input lines P0 to P7 of the counter. This number will only be loaded into the counter when line PE (pin 15) is at a logic-low state. The need to relate the programmable delay to the zero-crossing instant introduces the need for using the output of a zero-cross detection circuit to activate line PE of the counter. In this way, the counter reads the count and starts counting as close to the zero-crossing instant as possible without any intervention from the microcomputer. The latter, however, should load the value of the required delay prior to the zero-crossing instant. Assuming that the output of the zero-cross detector is at a logic-low level at all times other than when indicating a zero-crossing point, the circuit of Fig. 14.12 inverts the signal with an inverting gate. This is because line PE should be at a logic-high level at all times other than when loading the count.

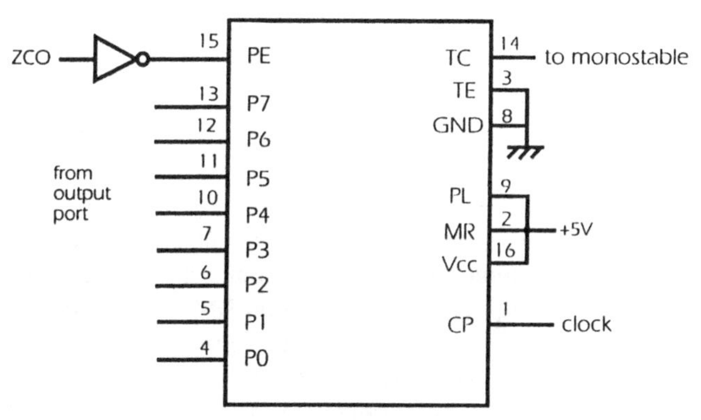

Fig. 14.12 Counter-based temperature controller.

During its operation, the counter decrements its count by one digit every time a clock pulse is received through line CP (pin 1). Output line TC (pin 14) is at a logic-high state during the counting-down process. When the count is decremented to 0, the state of line TC changes to a logic low for the duration of a single clock pulse. The change of state indicates the instant where the triac is to be turned on. As in the previous example, a monostable can be employed to widen the pulse and use it to trigger the triac.

Note that after producing the output pulse, the counter will continue to count down from 255. With the choice of an appropriate clock frequency, the count will not reach 0 twice in a half-cycle of the applied a.c. signal. Note that even if the clock frequency is high, the generation of a second triggering pulse should not cause a problem, since the triac will be conducting until the end of the half-cycle anyway. But if the frequency is too high, then there is a possibility of the occurrence of the second pulse close to the next zero-crossing instant. If the triggering pulse is wide, then conduction could start just after a zero-crossing instant. It is not difficult to stop the counting as soon as a triggering pulse is delivered by the counter. A simple modification is required to permit the delivery of a logic-high signal to line TE (pin 3) of the counter. Such a signal can be obtained by employing a flip-flop. The output of the zero-cross detector is connected to one input of the flip-flop. The occurrence of a zero-cross sends the output of the flip-flop to a logic-high state. The second input of the flip-flop is supplied from line TC. When the end of the count is reached, line TC is activated. It requests the triggering of the triac and resets the flip-flop (changing the state of its output line to a logic low). If the inverted output of the flip-flop is connected to input line TE of the counter, the counting process will stop between the instant of triggering the triac and the next zero-crossing point.

As in the previous example, the use of the counter allows the microcomputer to control the temperature with minimum intervention. This is possible because the microcomputer delivers the count once and it will be reloaded into the counter at every zero-crossing instant.

Let us now combine the ideas presented so far to solve the following example. As an exercise, try to solve it before looking at the solution.

Example 14.9: **You are given the task of designing a triac-based temperature controller. The circuit should be controlled from a microcomputer whose only interface lines are the output lines of an internal 10-bit up-counter which counts from 0 and resets internally by a zero-crossing signal. Means should be provided to permit changing the temperature from the keyboard of the microcomputer, as well as manually (with a potentiometer). The microcomputer should have the ability to turn the controller on and off.**

The potentiometer provides an analogue signal, while the microcomputer delivers a digital count. Since no additional interface lines are available, an interface circuit should be designed to combine the two signals. The digital output of the internal counter can be converted into the analogue form by a unipolar 10-bit D/A converter. An analogue comparator can be employed to compare the output of the converter with the setting of the potentiometer, as shown in Fig. 14.13(a).

Since the internal counter is reset at every zero-crossing instant, the output of the D/A converter is a ramp signal which increases from 0 and resets internally when the mains signal crosses the 0-V line. It is shown in Fig. 14.13(b) as signal X2 together with a d.c. level (signal X1) representing the setting of the potentiometer. The output of the analogue comparator is shown in Fig. 14.13(c) as signal X3. Let us assume that an operator alters the setting of the potentiometer to increase the d.c. voltage supplied to the inverting terminal of the analogue comparator. This alteration delays the instant of changing the state of the output of the comparator from low to high, thereby reducing the conduction period of the triac.

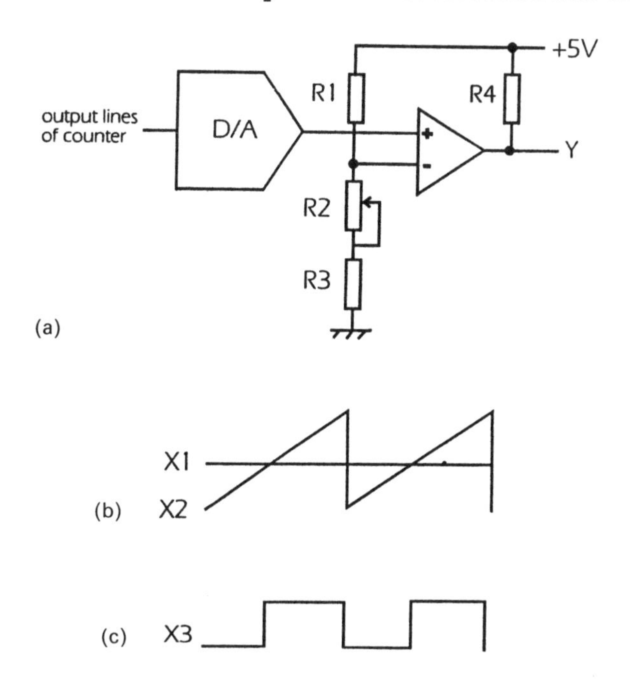

Fig. 14.13 Manual and programmable control of temperature: (a) circuit; (b) compared signals; (c) triggering signal.

The next step is to allow the microcomputer to change the temperature without altering the setting of the potentiometer. This can be achieved internally by changing the frequency of the clock operating the counter according to a demand level entered from the keyboard. Changing the frequency alters the rate of counting, which will in turn change the slope of the analogue ramp (signal X2 of Fig. 14.13(b)). The result is a corresponding change of the triggering angle of the triac. The allowable increase in the clock frequency is defined by the resolution of the counter. In other words, it is limited by the maximum count before the counter overflows to 0 (which is 1023 for a 10-bit counter). The overflow has the same effect as applying a reset signal for a very short interval and should only be activated by a zero-crossing signal.

The final requirement is to provide the microcomputer with the ability to turn the controller off. This is done with the help of resistor R3. It is included to limit the minimum d.c. voltage produced by setting the potentiometer. The microcomputer can either stop the delivery of clock pulses to the counter as soon as the zero-crossing signal resets the counter, or apply a logic signal to reset the counter at any instant. Either way, the output of the D/A converter will be too low to cause a change in the state of the output of the comparator.

The value of R3 should be high enough to ensure that no triggering pulses will be produced as a result of electrical noise.

14.6 Temperature conversion

Temperature can be presented in degrees Celsius (°C), degrees Fahrenheit (°F), or kelvins (K). The relationship between them is given by the following equations:

$$°C = (5/9) \times (F - 32) \tag{14.4}$$

$$°F = (9/5) \times C + 32 \tag{14.5}$$

$$K = °C + 273.15 \tag{14.6}$$

Solving these equations is easily performed by a microcomputer. In some cases, however, additional means are needed to perform the conversion, as shown in the following example.

Example 14.10: **A microcomputer delivers a regularly updated reading of the room temperature in degrees Celsius. The reading is supplied as an 8-bit number through an output port. How can the reading be converted to degrees Fahrenheit without altering the software of the microcomputer?**

Assume that the room temperature cannot exceed 99°F. The application requirement can be satisfied by a table to convert an 8-bit number into another 8-bit number. The table can be stored in an EPROM such as the one used in Fig. 12.2. The table should be presented so that a Fahrenheit value is stored as data at an address represented by a Celsius value. As soon as the microcomputer delivers an 8-bit value, the EPROM supplies the corresponding Fahrenheit reading through its data lines to be used by an external circuit.

14.7 Heat generation and dissipation

Heat is generated during the operation of microcomputers and the equipment they control. The majority of microcomputer-based applications are of relatively low power and generate a relatively low amount of heat. When using a microcomputer in power-control applications, the generated heat can be significant and is dependent on many factors. One source of heat comprises power-switching devices. When turning a switching device on, the voltage across it reduces, while the current through it increases. Similarly, when turning such a device off, the current decays to zero and the voltage increases. Such changes cause power losses in the device. These are not the only losses, since when conducting, a switching device possesses a small internal resistance which leads to a voltage drop across the device and causes the formation of heat. The amount of losses depends on many factors, such as the type of device used, the way it is driven, the switching frequency, and the type of operation. Section 10.3.2 illustrates that using a switching device in one region of its characteristics leads to different losses than if it is used in another region.

The losses lead to the formation of heat which is conducted through the internal structure of the device to the outside world either directly, or with the aid of a heat sink. During operation, the temperature of a device will rise until it reaches a steady state defined by the heat-dissipation method. In order to assist in the heat-transfer process, manufacturers of power-switching devices package such devices in a form that helps in the dissipation of heat. One side of the device (or the whole package) is suitable for connection to the heat sink. It is usually made of metal to ensure good transfer of heat. Depending on the device, the contact section may or may not be electrically isolated. The use of a non-isolated device means that the heat sink will be live and can be isolated from other areas it is in contact with. Alternatively, use can be made of electrically-isolating material (e.g. in the form of a washer or tape) through which heat will transfer from the device to the heat sink. Although such material has good heat-transfer characteristics, it adds a small resistance to the heat transfer. This is justified since it eliminates the need to isolate the heat sink.

Clearly, not all heat sinks dissipate the same amount of heat; they are classified according to their thermal resistance which is presented in degrees-per-watt units. The rate of exchange of heat through a heat sink depends on its material, shape and size. It also depends on the difference in temperature between the heat sink and ambient. Dissipating heat is more difficult when power devices are operated inside cabinets. Sufficient ventilation is required and, in many cases, forced ventilation is introduced by the use of electric fans. Such fans consume power and can introduce noticeable noise. Consequently, their number and positioning are important for the successful operation of the unit. In many very high-power applications, it becomes more practical to use water or oil cooling, where the liquid is pumped through special paths within heat sinks to absorb the heat.

Since heat transfers to the top, the best heat exchange can be achieved by placing the heat sink so that its fins are vertical, with no obstructions to the flow of air. Furthermore and in order to achieve efficient transfer of heat between the heat-generating device and the heat sink, one needs to ensure that contact surfaces are smooth and free from dirt. In practice, however, the surfaces of heat sinks may not be perfect. Consequently, the area of contact between the heat sink and each device can be covered with a thermal compound material in order to fill any gaps. Aluminium gives a good combination of thermal conductivity and light weight, making it a very attractive material to use in heat sinks.

Temperature sensors are widely used to measure the temperature inside enclosures and/or to provide a voltage level indicating over-temperature, as shown in Example 14.1. Commercially produced electronics equipment is tested in this way to ensure correct operation with sufficient heat transfer. After production, temperature sensors are also required within the unit, since temperature can increase due to an overload, a blockage of the air-flow path, or a failure of the cooling device.

The microcomputer may not sense the fault, e.g. due to a fault in the temperature-sensing device. Alternatively, the microcomputer may not be able to take an action, e.g. due to a component fault, or in the unlikely event that the microcomputer may be locked in an endless loop. The safety requirements of many applications introduce the need for the inclusion of hardware devices to detect unacceptable temperature levels and take an appropriate action, e.g. shutdown or sounding an alarm.

Example 14.11: **The safety requirement of an application imposes the need to modify the solution of Example 14.1 in order to introduce a further level of protection. This level should only be activated if after interrupting the microcomputer, the temperature continues to rise and increases beyond the safe-operating level. Design an additional circuit to sound an alarm and produce a shutdown signal.**

The detection of a violation of the safe temperature level can be achieved by the use of an additional analogue comparator connected in a similar way to that of Fig. 14.1. The state of its output changes from low to high when the temperature level is violated. That level can be used to activate a power-latch circuit, such as the one given in Fig. 14.14(a). It is similar to the crowbar circuit of Fig. 10.23, except that here the thyristor is triggered by the active-high over-temperature signal applied to point X. Since thyristor TH is operated from a d.c. supply, once it is turned on it will latch to conduction. The state of output line Y will change from high to low. This signal can therefore be used to activate a shutdown when in the low state.

As long as TH is in the conduction state, a d.c. current passes through the alarm AL. It will sound and continues to do so until the power is turned off. If the operation of the alarm requires a high voltage, then the low-power electronics can be isolated by the use of an optocoupler, as shown in Fig. 14.14(b). This approach introduces the need for an additional power supply (V2) to operate the switching device of the optocoupler.

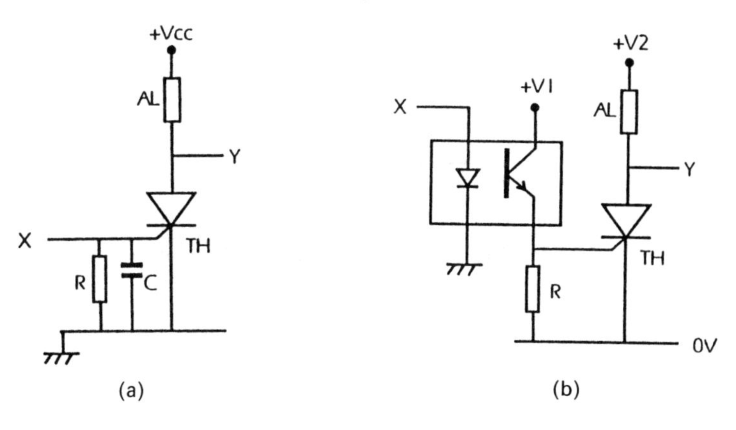

Fig. 14.14 Fault-latching circuit: (a) not-isolated; (b) isolated.

Chapter 15

Motor control

15.1 Introduction

Programmable motor controllers are designed by interfacing power-switching devices to microcomputers and implementing a suitable control algorithm. The number of devices, configuration and rating depend on several factors, such as the type of motor, the source of power (a.c. or d.c.) and the power levels. Microcomputer-based motor controllers can provide precise and stepless speed and position control. They make use of the microcomputer's fast decision-making capability to deliver high performance and fast fault detection. Sophisticated algorithms can be implemented by software means, or, alternatively, the microcomputer can take a supervisory role, leaving a digital or an analogue circuit to perform the required control.

Microcomputer-based motor controllers are often referred to as intelligent controllers and find uses in a variety of applications, such as wire-winding processes, printing machines and electric vehicles. Some controllers are designed to achieve precise position control, e.g. positioning of robot arms and automatic insertion of electronic components on PCBs. One of the main advantages of microcomputer-based controllers is programmability. A sequence of speeds and directions of rotation can, for example, be programmed in to a microcomputer, specifying the duration of each step. A fixed sequence can be programmed in to an EPROM, while a reprogrammable sequence can be programmed in to RAM, EEPROM or a disk. Sophisticated algorithms can be developed on an off-line computer, e.g. a desk-top PC, and simulated to observe the theoretical behaviour before transferring them to the microcomputer which controls the operation of the motor on line. Such an on-line microcomputer is likely to be designed specifically for the control of a specific type of motor within a specified range of power and speed. It may consist of a single board using an 8-bit or a 16-bit microcontroller plus peripherals and interface devices.

In an intelligent motor controller, the microcomputer is not only employed to control the rotational speed and direction, but it also monitors various parameters and communicates with the operator. The degree of communication varies from one system to another. Generally, one is likely to see better user communications in controllers designed for higher-power machines. Since the power components of such controllers are of a high cost, the cost of a more informative display will only add a small proportion to the total cost. In the last few years, however, the advances in optoelectronics technology have reduced the cost of displays (especially alphanumeric and graphical displays) and made them easier to interface and use. The result is the inclusion of more informative displays in lower-cost motor drives thereby increasing their user-friendliness and easing the monitoring and fault-finding tasks.

15.2 Control of d.c. motors

Low-power low-voltage d.c. motors are incorporated in low-power electronic circuits making use of the available d.c. voltage. Higher-power d.c. motors require rectifying the mains a.c. signal to a d.c. voltage before using it to operate the motor. This is not difficult to do, and with a few modifications the rectification circuit can be transferred to a speed changer operating under the control of a microcomputer. When compared to an a.c. induction motor of the same power rating, a d.c. motor is larger in size, costs more and needs more maintenance, but it offers a wide range of speed and is relatively easy to control.

Figure 15.1 shows an example of a d.c. motor. Its parts of interest here are the armature and the field. They are shown to be separately excited by V_a and V_f respectively, causing current to flow in the armature (I_a) and in the field (I_f). The latter generates a magnetic flux (M_f) which links with the armature and creates motion. The relationship between the voltage and current in the armature circuit is given by:

$$V_a = E_a + I_a R_a \tag{15.1}$$

where E_a is the back e.m.f. (electromotive force) generated as a result of rotating the machine, and R_a is the resistance of the armature.

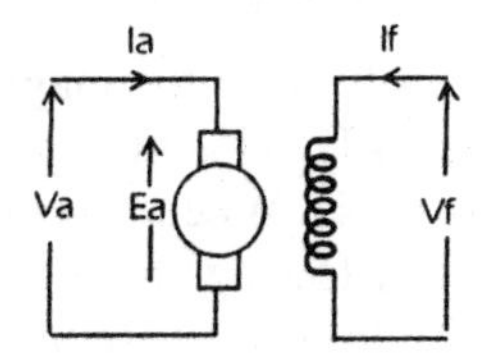

Fig. 15.1 Separately excited d.c. motor.

The power is given by:

$$V_a I_a = E_a I_a + I_a^2 R_a \qquad (15.2)$$

This represents: Applied power = Useful power + Armature losses.

The operation of the machine can be understood by considering the following equations:

$$E_a = K_1 N M_f \qquad (15.3)$$

$$T = K_2 I_a M_f \qquad (15.4)$$

$$M_f = K_3 I_f \qquad (15.5)$$

where N is the speed of rotation, and k1, k2 and k3 are constants.

Substituting Equation 15.3 into Equation 15.1:

$$V_a - I_a R_a = K_1 N M_f \qquad (15.6)$$

Substituting Equation 15.5 into Equation 15.6 and rearranging:

$$N = (V_a - I_a R_a)/(K I_f) \qquad (15.7)$$

where $K = K^1 K^3$

The method of controlling the speed of the motor can be derived from the relationship presented in Equation 15.7. It shows that for a constant field current, the speed of the motor is proportional to the applied voltage (less the voltage drop on the armature resistance). The speed can be reduced by connecting a resistor in series with the armature. Alternatively, the speed can be increased by decreasing the field current. Equation 15.3 shows that for a fixed field, the back e.m.f. is proportional to the speed. Reducing the field is a process known as 'field weakening'; it is implemented to increase the speed, but, according to Equation 15.4, the armature current needs to increase in order to meet the requirement of the load torque. This is less favourable and sets a limitation to using this approach.

Reversing the direction of rotation of the motor can be achieved either by reversing the direction of I_f, or by reversing the polarity of the voltage applied to the armature, as will be shown later in this chapter.

Rather than having a separately excited configuration, the motor can be connected as series or shunt, as shown in Fig. 15.2. In the former, the supply current (I_s) is given by:

$$I_s = I_a = I_f \qquad (15.8)$$

Since the flux is proportional to I_f up to the magnetic saturation, in a series motor the flux is proportional to I_a. This makes the torque proportional to the square of I_a (see Equation 15.4). When saturation is reached, the flux is almost constant. Using Equation 15.4 again reveals that the torque becomes proportional to I_a. A series motor is therefore suitable for applications requiring a high starting torque.

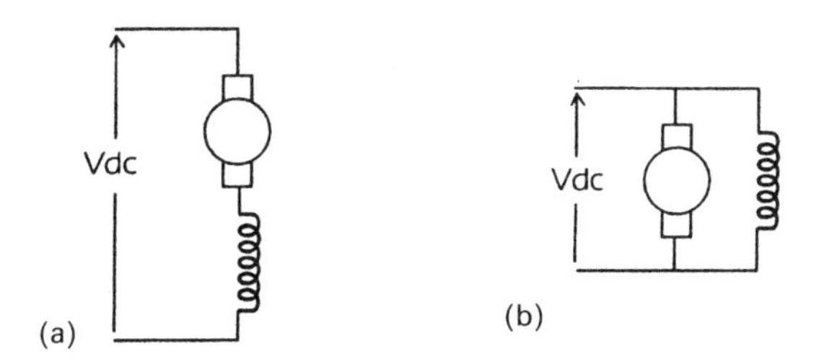

Fig. 15.2 Direct current motor connections: (a) series; (b) shunt.

The case is slightly different for a shunt motor, where the supply current is given by:

$$I_s = I_a + I_f \tag{15.9}$$

The flux and the speed are practically constant (there is a small change in speed when the load is increased). This implies that shunt motors are suitable for applications where constant speed is required.

15.2.1 Operating from a d.c. supply

The simple circuit of Fig. 15.3(a) is an example of a microcomputer-controlled driver of a d.c. motor. The microcomputer sends a logic-high signal from a digital-output port to drive the transistor to saturation and operate the d.c. motor at a fixed speed. The direction of rotation depends on the way the motor is connected. This implies that the microcomputer has no control over the rotational direction, since the simple circuit has no means of reversing the current through the motor. With this type of circuit, reversing can only be achieved by physically reversing the connections of the motor terminals. Microcomputer-controlled reversal of current can be introduced by the use of additional switching devices, e.g. transistors or relays. The transistor circuit shown in Fig. 10.15 is an example of a microcomputer-controlled current-reversal circuit. The direction of the current depends on which pair of transistors is switched on.

Rather than utilizing four switching devices and their associated interface and drive circuits, the configuration of Fig. 15.3(b) employs two switching devices connected across a split supply. The additional requirement here is the need for +V and −V power supplies, where only one of the two will be supplying the motor at one time, depending on which direction of rotation is required.

A microcomputer controls the operation of a d.c. motor in an open-loop or a closed-loop manner. Open-loop control can be based on supplying a d.c. voltage whose magnitude is defined by the microcomputer. The method assumes that the motor will run at a speed proportional to the average of

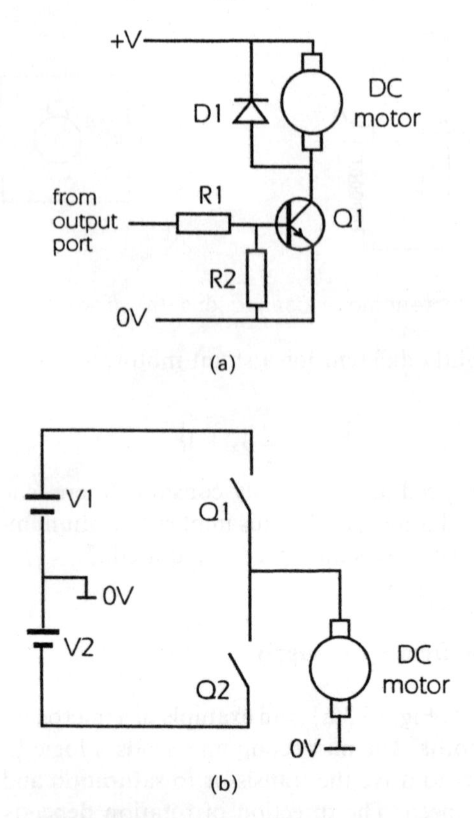

Fig. 15.3 Control of d.c. motor: (a) unidirectional; (b) bidirectional.

that voltage. The microcomputer has no means of knowing if the speed is altered due to such effects as a sharp change in load. This is not the case with a closed-loop control where the speed, the current and/or the voltage will be fed back to the microcomputer or to the interface circuit for testing and comparison with the demanded level. Regardless of which control method is adopted, the required level of voltage or speed may be programmed in the microcomputer, or supplied by the user through such means as a keyboard or a potentiometer. A user-supplied reading can, if required, be stored in an EEPROM, or on a disk within the microcomputer to permit its use after the power is turned off and on again.

Changing the speed is achieved by changing the average voltage. One way of implementing this is by the use of PWM (Pulse Width Modulation). With this method, the microcomputer supplies a series of pulses with a controllable mark-to-space ratio (adjustable duty cycle). A switching device, e.g. a transistor, can be utilized to allow the microcomputer to connect the full voltage to the motor when the transistor is on and 0 V when the transistor is off. The designer needs to make a decision on what frequency the pulses should be generated at. A frequency of 20 kHz may seem attractive, since it is higher than the audible range and may not be high enough to cause significant switching losses. This, however, is not always true, since the switching losses depend on both the switching device and its drive circuit. When driving a motor with pulses, however, the current will increase when the switching device is turned on and decay when it is turned off. The effect of the inductance limits the rate of change in current. The result is a ripple as the current changes between an upper and a lower limit. The use of high-frequency switching can reduce the ripple, since the next pulse is applied before the current decays to a low value.

Example 15.1: **Modify the circuit of Fig. 15.3(a) to allow a microcomputer to perform closed-loop control of a d.c. motor.**

Method of interfacing

A resistor of a very low value, e.g. 0.1 Ω, can be connected between the emitter of the transistor and the 0-V line. The motor current passes from the collector to the emitter of the transistor and then through the added resistor. This implies that the voltage drop across the resistor is proportional to the motor current, which in turn is proportional to the motor's torque. The reason for using a low-value resistor is to minimize its effect on the operation of the motor drive circuit. The voltage drop across the resistor can be fed back to the controlling microcomputer through an A/D converter. An operational amplifier, similar to that of Fig. 6.4, may need to be incorporated to amplify the analogue signal, thereby reducing the quantization error when converted to the digital form.

Designing the control algorithm

Let us choose a simple algorithm. The microcomputer has a demand signal, which is either fixed (programmed) or supplied by an operator. The microcomputer compares the demand signal with the digitized value of the feedback voltage using the following two equations and taking into consideration the sign of the error:

$$\text{error} = \text{demand} - \text{feedback} \tag{15.10}$$

$$\text{correction} = \text{error} \times \text{gain} \tag{15.11}$$

The gain can either be fixed or has a default value which can be altered by the user. Clearly, if precise control is required, then a more sophisticated control algorithm is needed. This increases the complexity of Equation 15.11 leading to an increase in the execution time, but can give an improved dynamic performance.

Implementation

Let us base our solution on Equations 15.10 and 15.11 and use the PWM approach to control the motor. The microcomputer reads the feedback signal, calculates the required correction, and uses it to alter the mark-to-space ratio of the pulses supplied to the base of the transistor. The maximum rate at which a change can take place depends on a few factors, e.g. the time needed for the A/D converter to digitize the analogue signal, the time needed for the microcomputer to calculate the required correction, and the frequency of the PWM pulses supplied to the motor drive circuit. Using a frequency of 20 kHz means that each PWM cycle has a duration of 50 µs. For many modern processors, this duration is sufficient to achieve an A/D conversion, calculate the required correction, and send the value to the drive circuit. Depending on the type and ability of the processor, such a rate of update may mean that the microcomputer is dedicated to the implementation of the control algorithm and has little, or nothing else, to do. This is not likely to be acceptable in practice and the microcomputer may need to perform the calculations at a lower rate, e.g. with a reduced PWM frequency, or to attend to the calculation once in every fifth pulse. The latter implies that the microcomputer will perform the calculations once every 250 µs which corresponds to an update rate of 4 kHz.

Selecting the instant of sampling

The instant of starting the A/D converter has a significant effect on the performance of the overall system. Let us consider three alternative approaches where the choice depends on the type of application, the method of control and the conversion time of the A/D converter:

- Starting the A/D converter at the beginning of the interrupt-servicing routine. This has the advantage of regular sampling at fixed intervals and the result will be used as soon as it is produced. During the conversion time, the processor will either have to wait or perform other useful tasks. The use of such a method involves the execution of additional instructions to start the converter and test for the end of conversion.
- Rather than having to wait during the conversion time, the A/D converter can be started as soon as the digitized value is read. This approach permits the digitization process to take place while the processor is calculating the required correction. This saves time, but there is a time interval between the end of conversion and using the digitized value. This introduces a phase lag whose effect on the operation depends on the method of control. Since the interval between successive samples is longer than the conversion time, the use of this method removes the need to test for the end of conversion.

- Instead of consuming time starting the A/D converter and testing its EOC flag, the A/D converter can be operated in the continuous-conversion mode (as explained in Chapter 7). In this way, the A/D converter stores its result in a regularly updated latch. The microcomputer reads this latch as it reads any memory location. The method can save a considerable amount of time, especially at a high rate of sampling. If, for instance, the start of conversion and the test of the EOC flag take 5 μs and the sampling interval is 50 μs, then not performing these two tasks saves 10% of the execution time. If, on the other hand, the sampling interval is increased to 250 μs, then the saving is only 2%.

Alternative approach

Rather than using the PWM approach, the control may be applied by supplying the base of the transistor with an analogue signal from a D/A converter. The amplitude of the signal will determine the base current. The current flowing through the motor is the collector current of the transistor, which is proportional to the base current. This gives direct control, but implies operating the transistor in its active region where the losses are higher.

Example 15.2: **Rather than using the microcomputer to perform the closed-loop control, how can hardware be utilized to perform such control and save execution time? The microcomputer controls the operation by supplying the demand signal.** The circuit of Fig. 15.4 is one way of satisfying the requirement. It uses a low-value (0.1-Ω) resistor R3 connected in a similar way to the previous example. Rather than reading the voltage drop across R3 by the microcomputer, this circuit incorporates an analogue comparator to compare that signal to

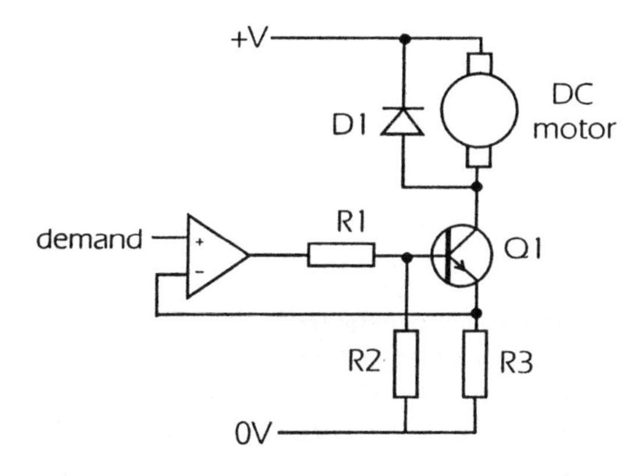

Fig. 15.4 Unidirectional closed-loop speed control.

the demand signal in an analogue form. The microcomputer supplies the demand signal in the form of a d.c. level generated through a D/A converter. It is connected to the positive terminal of the analogue comparator whose negative input line is connected to the feedback signal. If the motor current is less than the demand, then the output of the comparator is at a high level, causing the transistor to conduct. Otherwise, it will be at a low level and turns the transistor off. The result is therefore a controlled series of pulses.

Measuring the current in the last example is achieved by measuring the voltage drop across resistor R3. Since one end of the resistor is connected to the 0-V line, the voltage measurement is obtained from a single point. This is not always possible and the measurement may need to be obtained away from the 0-V line. In such a situation, the voltage drop across the resistor is represented by the voltage difference between its two terminals. A differential amplifier can be employed, as shown in Fig. 15.5. The value of the resistors depends on the value of both the supply voltage +V and the current I. The voltage is applied across the resistors forming a voltage divider, while the value of the current causes a voltage drop across R_s which will be represented at the output of the amplifier.

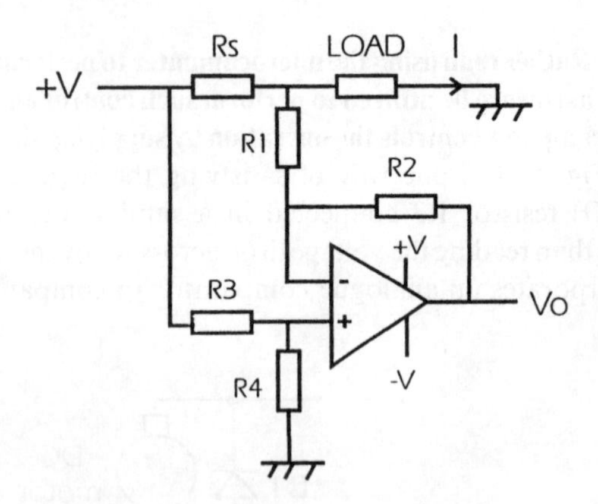

Fig. 15.5 Current measurement.

Example 15.3: **Design a microcomputer-based controller to operate a d.c. motor with a programmable speed and direction.**

The method described with the aid of Fig. 15.3(b) can be applied as shown in the interface circuit of Fig. 15.6. A triangular wave can be generated (as shown in Chapter 12) and supplied to point W. It is compared to the signal at terminal X which is a microcomputer-generated d.c. level (produced via

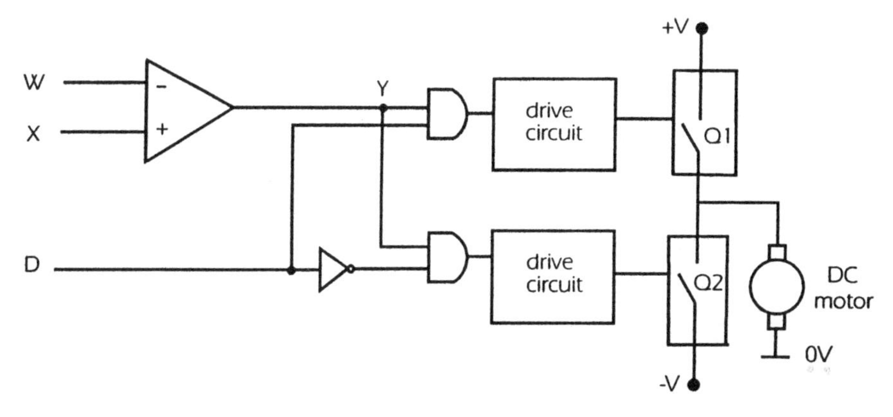

Fig. 15.6 Interfacing to a bidirectional motor drive.

a D/A converter) representing the speed demand. As in the previous example, the output of the comparator is a series of pulses whose width is proportional to the demand signal.

The circuit also allows the microcomputer to set the direction of rotation through a single digital-output line D. A logic-high signal on that line enables the upper AND gate to deliver the pulses from point Y to the upper switching device Q1, which will turn on and off according to the mark-to-space ratio of these pulses. This operates the motor from the positive supply rail. Reversing the direction of rotation is accomplished by the microcomputer changing the state of line D to a logic low. This disables the upper AND gate and enables the lower one. The speed-control pulses will then pass from point Y to the lower switching device Q2. This action leads to operating the motor from the negative rail, thereby reversing the direction of rotation.

Care should be taken when changing from one direction of rotation to the other, since a short delay is required to ensure that the current through a switching device has decayed to zero before attempting to turn the other switching device on (Section 10.6).

Direct current motors can also be employed as generators. One method is shown in the following example.

Example 15.4: **Define a configuration which will enable a d.c. motor to be used in the motoring and in the generating modes under the control of a microcomputer.**

The circuit of Fig. 15.7 illustrates the use of two MOSFETs to drive a d.c. motor. Only one MOSFET will be turned on at any instant of time. The voltage delivered to the motor is controlled by the microcomputer, which

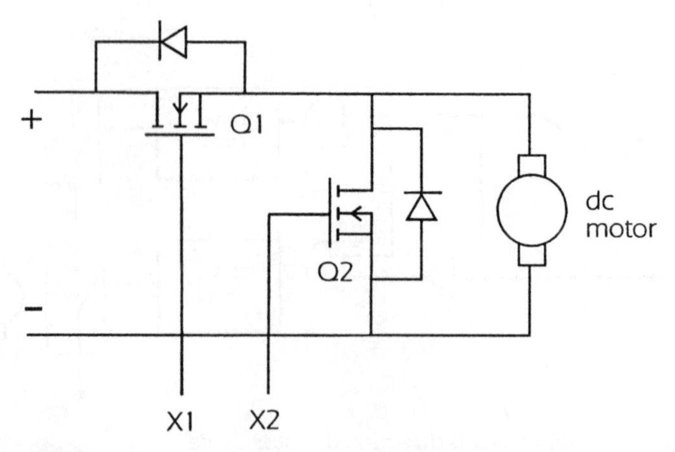

Fig. 15.7 Controlled motoring and generating modes.

supplies pulses through a digital-output line. Pulsing line X1 switches device Q1 on and off while Q2 is off. Switching Q1 on delivers current to the motor, while switching it off permits the flow of the motor current through the internal diode of Q2. The motor's current is therefore a d.c. level with a ripple which depends on the frequency and the mark-to-space ratio of the applied pulses.

In order to use the motor as a generator, the microcomputer keeps MOSFET Q1 off while delivering pulses to switch Q2 on and off through the digital-output line X2. This causes current to pass through Q2 when it is turned on. The current path changes every time Q2 is turned off, as it passes to the d.c. supply through the internal diode of Q1. As in the motoring mode of operation, the motor current possesses a ripple. It is, however, in the opposite direction.

15.2.2 Operating from an a.c. supply

The availability of mains power in the a.c. form means that it will need to be converted to d.c. before using it to operate a d.c. motor. The process is called 'rectification' and is explained in Section 10.2, where a diode bridge is employed to supply a d.c. voltage proportional to the a.c. voltage applied to it. The circuit of Fig. 15.8 incorporates two diode bridges to supply d.c. voltage to both the armature and the field of the motor. This connection leads to uncontrolled operation. The circuit can be modified to allow a microcomputer to control the output of the bridge, which will in turn define the speed and direction of rotation of the d.c. motor, as shown in the following examples.

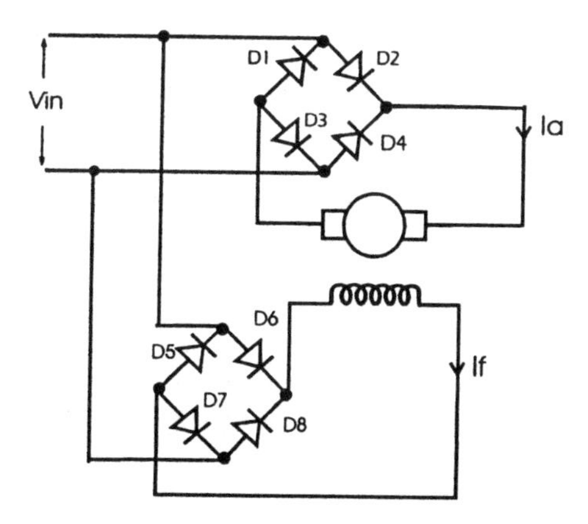

Fig. 15.8 Operation from an a.c. supply.

Example 15.5: **Modify the circuit of Fig. 15.8 to allow a microcomputer to change the speed of the motor.**
Since the variation of the speed of a d.c. motor can be achieved by the variation of the average voltage applied to it, the uncontrolled rectifier bridge consisting of D1, D2, D3 and D4 can be modified to make it a controlled rectifier by using thyristors instead of diodes. The control that the microcomputer has is the phase angle of triggering the thyristors. This covers half a cycle of the input a.c. signal, which is 10 ms for a 50-Hz supply (8.33 ms for a 60-Hz supply). In other words, operation from a 50-Hz supply limits the maximum rate of change of speed to 100 Hz, which is much lower than the rate achieved with PWM when operating from a d.c. supply.

The implementation of the control can be relatively simple, especially if the microcomputer possesses an internal timer/counter. It requires a zero-cross detector (such as the one described in Section 10.9) to supply the microcomputer with successive pulses. Depending on the type of detector used, a pulse will be supplied either once every cycle, or once every half-cycle. Either way, the microcomputer measures the duration between the pulses in order to calculate the maximum allowable phase angle (half a cycle). Assuming that a simple closed-loop control is required, the microcomputer will introduce a delay before supplying triggering pulses to the thyristors. It then reads the speed of the motor (e.g. through the shaft encoder described in Section 11.8). An error signal is then generated by software by comparing the feedback signal with the demand. Depending on the sign and magnitude of the error, the microcomputer alters the delay in order to change the phase angle and minimize the difference (more sophisticated algorithms can be implemented by the use of more than one feedback loop).

The above examples deal with controlling the motor from a single-phase a.c. supply. Direct current motors, particularly those of high power, are operated from a 3-phase a.c. supply. The 3-phase voltage is rectified by the use of a 3-phase bridge rectifier similar to the one shown in Fig. 10.3. The diodes of the bridge are normally replaced by thyristors to permit a microcomputer to adjust the speed of the motor. The resulting configuration is known as the 'fully controlled bridge rectifier'.

Example 15.6: **Design a circuit to allow a microcomputer to operate a 6-thyristor fully controlled 3-phase bridge rectifier using the minimum number of interface lines.**

The microcomputer needs to send a triggering pulse to each of the six thyristors of the bridge rectifier. The output of the bridge rectifier is adjusted by the microcomputer through controlling the instants of sending these pulses. In a closed-loop system, the adjustment is repeated regularly in order to minimize the difference between a reference and a feedback signal. On the other hand, the operation of an open-loop system relies on setting the triggering instants according to a preprogrammed or a manually settable value. In either case, the interface circuit takes the form of the block diagram presented in Fig. 15.9. The low-power control electronics are isolated from both the input power supply and the power-switching devices. The isolated signal is supplied to a zero-cross detection circuit (Section 10.9) to generate pulses at the zero-crossing instants. The operation of the 3-phase controlled rectifier needs six pulses per cycle, which can be delayed to produce the required voltage. Let us consider two ways of applying the control.

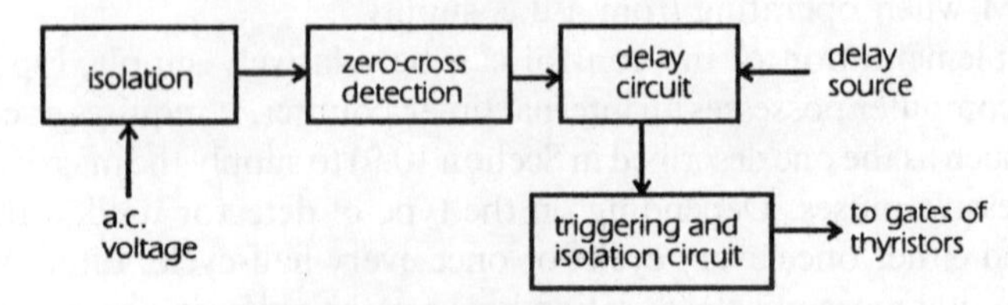

Fig. 15.9 Control of a 3-phase bridge rectifier.

Solution 1 Microcomputer-controlled delay

The output of the bridge rectifier is set according to the time between a zero-crossing point and the instant of delivering the triggering pulse to the corresponding thyristor. Since the applied 3-phase a.c. signals possess the same frequency and are spaced 120 electrical degrees apart, only one of the phases can be tested for a zero-crossing point. The microcomputer can then deduce the zero-crossing instants of the other two phases. Using a single zero-cross detector and a counter, the microcomputer can measure the time between two successive zero-cross signals. This gives a measure of the period of a single cycle of one of the phases of the input signal. The microcomputer can then divide the measured period by 6 and load the result to a continuously running timer to produce six equally spaced pulses per cycle. The time delay of each pulse is set digitally according to a number written to the timer by the microcomputer. The output of the timer will then be supplied to the point marked 'input' in the circuit of Fig. 15.10(a).

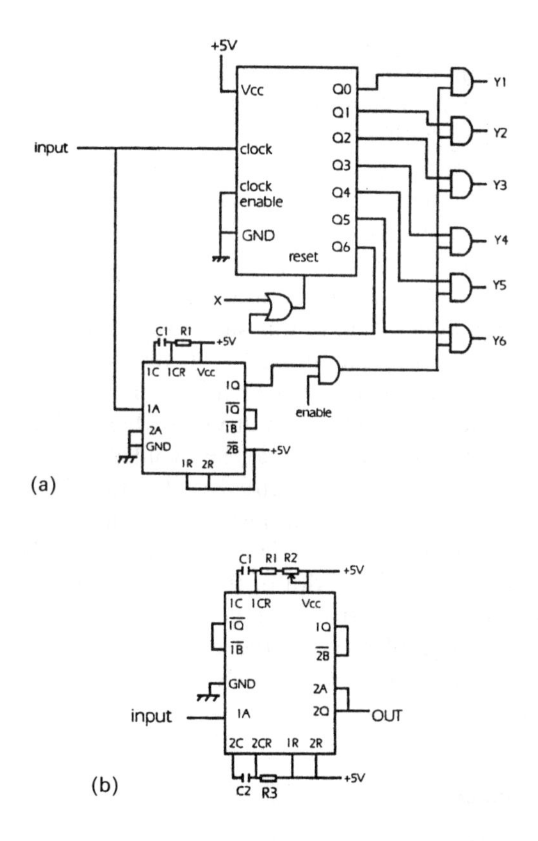

Fig. 15.10 Interface circuit of a 3-phase bridge rectifier: (a) programmable delay; (b) introducing a manual delay.

The figure shows the input signal supplied to two ICs. The first is the 74HCT4017. It receives the pulses through its clock input line and distributes them to six sequentially activated lines (Q0 to Q5) before resetting to repeat the sequence. The reset of this IC is achieved by connecting its output line Q6 to the reset pin via an OR gate. The second input of this gate is connected to a digital-output line of the microcomputer (line X). This line is utilized to give the microcomputer the ability to reset the IC, thereby synchronizing the operation and delivering the pulses to the correct thyristors.

The second IC is the 74HCT4538 dual retriggerable monostable multivibrator (this IC is also used in Section 9.5.1). It receives the input pulses through line 1A and generates an output pulse whose width is determined according to the relationship T = 0.7(R1)C1. It is configured for leading-edge triggering and its function is to set the pulsewidth of the output pulses delivered to the thyristors. Figure 15.10(a) shows the output of the monostable (line 1Q) connected to a single AND gate. This gate is incorporated in order to provide an additional means of stopping the supply of triggering pulses to the thyristors through the enable line. This line can be controlled by fast-detection circuits to provide a hardware protection of the power-switching devices. Analogue comparators can, for example, be employed to detect fault conditions, such as over-current, over-voltage and over-temperature. An alternative use of the enable line is to provide an operator with a manual on/off control of the bridge rectifier. In a closed-loop control system, it is important to send the external enable/disable command to the microcomputer. This is because the application of this signal stops the bridge rectifier from generating an output. This results in a large difference when the microcomputer compares the feedback signal with the demand. The result is the delivery of a request for the highest possible output value which will be delivered when the enable signal is reapplied. This can be avoided if the command is also delivered to the microcomputer through a digital-output line for regular testing.

Once enabled, the AND gate will permit the transfer of signals from the output of the monostable to six AND gates. The width of the monostable's output pulses determines the width of the signals delivered to output lines Y1 to Y6. The instants of starting the delivery of any of the six pulses are defined by the change of state of the corresponding output line of the 74HCT4017 IC. A triggering circuit (such as that of Fig. 10.24) is needed to enable each of output signals Y1 to Y6 to drive one of the six thyristors.

Solution 2 Manual delay

A potentiometer can be incorporated as a means of providing manual speed setting. It can be connected to alter the pulsewidth of a leading-edge triggered monostable, as shown in Fig. 15.10(b). The circuit utilizes the same monostable IC employed in Solution 1 of this example, but it uses both of the monostables within that IC and adds potentiometer R2 to provide the manual control.

During operation, the microcomputer generates the six pulses without a delay and supplies them to the input line of monostable 1 (in the same way as in Solution 1). Without a delay, these pulses represent the six zero-crossing instants within a single cycle. Monostable 1 is configured to detect the leading edge of the pulses. Its output (line 1Q) is a series of pulses whose width is set according to the value of C1 and the sum of R1 and the setting of R2. The width of the pulses represents the delay in triggering the thyristors. Consequently, the values of C1, R1 and R2 are determined by the minimum and maximum delay required from the configuration.

The second monostable is configured for falling-edge triggering and is employed to detect the end of the time delay introduced by the first monostable. Its function is to set the width of the triggering pulses and deliver them from its output line (2Q) which is connected to the six AND gates (in the same way as in Solution 1). The enable control line can also be included by the use of the additional AND gate of Fig. 15.10(a).

15.3 Control of a.c. motors

The induction motor is an example of an a.c. motor which is very attractive to use. It is simple, rugged and very reliable. It needs minimum maintenance, has a relatively good power factor and efficiency, and is cheaper than a d.c. motor of the same power. The motor operates at its rated speed when connected directly to the a.c. mains. Controlling its speed efficiently over a wide speed range is, however, more difficult and costly than controlling a d.c. motor. The advances in microcomputer and switching-device technology have provided higher processing power and faster switching capability at low prices. Microcontrollers are normally utilized in the design of motor drive controllers. They not only offer high processing power, but they also include many of the required peripherals on a single IC. This has led to the development of smaller and more efficient motor drives at lower cost. When deciding which machine to use, one of the main consideration is cost. Direct current machines are more expensive than induction machines, but the former require a simpler, lower-cost controller. The lower cost of controllers for induction machines makes them more favourable in a large number of applications. An induction machine does not need the maintenance required by a d.c. machine, but one should pay attention to the motor drive characteristics and the efficiency of the controller. Harmonics can be generated by some a.c. drives which increase the machine losses and affect the smoothness of rotation.

15.3.1 Operation of induction machines

An induction machine consists of two main parts; they are the stator and the rotor. The stator of a 3-phase induction machine contains a 3-phase winding which, when connected to a 3-phase supply, generates a magnetic flux rotating at 'synchronous speed' defined by:

$$N_s = 60 \times f_s/p \tag{15.12}$$

where f_s is the supply frequency, p is the number of pole pairs, and N_s is the synchronous speed given in rev/min.

The rotating flux intersects the rotor conductors, causing current to flow. The rotor will turn at a lower speed than the synchronous speed. The difference is represented by the slip (s) which is given by:

$$s = (N_s - N_r)/N_s \tag{15.13}$$

and

$$f_r = s \times f_s \tag{15.14}$$

where N_r and f_r are the speed and frequency of the rotor respectively.

The torque/speed characteristics of an a.c. induction motor are shown in Fig. 15.11(a), where the stable operating region is between point X and the point at which the curve intersects the zero-torque line. The speed of the motor can be altered by changing the applied voltage or frequency. Figure 15.11(b) shows the effect of changing the voltage, while the effect of changing

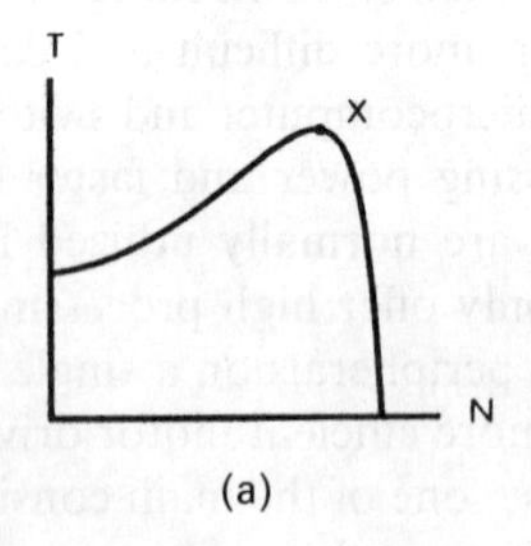

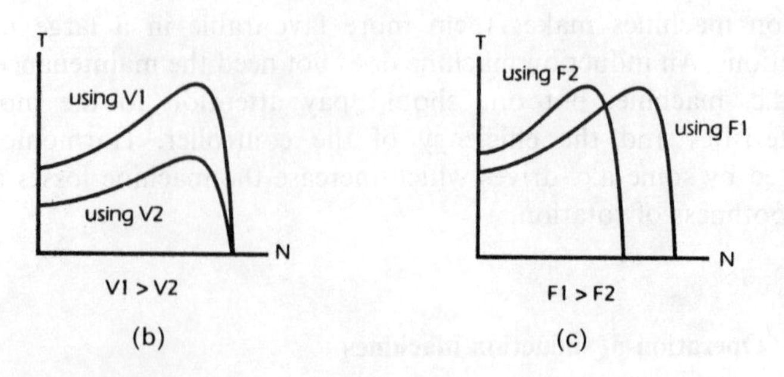

Fig. 15.11 Characteristics of an induction motor: (a) torque to speed; (b) effect of changing voltage; (c) effect of changing frequency.

the applied frequency is illustrated in Fig. 15.11(c). Changing the voltage is simpler to do, but is less efficient. Let us compare the two curves of Fig. 15.11(b). The most obvious difference is the reduction in the peak point of the characteristics. It corresponds to the maximum torque that the motor can handle. Assume that the motor will operate under a constant torque T which is less than the maximum torque. This can be represented by drawing a horizontal line intersecting both curves. The intersection points show that a slower rotation speed can result from reducing the applied voltage. Using Equation 15.13, one can conclude that reducing the speed by reducing the voltage corresponds to an increase in the slip. This leads to higher losses and therefore a lower efficiency. Figure 15.11(c), on the other hand, shows that changing the speed by altering the frequency does not increase the slip, as it changes the synchronous speed of the motor.

15.3.2 Changing speed and direction

Let us consider examples of using a microcomputer to control the operation of a.c. motors. We start by considering the adoption of the simpler voltage-control approach, and then proceed to illustrate the implementation of the more efficient frequency-control approach.

Example 15.7: **Select switching devices and a suitable configuration to enable a microcomputer to control the speed of a 3-phase induction motor using the voltage-controlled approach. How can the direction of rotation be reversed?**
The requirement can be satisfied by using triacs as the power-switching devices. A single triac is incorporated to control the voltage of each of the three phases, as shown in Fig. 15.12(a). Each of the three triacs (T1, T2 and T3) is interfaced to the microcomputer through a triggering circuit; such a circuit is presented in Fig. 10.24. The required speed is set by adjusting the phase angle of triggering the triacs according to a demand signal programmed into the microcomputer or set manually by a potentiometer.

This is an open-loop control which satisfies the requirements of a vast number of applications, e.g. controlling the speed of a fan. The control scheme can be converted to a closed-loop form by using a speed-measuring device, e.g. the optical encoder described in Section 11.8. The speed signal can be fed back to the microcomputer to be compared to the required speed by software. The microcomputer will then adjust the phase angle in order to minimize the difference.

Reversing the direction of rotation of the motor can be achieved by reversing the sequence of the phases. Rather than performing this manually, Figure 15.12(b) provides a microcomputer-based solution. Triacs T4 and T5 are incorporated to provide alternative paths to those offered by triacs T1

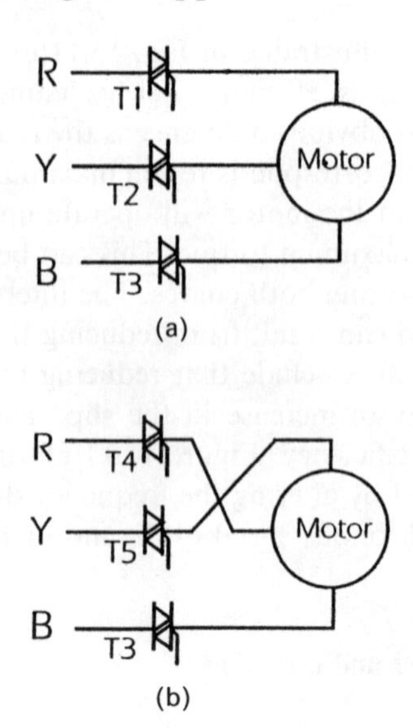

Fig. 15.12 Voltage-controlled operation: (a) forward rotation; (b) reverse rotation.

and T2. This method implies that a speed controller can be provided with the ability to reverse the direction of rotation by employing five triacs. The microcomputer can be programmed to turn on the three triacs that will give the required direction of rotation.

Rather than using the voltage-controlled approach, one can adopt the more attractive frequency-controlled approach. The change of frequency is accomplished by rectifying an a.c. voltage (e.g. the 50-Hz or 60-Hz mains) into a d.c. voltage which will then be changed by an inverter to an a.c. signal at the required frequency. A typical variable-speed drive system is represented by the block diagram of Fig. 15.13. The resulting shape (and hence the harmonic contents) of the a.c. waveform depends on the inverter topology. In general, a 3-phase inverter consists of six switching devices operated repeatedly and sequentially under digital or analogue means of control. A microcomputer-controlled inverter can control an a.c. motor efficiently, providing a stepless variation of speed associated with a fast on-line

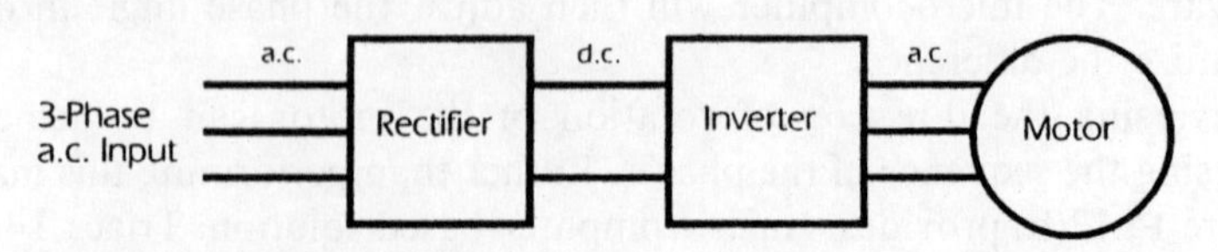

Fig. 15.13 Three-phase variable-speed drive.

decision-making capability. The operation of many microcomputer-controlled variable-speed drives is based on programming the microcomputer to deliver pulses to each of the switching devices at high frequency. At the same time, the microcomputer examines feedback signals, such as the speed of the motor, and adjusts the inverter's output signal. The microcomputer will also examine other signals to identify faults and take the required action quickly.

Functionally, the inverter can either be a frequency changer or a frequency-and-voltage changer. In the first case, the magnitude of the voltage is controlled before being delivered to the inverter. One commonly used approach is to alter the voltage within the rectification stage of the configuration of Fig. 15.13. This is performed by utilizing a fully controlled 3-phase bridge rectifier rather than a diode bridge. With this approach, the magnitude of the d.c. voltage is controlled and will therefore define the amplitude of the pulses supplied from the inverter to the motor at a frequency set through the inverter. The use of a fully controlled bridge rectifier introduces two main limitations. First, it injects harmonics into the mains, and second, it limits the rate of changing the voltage to six times per cycle, as explained in Example 15.6. This leads many designers to favour the use of a 3-phase diode bridge (uncontrolled) at the front end followed by a d.c. chopper to change the magnitude of the rectified signal by producing pulses with an adjustable mark-to-space ratio. This ratio is set by a microcomputer, or by a dedicated driver IC. This approach uses additional components, but can increase the speed of response to a much higher rate. This is because the PWM pulses supplied to the chopper are at a high frequency and a decision can be taken within one pulse (50 μs for a 20-kHz chopping frequency). Rather than employing separate devices to control the voltage, the inverter can be operated to control both the voltage and the frequency. This can be achieved by using the pulse width modulation (PWM) approach explained in the next section. One should keep in mind, however, that the use of a fully controlled rectifier allows the controller to limit the magnitude of the rectified voltage, or terminate the rectification process if the input a.c. voltage is increased to excessive levels. This is a very useful means of protection.

Example 15.8: **Design a microcomputer-controlled circuit to control the speed of a 3-phase induction motor using the frequency-controlled approach.**

Power transistors can be incorporated to form a 3-phase inverter. It consists of three identical legs, one of which is shown in Fig. 15.14(a), where a resistor, a capacitor and a diode are connected in parallel with each transistor to form a snubber circuit for protection. Each of the other two diodes (D1 and D2) is connected as a flywheeling diode to provide a path for the flow of power when the transistor turns off. The transistors are incorporated as microcomputer-controlled switches which are operated either in the cut-off or the saturation region of their characteristics. When conducting, the upper transistor connects the upper rail of the d.c. bus to the output. On the other hand, switching on the lower transistor connects the negative rail of the d.c. bus to the output. A dead time is required to prevent a short-circuit across the d.c. supply. The connection of the three phases is shown in Fig. 15.14(b).

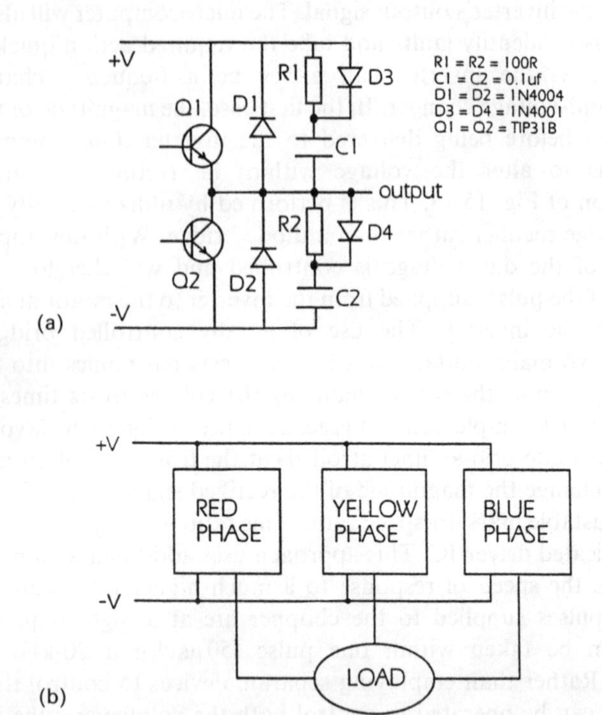

Fig. 15.14 Three-phase transistor inverter: (a) one of the three legs; (b) block diagram.

There are many ways of interfacing the microcomputer to the six power transistors of the inverter. They vary in their complexity and in the degree of control given to the microcomputer. Let us select a relatively simple interface circuit, such as the one presented in Fig. 15.15. Rather than using six lines from the microcomputer, the circuit only needs a single digital-output line (marked 'input'). The microcomputer supplies pulses at the required frequency through the input line to the clock pin of a 74HCT4017 IC. These pulses will be distributed between six channels (Q0 to Q5) sequentially before the 74HCT4017 is reset by output line Q6. Each three output lines of the 74HCT4017 are ORed together to produce a square wave with equal mark and space. The result is six waves generated sequentially, one every 60 electrical degrees ($\frac{1}{6}$th of a cycle), as shown in Fig. 15.16. After being delayed

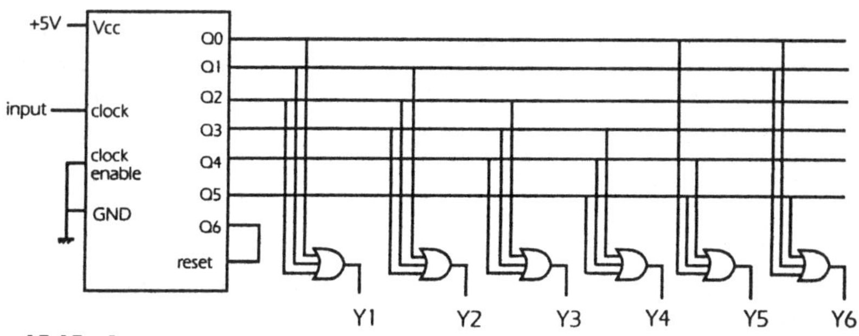

Fig. 15.15 Inverter-control circuit.

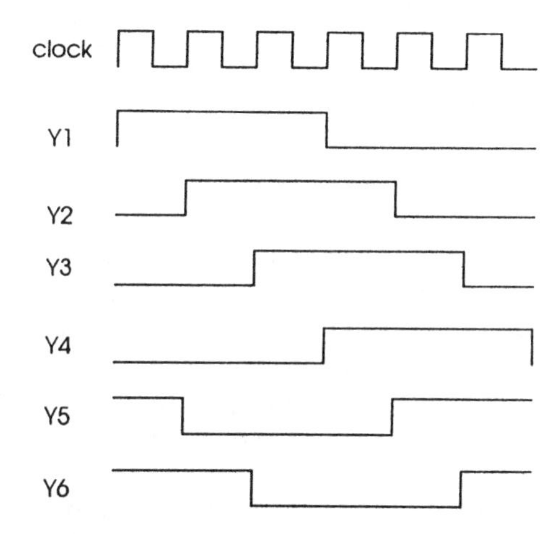

Fig. 15.16 Three-phase drive waveforms.

(to provide dead time), each of these signals is isolated and amplified before being used to drive one of the six transistors of the inverter. The sequence of turning on the transistors is the same as that given in Section 4.6.3, where it is deduced from the waveforms of Fig. 4.19.

By using this approach, the function of the microcomputer is reduced to a programmable frequency changer. Its function can be performed as described in Chapter 9. The method reserves a small proportion of the processing power, leaving the microcomputer with sufficient power to communicate with the user and to monitor the operation and perform additional tasks.

Example 15.9: **How can a microcomputer use the open-loop approach to control the speed of two inverter-driven induction motors where one of them needs to run at four times the speed of the other? Assume that the motor contains a single programmable timer.**

The speed of an induction motor is proportional to the frequency of the driving signal. The microcomputer's timer can be employed to generate the frequency for the faster motor. A simple circuit can then be designed to derive the frequency for the second motor from that of the first. Such a circuit is a divide-by-4 counter such as the one shown in Fig. 9.2.

Changing frequency by a microcomputer can be achieved by the use of counters and timers (Section 9.4). A counter in the last example performs the frequency change and will keep the ratio of the two speeds fixed. Flexibility can be provided if the microcomputer has the ability to alter the ratio of the speeds of the two motors. This is needed in many applications. Consider, for example, transferring paper from a large roll to a smaller one. One turn of the large roll corresponds to several turns of the small roll. As the rolling process continues, the large roll becomes smaller and the small roll becomes bigger. If each roll is driven by a motor, then the ratio of the two speeds will need to be altered regularly. In a digital system, the change of the speed ratio is performed in steps whose number and size depend on the resolution of the controller and the interface circuit. Such a task can be achieved by a microcomputer-based controller if it includes one or more programmable timers/counters. Such devices are available either in IC form or integrated within a microcontroller IC. Its operation is mentioned in Section 9.4.6. The use of such a device gives the microcomputer the ability to alter the speed of an induction motor by writing information to two registers. One register sets the frequency of the pulses and hence the speed of the motor, while the other register sets the mark-to-space ratio, thereby defining the output voltage. The use of such a device will leave further time for the microcomputer to execute control algorithms, monitor various signals, and communicate with the user.

Rather than using programmable timers/counters, Example 9.12 performs the alteration of both the frequency and the mark-to-space ratio by incorporating external hardware in the form of two digital comparators and a counter. All the microcomputer needs to do is write to two output ports identifying the required 'mark' and 'period'.

15.3.3 The pulse-width-modulation approach

When designing a motor controller for a 3-phase induction motor, the frequency-controlled approach is often used. The solution of the last example gives the inverter the task of changing the frequency and leaves the setting of the required voltage to be performed by the rectifier or by a d.c. chopper. Rather than introducing additional measures for changing the voltage, means can be introduced to allow the inverter to perform the task of changing the voltage as well as the frequency.

There are several methods of satisfying this requirement. One of the most widely used is pulsewidth modulation (PWM), which is based on altering the width of the pulses delivered to the motor in order to meet two requirements:

1 To change the mark-to-space ratio of the pulses sequentially in order to generate a sine wave plus a minimum number of harmonics at the input terminals of the motor. This requirement sets the relationship between one pulse and another.
2 To alter the mark-to-space ratio of the pulses in order to make the generated sine wave meet the voltage set point. This requirement can be met by scaling the mark of all the pulses, but keeping their relative relationship constant. In other words, each mark of a pulse is multiplied by a scaling factor which defines the required output voltage.

Most of the commercially available speed and direction controllers of induction motors use the PWM approach. In such controllers, the number of pulses per cycle is optimized in order to achieve a balance between noise, losses and the harmonic contents in the output waveform. A higher number of pulses gives a better representation, but leads to higher switching losses when the fundamental frequency is high. The effect of the harmonics, however, becomes less noticeable as the fundamental frequency is increased. This means that the number of PWM pulses per cycle of the fundamental frequency can be decreased when operating at a high fundamental frequency and increased when the frequency is reduced. The utilization of a microcomputer to perform such a task simplifies the design. This is because the generation of pulses can be performed by the use of interrupts. Changing the number of pulses per cycle means altering the rate of occurrence of the interrupt. If the required mark of each pulse is read from a table, then the alteration of the number of interrupts should be accompanied by reading the value of each mark from an alternative table.

There are several methods of producing PWM pulses and interfacing them to the inverter by using a suitable drive circuit. A software-based method requires the involvement of the microcomputer with the generation of each pulse. Such a method requires off-line calculations to produce tables containing the required switching instances (or pulsewidths). A signal of a regular frequency needs to be employed to interrupt the microcomputer. The interrupt-service routine increments a pointer, reads a value from a table, processes it and loads it into a timer or a counter. The amount of processing that the microcomputer does on the value depends on the adopted algorithm. Internal timers and/or external hardware is utilized to assist in achieving the goals efficiently. A circuit such as that of Example 9.12 can be used where signal X2 can be utilized to interrupt the microcomputer, which will deliver a digital value corresponding to the next mark. If the operation mode was open-loop, then this value can be delivered to the output port immediately after being read from the table. On the other hand, a closed-loop mode requires reading a reference from a table, comparing it to a feedback signal and altering the digital number delivered to the output port accordingly. This alters the mark-to-space ratio of the pulses, which will in turn alter the instantaneous amplitude of the sine wave in order to reduce the difference between the reference and the generated waveform.

An alternative method is to generate the switching instances by hardware means. A widely used approach is based on comparing two types of waveforms, a sine wave at the required fundamental frequency, and a triangular wave at the required PWM frequency. The latter is referred to as the carrier wave and is normally at a much higher frequency than the fundamental. When using the hardware approach, one needs to decide on the degree of control given to the microcomputer.

Example 15.10: **How can a microcomputer be used to control the speed of a 3-phase induction motor from a 3-phase inverter using the PWM approach?**
A 3-phase inverter, such as that presented in Example 15.8 and shown in Fig. 15.14, may be utilized to satisfy the requirement. One can use the software-based method and implement it by modifying the solution of Example 9.12 in the manner explained above. The requirement can also be met by using an analogue method based on the use of sine and triangular waves. Figure 15.17 represents the circuit required to satisfy the requirement. The microcomputer generates three sine waves shifted by 120°. They are compared by three analogue comparators to a high-frequency triangular wave (the carrier wave). When the amplitude of the sine wave exceeds that of the triangular wave, the output line of the comparator changes to a high

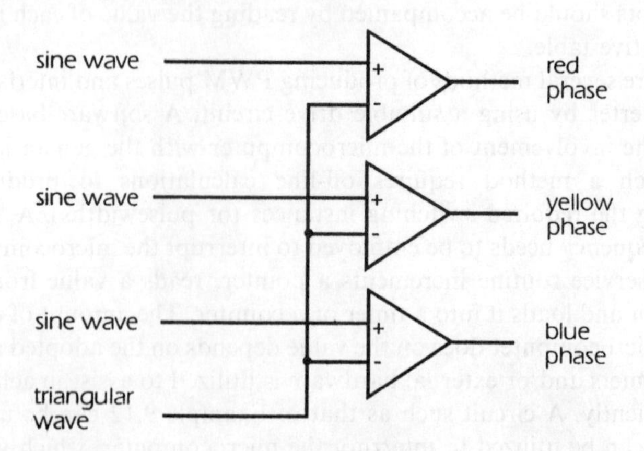

Fig. 15.17 Using the sine/triangular method.

state. Otherwise, it is at a low state. This represents the mark and space of the PWM pulses of one phase and determines the switching times of the two switching devices (power transistors in the case of the circuit of Fig. 15.14) within one leg of the inverter. Clearly, a dead time is required between switching the power devices within each leg of the inverter (Section 10.6). The dead time can be reduced by replacing the power transistors with faster switching devices, e.g. MOSFETs or IGBTs.

As explained above, generating the sine waves by the microcomputer requires a regular interrupt to instruct the microcomputer to read a value from a table. The length of the table is limited by the number of times the microcomputer is allowed to be interrupted within one cycle of the fundamental. For instance, the duration of each cycle of a 50-Hz fundamental is 20 ms. If the requirement is to have 180 PWM steps within a single cycle, then an interrupt is required once every 111 μs.

Let us first assume that a single sine wave is required to be produced from a single table using open-loop control. This implies that the interrupt-service routine will be as simple as transferring a value from a table to an output port before incrementing a pointer. The routine also needs to check whether or not the pointer has reached the end of the table. If it has, then the microcomputer needs to move the pointer to the first parameter of the table.

Generating the three sine waves can still be accomplished by the use of a single table. Three pointers are required and are placed 120° apart, i.e. one-third of the table apart. The interrupt-service routine will then need to read and send three values, increment three pointers, and adjust any pointer which has reached the end of the table. This has tripled the amount of work needed within the interrupt and will set a limit to the maximum allowable sampling frequency.

The design is not yet complete, since the pulsewidths should be adjusted in order to produce the required output voltage. This can be achieved by scaling each value read from the table. A multiplication factor needs to be set to a default value corresponding to an output voltage. This value can then be altered according to a command from the operator, or in order to minimize the difference between a demand and a feedback signal in a closed-loop control system.

One has to keep in mind that real-time control systems normally use integer calculations in order to reduce the execution time (floating point calculations take a much longer time and require the use of libraries which reserve a large area of memory). Representing a multiplication factor by integer calculations, however, can limit the resolution. A better approach is to use a multiplication-and-division factor. This means that the value read from the table can be multiplied by a variable before being divided by a fixed number, e.g. 100. This variable represents the required output voltage. This approach helps to achieve finer control. For instance, giving the multiplication variable a value of 100 corresponds to a scaling factor of 1

(100%), while giving that variable a value of 3 corresponds to a scaling factor of 3%. When performing integer calculations, one should also keep in mind that performing the multiplication before the division leads to a more accurate result. For example, the result obtained from performing integer calculations on (5 × 5)/3 is closer to the exact value than performing (5/3) × 5.

Example 15.11: **A test circuit is required in a laboratory to operate a PWM inverter at a fixed voltage and frequency. The frequency is set by a microcomputer through a series of clock pulses of a fixed frequency delivered through a single output line. Design the circuit assuming that there is no excess to any of the microcomputer's other interface lines.**
A fixed voltage and frequency mean a single table containing the required pulsewidths. Parameters within the table will be read sequentially and repeated continuously. The table can be stored in an EPROM whose address lines are driven from an up-counter in a similar way to that explained in Example 12.3. The counter is clocked by the fixed-frequency pulses supplied by the microcomputer. The contents of the table are stored in sequential locations within the EPROM.

If the contents of the table correspond to the required mark of each pulse, then a D/A converter is required to convert that number to an analogue signal to be compared with a triangular wave in a similar manner to that of the previous example. This, however, requires several components. A simpler and a lower cost solution can be reached if the contents of the table are modified.

The EPROM possesses eight data lines. Six of these lines can be utilized as the output lines of this simple interface circuit. Each line supplies the on/off states of a single switching device of the inverter. If the output of the inverter is to be a square wave, or a quasi-square wave, then the table within the EPROM will consist of a few states. Generating a sine wave is more difficult, since the mark-to-space ratio of each pulse is different from the others. Can you think of a way of arranging the table?

The answer is based on dividing the table into equal sections, each corresponding to a single PWM pulse. Let us assume that we want 90 pulses of PWM per cycle of the fundamental. Let us divide each pulse into 100 steps (remember that each output pulse is represented by a single bit within the EPROM). This means that in order to supply switching device Q1 with an equal mark and space, 50 successive locations should contain a 1 in their lowest bit followed by 50 locations containing 0 in their lowest bit. In other words, the number of successive 1's in the table corresponds to the percentage of the mark of each of the successive pulses.

This arrangement increases the number of bytes reserved by a single table. The arrangement under consideration reserves 9000 bytes (90 pulses × 100 steps per pulse). An EPROM such as the 27256 has a 32-kilobyte storage capacity and can (if required) include three such tables. Alternatively, the extra space can be used to increase the resolution or the number of pulses per cycle of the fundamental. There are, however, limitations to these values, e.g. the switching capability of the power devices and their switching losses as well as the upper frequency that can be supplied by the microcomputer. The latter can be calculated as ($9000f_s$), where f_s is the fundamental frequency to be generated by the inverter and the 9000 is the number of locations per cycle that we selected. This implies that in order to generate a 50-Hz output signal, the clock should be 450 kHz.

Chapter 16
Miscellaneous applications

16.1 Introduction

This chapter starts by giving alternative solutions to interfacing keyboards to microcomputers. It proceeds to illustrate methods of exchanging information between microcomputers and add-ons. Section 16.4 gives a brief description of the use of microcomputers as testing devices and considers a few interfacing applications. Any interface between a microcomputer and an add-on may need to be modified at a later date to suit changes in requirements. This may not be possible to achieve. Even if it was possible, it may not be easy to implement and requires a careful consideration in order to satisfy the requirements efficiently. This is described in Section 16.5. Limitations may also be imposed on interfacing applications in respect of the power requirement. Section 16.6 explains that the solution depends on the application and may be satisfied by the internal power supply of the microcomputer, or by the introduction of an external power supply.

Noise glitches can be generated from various sources and affect microcomputer-controlled systems. They may lead to failures due to hardware malfunctions or software upsets. They can, for example, cause a program jump to a random location in memory leading to unpredictable results. The design can be optimized to minimize the noise effects without considerable increase in cost and without sacrificing considerable computation time. The designer needs to know the expected noise level in the system and then decide on the most economical level of protection. The last section of this chapter explains causes of noise glitches and presents hardware and software solutions to minimize noise effects on the operation of the microcomputer.

16.2 Interfacing keyboards

Keyboards and keypads consist of a group of switches which can be interfaced to a microcomputer through input lines. Figure 4.1 shows how individual keys can be interfaced to a microcomputer together with methods of debouncing. In many applications, however, a large number of keys is required to be interfaced to the microcomputer. For instance, allowing an operator to enter decimal numbers requires 10 numerical keys. The control pad is not likely to be complete without a few additional function keys. A 12-key or a 16-key arrangement is widely used. The 16-key arrangement can also be configured as a hexadecimal keyboard, as shown in Fig. 16.1. Rather than reserving 16 input lines to interface a 16-switch keyboard to a microcomputer, the switches of Fig. 16.1 are arranged in a matrix form. The arrangement only uses eight lines and its operation is described in the following example.

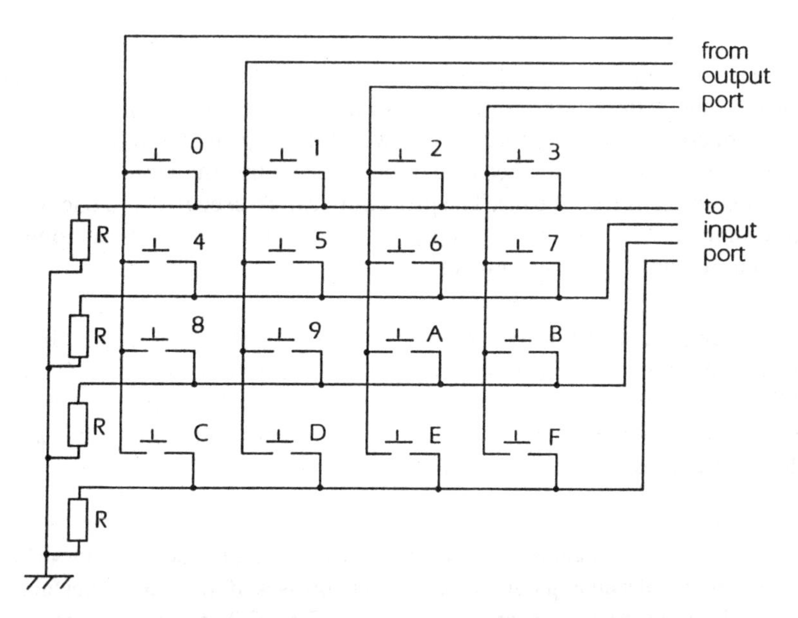

Fig. 16.1 Interfacing a keyboard to a microcomputer.

Example 16.1: **How can a microcomputer operate the 16-switch keyboard shown in Fig. 16.1?**

The 16 switches are interfaced to the microcomputer through four digital-input lines and four digital-output lines. The method of operation is based on enabling each of the output lines in sequence and reading the input lines to identify the state of the switches. This is shown in the following procedure, where 'y' is a byte and '&' is a software AND.

```
1  SET OUTPUT LINE Q0 HIGH
2  READ INPUT PORT AND STORE IN 'y'
3  y = y & 00001111
4  IF (y > 0), THEN
        IF        y = 1, THEN SWITCH 0 IS PRESSED
        ELSE IF   y = 2, THEN SWITCH 4 IS PRESSED
        ELSE IF   y = 4, THEN SWITCH 8 IS PRESSED
        ELSE IF   y = 8, THEN SWITCH C IS PRESSED
```

Step 1 in the procedure sets one line by clearing the lowest four bits of the output port (using AND 11110000), and then ORing it with a number corresponding to that line. Line Q0 of the port is set by ORing with 00000001, as shown in the previous chapters. By using lines Q0 to Q3 of the input port, the procedure extracts and tests the four least significant bits in steps 3 and 4.

By setting a single output line, the procedure can only test 4 of the 16 switches. Software will therefore need to be written to repeat the procedure for each of the four output lines, thereby testing the state of the 16 switches. Clearly, the process takes many steps and can consume a large percentage of the microcomputer's processing power if it is repeated at a high rate. Key presses, however, do not occur at a high speed and the keys can be tested, say once every 100 or 200 ms. The procedure can be executed as a background task, since it does not need to occur at precise time intervals, as long as it is done within a specified interval.

The arrangement of Fig. 16.1 reduces the number of interface lines. In many microcomputer-controlled applications, input/output lines are used for different functions, leaving a few lines available for the interface of switches. Rather than introducing additional ports, designers often find ways around the problem, as shown in the following example.

Example 16.2: **An application requires the use of four push-to-make switches. How can a microcomputer detect switch presses if the only interface line available is an input to a unipolar A/D converter? The voltage range of the A/D converter is 0 to +5 V.**

The four switches need to be interfaced to the analogue-input line in such a way that the closure of each of them supplies a unique voltage level. The arrangement of Fig. 16.2 can be used where a series of resistors is incorporated to provide different voltage levels. Let us assume that all the resistors are of equal value.

The resistors form a voltage divider whose divide ratio depends on which of the switches is closed, if any. When all the switches are open, the

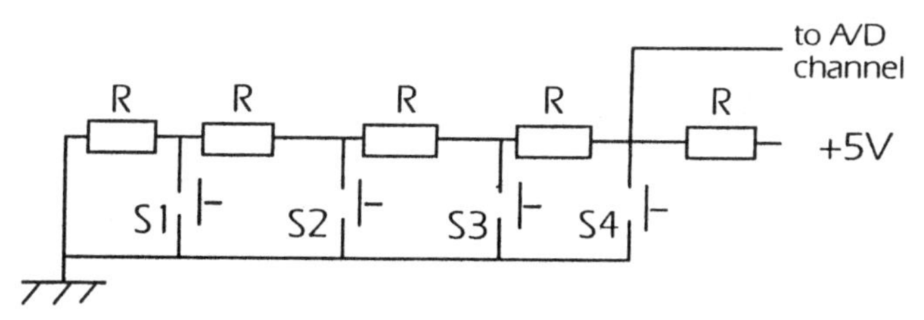

Fig. 16.2 Interfacing switches through an analogue-input line.

microcomputer reads a voltage of:

$$5\,V \times (4R/5R) = 4\,V$$

If switch 1 is closed, then the microcomputer reads:

$$5\,V \times (3R/4R) = 3.75\,V$$

Similarly, closing switches 2, 3 and 4 gives 3.3, 2.5 and 0 V respectively. The microcomputer reads the A/D converter, compares the reading with pre-programmed levels, and decides which of the switches is closed.

Using resistors of equal values can reduce the assembly time of the product and can reduce space. This is because a resistor network can be employed where the resistors are included in a DIL IC, or in a single-in-line package. The drawback of using resistors of equal values in the circuit of Fig. 16.2, however, is the non-linearity of the voltages supplied by the closure of different switches. This is not a problem if the number of switches is low. It may lead to confusion if a high number of switches is required to be interfaced in this way, especially for operation in a noisy environment. With a simple calculation, one can alter the value of the resistors to have almost equal voltage steps between the switches.

As an exercise, select resistor values to provide equal voltage steps. What changes are needed if the number of switches is increased to 6?

16.3 Alternative solutions to a data-transfer problem

There exists a few methods of transferring information between a microcomputer and add-ons. Such a transfer is normally accomplished through the data bus by using 3-state buffers. This section considers four methods.

16.3.1 The program-controlled approach

This approach is used in solving many of the examples presented in the previous chapters, e.g. to transfer a single value from an A/D converter to a microcomputer. Let us consider its suitability for the transfer of blocks of data. It is based on dividing the processing time of a microcomputer to execute tasks in sequence, where the microcomputer decides on the instant of attending to each task. In the case of executing a task involving the transfer of information from the microcomputer to an add-on, the former fetches the information from memory, transfers it to the add-on and increments a pointer to point to the next item in the list. Similarly, when information is required to be transferred from an add-on to a microcomputer, the latter reads the information, stores it in memory and increments a pointer. Although these steps take a very small portion of the execution time, their use can, however, add up to a long time when the transferred information is in the form of a very long message. Performing the transfer by the program-controlled approach is therefore best suited for short messages.

16.3.2 The interrupt-controlled approach

With this approach, the processing of tasks is accomplished as a result of responding to a demand either from an external source requesting to be serviced, or from a timing device asking the microcomputer to communicate with an add-on. Like the program-controlled approach, this approach has been used by many of the examples presented in the previous chapters. It removes the need for executing software timing loops and performs the transfer of data in single values. It is therefore best suited for the transfer of short messages.

16.3.3 Direct Memory Access (DMA) approach

The approaches presented above are widely used, simple to apply, and best suited for the transfer of short messages. There are, however, applications where minimum time needs to be spent on the transfer of long blocks of data. Fast transfer of information from a specified source (e.g. memory or an input device) to a specified destination (e.g. memory or an output device) can be achieved by the use of the DMA (Direct Memory Access) approach. The transfer takes place through the microcomputer bus, but without passing through the microprocessor. During the transfer, the microprocessor will not use the bus, as it will transfer the control of the bus to a new master. This type of information transfer is used by some microcomputers and is achieved by a DMA controller which, in many microcomputers, is external to the microprocessor. However, in certain microprocessors, such as the high-integration Intel 80186, DMA channels are included within the microprocessor IC. When using DMA, there is a need to specify the length of a block of data as well the location of its source and destination. Due to the associated complexity and higher cost, DMA is mainly used for situations requiring fast transfer of a large block of data to, or from, memory where the program-controlled and the interrupt-controlled approaches are too slow.

A processor can instruct a DMA controller to begin a transfer of data from a given address and for a specified length. When an add-on has data ready for DMA transfer, it supplies a DMA request to the processor, which will then reply by sending a signal acknowledging the request and informing its originator that the bus is available for DMA transfer. The transfer then takes place for a block of data where the address is incremented and the data count is decremented after each transfer. This continues until the transmission of all the required data (the data count becomes equal to 0) is completed. The device then informs the processor that the transfer is accomplished and releases the bus.

In the Z80 microprocessor, the bus-request signal is supplied to the BUSRQ (BUS ReQuest) line of the processor, which in turn acknowledges the request by activating the BUSAK (BUS AcKnowledge) line. Both of these lines are active-low. The 68000 microprocessor uses three active-low lines. It receives the bus-request signal through a line named the BR (Bus Request) line. It acknowledges the reception and indicates that it is releasing the bus by activating the BG (Bus Grant) line. The originator of the request then activates a third line called the BGACK (Bus Grant ACKnowledgment) line to indicate that it is the new master of the bus. When the 68000 is not using the bus, both the AS (Address Strobe) and the DTACK (Data Transfer ACKnowledge) lines are not active, i.e. at a logic 1 state (both the AS and the DTACK lines are used in Example 2.3). When no master is in control of the bus, the BGACK line is at a logic 1 state.

Many microcomputers use a DMA controller and offer DMA request and DMA acknowledge lines within their expansion connector. In addition to such lines, the IBM XT and compatible microcomputers provide a signal called the AEN (Address ENable) signal. When at a logic 1, it indicates that a DMA is taking place. When at a logic 0, it indicates normal operation, i.e. the bus is under the control of the processor. This signal is used by interfacing circuits (see Example 3.5).

16.3.4 The handshaking approach

This approach differs from the ones considered in Sections 16.3.1 and 16.3.2 in that, before commencing transmission, the sender checks that the destination is ready to receive. It can be implemented by using digital lines to transfer status flags which permit the sender to request transmission, and allow the destination to indicate its readiness to receive information. This is known as *handshaking*. The advantage of this approach is that it enables a device to transmit data only when the destination is ready to receive them.

It therefore minimizes the possibility of losing information. However, in addition to its use of additional digital lines, this approach requires additional processing for setting and examining the flags before establishing connection. If not correctly organized, this approach could lead to unnecessary delays in either device.

Consider an application where data need to be transmitted from a microcomputer to an add-on containing an input buffer into which digital values need to be loaded from the microcomputer in sequence. When the microcomputer is ready to send data, it activates the RTS (Ready To Send) line, to inform the add-on of its readiness to transmit. When the add-on is ready to accept data, activates the RTR (Ready To Receive) line. Another line is also needed to allow the microcomputer to inform the add-on of the availability of data at its input, the data-available (DA) line. A fourth line can also be used, the acknowledgement (ACK) line. It is activated by the add-on to inform the microcomputer of the successful reception of a data value. The connection between the microcomputer and the device is shown in Fig. 16.3(a).

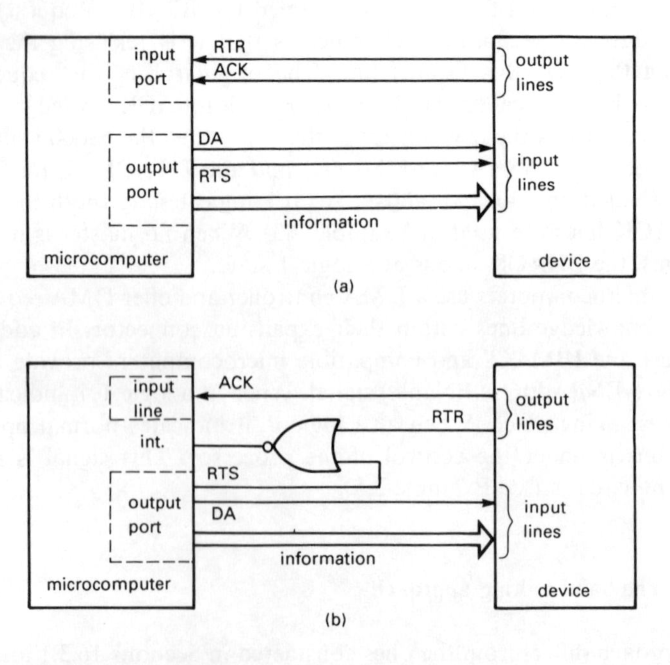

Fig. 16.3 Using the handshaking approach: (a) using the program-controlled method; (b) using the interrupt-controlled method.

The activation of the RTS and the RTR lines can take place at any instant of time. The transmission of data, however, can only take place when both the sender and the receiver are ready. When this is the case, the microcomputer sends the data, one value at a time. With each transmission, the microcomputer pulses the DA line. The add-on detects the change of state of that line, reads the data and pulses the ACK line to inform the microcomputer of the success of the communication step. The latter can then send the next byte. The designer needs to decide on what to do if the destination has not pulsed the ACK line. He or she can program the microcomputer to wait for the change to occur within a specified time interval. If such a change is not detected, then the microcomputer can repeat the transmission once more and retests the line. If the fault persists, then the transmission can be abandoned, a flag set, and the operator informed.

The transmission normally takes place through the data bus, which sets a limit to the number of bits representing each of the transmitted values. If the information is represented by a group of bytes, then they can be transmitted as a series of single bytes in an 8-bit microcomputer, and as a series of 2-byte words in a microcomputer with a 16-bit data bus. The latter type of transmission is not always adopted, as the receiver device may only be capable of receiving bytes.

Let us now consider an application where an add-on, which uses the handshaking approach, is replaced by another which does not operate on the handshaking principle. Instead, the new add-on is continuously ready to receive information. In this case, a decision is needed on what changes should take place to operate the new add-on. One solution is to alter the software so that it will ignore the handshaking part. This, however, is not always possible. An easy way to solve the problem is to feed back the microcomputer-generated RTS signal to the digital-input line of the microcomputer at which the RTR signal is expected. Similarly, the DA signal is fed back to the microcomputer to act as the ACK signal from the add-on. In this way, the software does not need to be altered and the microcomputer achieves the transmission without delay.

Let us consider two ways of implementing the handshaking approach in transferring data from the microcomputer to an add-on.

(1) Program-controlled handshaking

With this approach, the microcomputer requests to transmit by setting the RTS line and will not commence transmission until it detects an active RTR signal. The implementation depends on whether the microcomputer should wait for the RTR line to be activated, or carry on with other activities and monitor the state of the RTR line in between attending to other tasks. The second approach is usually preferred, as the first approach can prove costly in terms of processing time. The instant of commencing the transfer is therefore decided by the microcomputer and not by the change of state of the RTR line. The transmission, however, can only take place when both the microcomputer and the add-on are ready.

When considering the use of this method, one should take into consideration the processing time spent on monitoring the RTR line. If correctly applied, the implementation of this method takes a small proportion of the processing time, particularly in applications where the transmission is not repeated at a high rate.

(2) Interrupt-controlled handshaking

In order to reduce the time spent on executing the data-transfer task, testing the RTR signal can be replaced by the use of the interrupt-controlled input method. In this case, the RTR signal is employed to interrupt the microcomputer whenever the add-on is ready to receive data. A problem arises when an interrupt is generated and the microcomputer is not ready to send. One way of avoiding this problem is to design the add-on so that it will not activate the RTR signal until the microcomputer activates its RTR line. This, however, is not always possible, especially when using a ready-made add-on. An alternative solution can therefore be used. It is based on employing a NOR gate as shown in Fig. 16.3(b). It allows the RTR signal to interrupt the microcomputer only when the latter is ready to transfer data (when the RTS line is active). The configuration assumes that the RTR and the RTS signals are active-low, and operates as follows.

When the microcomputer wants to transfer data, it changes the state of the RTS line to a logic 0. If, at that instant, the external device is ready to receive, then the RTR signal is also at a logic 0 state. This results in producing a logic 1 signal at the output of the NOR gate which will in turn activate the interrupt and start the execution of the interrupt-service routine. This routine will reset the RTS line (to a logic 1) and transmit the data to the add-on (in the same manner as that of the program-controlled approach) before returning to the main program. Note that the configuration permits the microcomputer to be interrupted ony when both the add-on and the microcomputer are ready to communicate. This implies that the interrupt has almost the same effect as the request line in the program-controlled approach; the difference being that the microcomputer will not spend time monitoring the RTR line.

16.4 Using microcomputers as testing devices

Microcomputers represent a cost-effective way of designing test and measurement systems. They possess intelligent decision-making capability and high processing power. Many testing applications use a microcomputer to continuously monitor one or more real-time signals. If the signal is analogue, then an A/D converter is utilized to convert samples of the signal to a digital form. In other testing applications, a microcomputer is employed to supply signals to a circuit or a system and examine its behaviour by monitoring its output signals. Such testers can be produced by providing a microcomputer with digital and/or analogue input and output lines as shown in the previous chapters. Some testers need to provide high-power switching by the use of relays or transistors. The interface of these devices to a microcomputer is given in Chapter 10. Although a standard microcomputer is suitable for many testing applications, certain requirements (such as very high sampling rate) are sometimes imposed and cannot be met economically without introducing additional hardware. An example of fulfilling such a requirement is given in Section 3.5.2.

When designing computerized testing equipment, it is important to decide on whether the tester is to be tailored for a single application, or to include means to permit its use in a variety of applications. The decision is needed because the required resolution, speed of conversion, sampling rate, memory area, etc. vary from one application to another. After defining the exact object of the tester and selecting a microcomputer, a designer can introduce suitable input and output devices and coordinate the implementation of tasks. These can be arranged to produce the optimum solution which will satisfy the test and financial requirements efficiently. Certain tests impose restrictions; for instance, tests which require the collection of a large number of samples may introduce the requirement for additional memory area. A decision is needed on whether to expand the memory, or to introduce techniques such as those presented in Chapter 17.

During a test or a data-collection process, there may exist intervals where the microcomputer is waiting for results to be produced, e.g. waiting for an A/D device to complete its conversion. Such intervals can be used for carrying out useful tasks such as error detection. This is very useful, as detection of an error can be employed to automatically repeat a process, alert the operator, stop an operation, or shut down the system as soon as the reception of an unacceptable result is confirmed. After the successful completion of a data-collection process, the results can be presented to the operator. Some results, such as those produced by an A/D converter or a counter, are in a binary form, which may give the operator difficulty in relating them to the actual magnitude of the measured signal. The computation capability of the microcomputer can therefore be employed to present these numbers as values in the required units (e.g. degrees, centimetres, etc.). Such a presentation can, if required, take place off-line after completing the data-collection process.

16.4.1 A simple tester

An example of a simple low-cost manual tester is shown in Fig. 16.4. It allows a test engineer to place a probe at a single test point and examine each of four input analogue channels. The signal selection is achieved

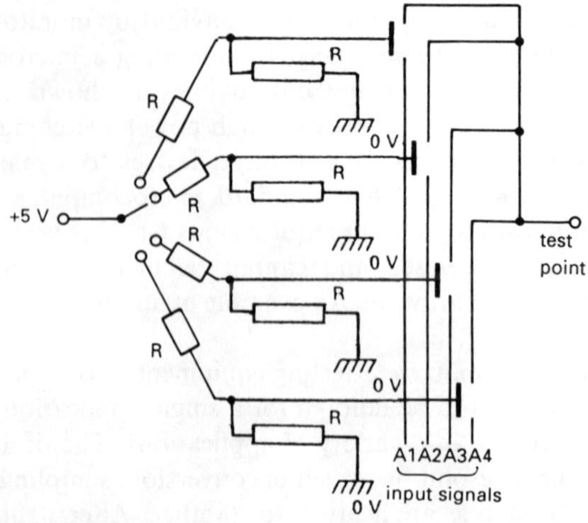

Fig. 16.4 Selecting signals by a rotary switch.

manually by using a rotary switch and four analogue switches. The enable line of each analogue switch is connected to the 0-V rail through a resistor to keep the contacts of the analogue switch open when not activated by the rotary switch. The latter supplies +5 V to the enable line of the selected channel. This results in closing the analogue switch and connecting the input signal to the test point. A test engineer places a probe at the testing point and examines each of the manually selected channels.

16.4.2 Employing a microcomputer

A simple way of utilizing a microcomputer is to use it as a data-collection device by interfacing it to the test point through an A/D converter. During operation, the operator selects the channel manually and instructs the microcomputer to commence the measurement and to save the results in a specified area in memory. This area can be divided into four regions, each containing the result of testing one of the four channels. With the aid of software packages, the results can be presented in an easy-to-examine form, e.g. as a graph.

16.4.3 Further modification

A microcomputer employed only as a data-collection device does not possess the ability to detect which of the input channels they correspond to. The tester can therefore be modified to allow a microcomputer to perform the channel selection as well as the testing. This can be accomplished in several ways. One way is to utilize the same hardware as the manual tester and connecting four digital-output lines from the microcomputer to the enable lines of the analogue switches. Alternatively, the analogue switches can be replaced by the use of a 4-channel multiplexer. If the tester needs to be provided with the ability to examine the channels simultaneously, then four individual A/D converters can be used.

16.4.4 Responding to commands

When performing the channel selection by a microcomputer, a test engineer needs to be provided with the ability to specify to the microcomputer the channels to be tested, the duration of each test, etc. This can be achieved by the use of the keyboard or external switches. During operation, the microcomputer examines the command-input source and performs the test accordingly. If required, the microcomputer can be programmed to display the selected channel number. This prevents misrepresentation, as the operator will only start the test when he or she is satisfied that the microcomputer has been provided with the required information.

16.4.5 Automatic testing

Rather than performing the selection by a command from an operator, the microcomputer can be programmed to test the four channels sequentially. The program needs to specify the required length of data, or the duration of each test. This approach is attractive to many applications, particularly where the test is repetitive and boring. The use of the microcomputer as the controller of the test reduces the intervention of the operator, whose function becomes commanding the start of the automatic-testing program and reading the results when the test is complete.

16.4.6 An input-output tester

Let us now consider a different method of testing where the microcomputer is employed to provide a circuit with a demand signal and test its output. With this approach, the test can take place without intervention from the operator. The test can be repeated and the results can be saved as a series of values in RAM, or even on a disk for future retrieval and examination. This approach simplifies the task of obtaining a hard copy of the results, as the contents of memory can be easily supplied to a printer or a plotter. The procedure depends on the type of signals and can take the form of repeatedly sending a value (digital or analogue) to the input of the circuit under test and reading its output.

16.4.7 User interface

In many applications where a microcomputer is employed as a tester, or as a controller, several programs can be stored in its memory, or on a disk. It is always attractive to adopt a user-friendly approach for the selection and execution of one or more of these programs. An on-screen graphical menu-based user interface is widely used to provide fast and simple access to programs. However, economic reasons may not permit the use of a suitable display, or the software overhead for generating such menus may be considered unacceptable. Furthermore, the associated software takes memory space which may be needed for implementing other tasks. In many interrupt-driven systems, the processing time needed to implement the menu structure is considered not significant, as it can be run as a background task.

Instead of using menus, full-size keyboards or monitors, the exchange of information in many microcomputer-based equipment is performed by other means which vary in cost, complexity and user-friendliness. Example 11.2 explains how a 7-segment display can be interfaced to a microcomputer. Example 6.2 illustrates supplying information to the microcomputer through external switches.

16.5 Modifying an existing interface

An interface designed for a particular application may not be suitable for another. Consequently, many designers plan their interfaces with future alterations in mind. For example, Section 3.7 gives an example of providing an interface with some degree of flexibility. In many situations, changing the specifications produces the need for designing a new interface. This, however, is not always the case, as a few modifications or some additional circuitry can be introduced to tailor an interface to meet the new specifications.

The amount of work involved in introducing the modification depends on many factors. If, for example, a single-channel A/D converter is interfaced to a microcomputer and a multichannel converter is needed, then a multiplexer can be introduced to allow the time-sharing of the A/D converter. Digital-output lines are obviously required to allow the microcomputer to select the channel of interest. If such lines are not available, then an output port will need to be introduced (as shown in Chapter 2). Furthermore, a device-selection signal is required to allow the microcomputer to address that port. If such a signal is not available, then it needs to be derived as shown in Chapters 2 and 3. Before deriving that signal, one needs to examine the memory map of the microcomputer to select a suitable unused address. Clearly, the additional circuitry can only be introduced if physical space is available. In many situations, this is not the case and one is forced to design the interface board even when the modification is simple.

Consider a situation where the hardware interrupt-request lines of a microcomputer are employed by an existing interface. The operation needs to be modified by the use of an additional interrupt. Example 1.3 illustrates a method of meeting the requirements easily. It is based on multiplexing an interrupt line and allowing the interrupt-service routine to detect the source of the interrupt.

Example 16.3: **Consider an application where the microcomputer of Fig. 6.1(a) is utilized to control an industrial process. It receives information through six digital lines and supplies commands through five digital-output lines. How can the system be expanded to receive information from two 8-bit A/D converters, supply an analogue signal, and deliver active-low pulses to command the start and stop of two independent external devices? The interface has to be implemented through the four ports of the microcomputer.**

In this case, an interface already exists and reserves some of the lines. The new requirement involves interfacing two individual A/D converters and a single-channel D/A converter as well as delivering four digital-output signals without affecting the existing interface. This implies that the expansion needs to be implemented by the use of 10 digital-input lines and 11 digital-output lines. Let us assume that each A/D converter accepts an active-low SC signal and generates an active-high EOC signal.

- *Input channels*
  Since we only have 10 input lines, the digital-output lines of the two A/D converters can be multiplexed and supplied to input port 3. The multiplexing is implemented by the use of two 3-state buffer ICs (74LS244) as shown in Fig. 16.5. This reserves eight out of the 10 available input lines, leaving two lines to use in monitoring the state of the two EOC lines.
- *Output channels*
  Assume that the requirement of the analogue output can be satisfied by the use of an 8-bit D/A converter. It can be supplied from output port 2. Since eight lines are reserved for operating the D/A converter, we have three output lines to supply the following: four lines to command the start and stop of the external devices, two SC signals to the A/Ds, and two lines to activate each of the 74LS244 ICs.
- *Satisfying the output requirement*
  We need to produce eight individual signals from three digital outputs of the microcomputer. This can be accomplished by employing a 74LS138

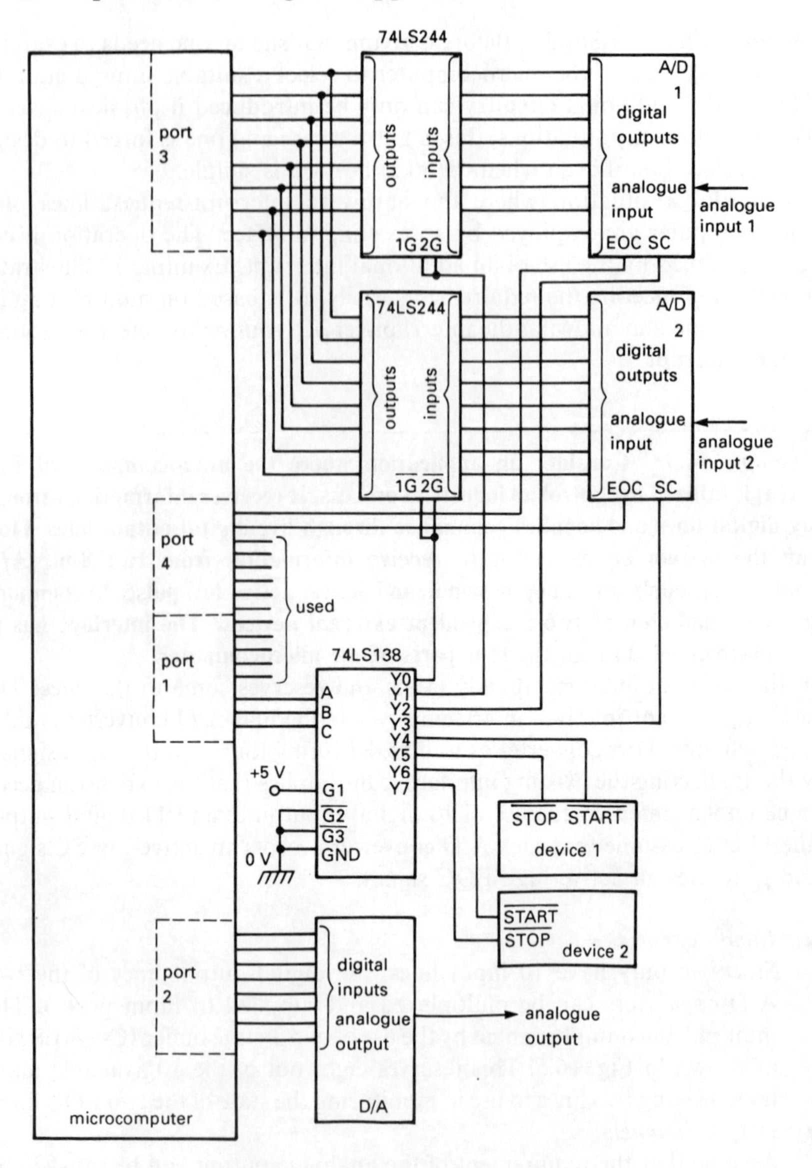

Fig. 16.5 Modifying an existing interface.

decoder IC (described in Section 2.2.5) as shown in Fig. 16.5. The decoder is connected to be always enabled (lines G1 = 1, GS = 0 and G3 = 0).

- *Operation*
  The microcomputer controls the operation of the interface circuit through the decoder. An output line of the decoder will be activated (logic 0) as long as its address is supplied to the three input lines of the decoder.

The procedure of supplying an address is in the form of performing a software AND followed by a software OR (as in Example 6.1).

To digitize channel 1, the microcomputer supplies the decoder with a value 02 followed by 01. This delivers a SC pulse to the corresponding A/D converter and activates the associated 74LS244 buffer IC. Assuming that the conversion time is short, the microcomputer can be programmed to monitor the state of the EOC line until it changes state. The microcomputer obtains the digital result by reading the state of port 3. Digitizing analogue input 2 is performed in the same way through outputs 3 and 0 of the decoder.

To update the analogue-output channel, the microcomputer delivers the value from output port 2 to the D/A converter without the need for a command from the decoder. Delivering the start and the stop pulses is performed through outputs 4, 5, 6 and 7 of the decoder. Therefore the ulitization of the decoder satisfies the requirement.

Exercise

As an exercise, write a procedure which begins by commanding the start of the two external devices. It continues to permit the microcomputer to read the two analogue-input channels and the six digital-input lines of port 4 in sequence. Following this, the microcomputer updates the analogue output before updating the five digital-output lines. Can this procedure be arranged differently to save execution time?

There are therefore many factors which dictate whether an interface circuit can be modified to meet new requirements or not. An important factor is the experience of the designer. In many situations, the redesign of a circuit can be avoided if the designer has carefully thought of how the existing circuit can be modified.

16.6 Additional power supply

In a large number of applications, the power needed to drive add-on circuits can be supplied by the internal power supply of the microcomputer. Clearly, the available power is limited and is stated in the manual of the microcomputer. If the available power is not sufficient, or if different voltage levels are required, then one can either design a power supply or purchase a ready-made power supply.

16.6.1 Important reminders

1 Care has to be taken when using the internal power supply, or connecting it to an external power supply, since wrong connections could damage the internal parts of the microcomputer and/or the add-on circuit.
2 In many applications, it is recommended to isolate the input and output lines of the microcomputer, e.g. by the use of optocouplers. In this way, the power supply within the microcomputer is isolated from the additional power supply and the interface circuit.
3 Reference points (normally the 0 V) of all power supplies need to be joined (unless the circuits are to be isolated from each other).

Figure 16.6 shows two linear power supplies; one generates a positive voltage, while the other produces two voltages, one positive and one negative. Each power supply consists of a stepdown transformer, a full-wave rectifier, a smoothing capacitor and a voltage regulator.

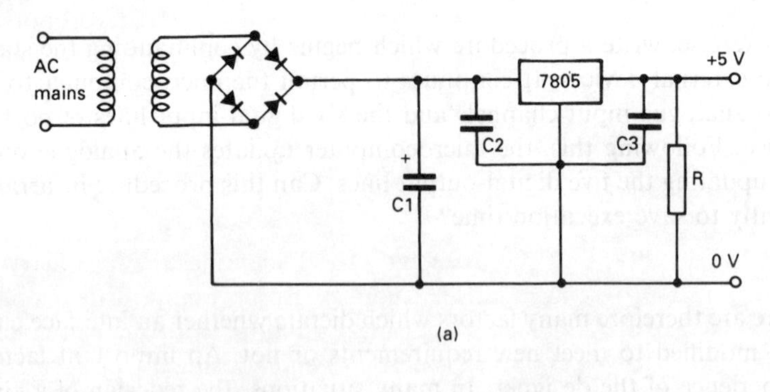

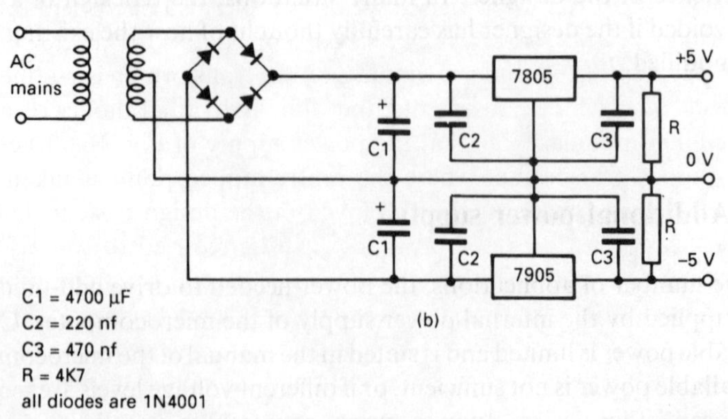

Fig. 16.6 Power supplied: (a) +5 V; (b) +5 V/−5 V.

16.6.2 Voltage regulator

This is the part that generates the voltage seen by the load and should be selected according to the load requirements. There are several types of voltage regulator. For example, the 7805, 7812, 7905 and 7912 are +5-V, +12-V, −5-V and −12-V 3-terminal 1 A voltage regulators respectively. Types of different current ratings are available and are identified by an additional letter, e.g. the 78L05 and the 78S05 are 0.1-A and 2-A +5-V regulators respectively. A designer can therefore select a regulator that satisfies the voltage and current requirements. It is a good idea to allow a spare current capability for future expansion of the circuit. The power supplies of Fig. 16.6 employ the 7805 and 7905 voltage regulators. Capacitors C2 and C3 are decoupling capacitors which need to be mounted as close to the regulator as possible.

16.6.3 Transformer

A stepdown transformer is selected to reduce the a.c. input voltage to a suitable level and to provide isolation. The reduction in voltage is proportional to the turns ratio of the transformer. Losses within the transformer are very low and can be determined by measuring the current into the primary side when the secondary side is open circuit. Since the losses are low, then one can say that:

$$V1\ I1 = V2\ I2 \tag{16.1}$$

where a '1' identifies the primary side and a '2' identifies the secondary side. Since transformers in linear power supplies are normally of the stepdown type, then V2 is a fraction of V1. Using Equation 16.1, one can conclude that I2 is a multiple of I1. Consequently, when choosing a transformer, one should not only select the voltage ratio, but should also pay attention to the power rating. In the power supply under consideration, the a.c. voltage appearing across the secondary winding of the transformer is rectified by a full-wave rectifier which consists of either four individual diodes or a single bridge-rectifier module. Note that the power supply of Fig. 16.6(b) employs a centre-tapped transformer where the centre-tapped point is taken as the 0-V line. If such a transformer is employed in the design of a single-output power supply, then the number of diodes can be reduced to two, as shown in Fig. 16.7(a). In many applications, however, several isolated outputs are required. For instance, the 3-phase inverter described in Section 15.3 includes six power transistors. Each transistor and its associated drive circuit may need to be isolated from the rest of the circuit by an optocoupler. An isolated power supply terminal will then be required to power the secondary side of each optocoupler. This and many other applications create the need for a transformer with a single primary winding and a few secondary windings. Figure 16.7(b) shows an example of a transformer with two individual secondary windings.

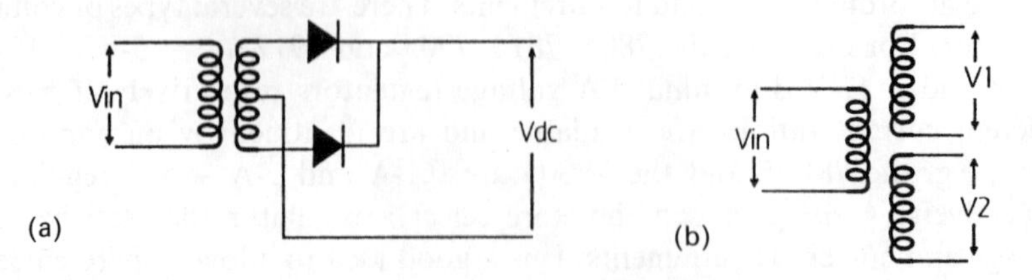

Fig. 16.7 Alternative configurations of transformers: (a) centre tapped; (b) two secondary windings.

16.6.4 Rectifier and capacitor

The rectified a.c. signal is smoothed by the use of a capacitor, (e.g. C1 of Fig. 16.6). The amount of the required smoothing depends on the load current I, the frequency of the supply f and the capacitance C of the smoothing capacitor. Consequently, for a given load current and supply frequency, the ripple dV available within the smoothed voltage depends on the capacitance of C1. The relationship between the voltage and current across the capacitor is given by:

$$I = C(dV/dt) \tag{16.2}$$

For a full-wave rectified signal, the value of dt corresponds to the duration of half a cycle of the supplied a.c. signal (which is the mains signal in this case). Equation 16.2 can then be written to present the ripple in terms of the frequency as:

$$dV = I/(2fC) \tag{16.3}$$

This equation shows that the higher the load current, the higher will be the ripple. Increasing the capacitance will, on the other hand, reduce the ripple. One can also see that for a given value of load current and capacitance, the ripple is inversely proportional to the frequency. Consequently, the ripple is higher for operation from a 50-Hz mains than from a 60-Hz mains. The presence of ripple means that the instantaneous value of the voltage across the capacitor varies with time. The capacitor will only be charged for the time where the input voltage is higher than that stored in the capacitor. Once charged, the capacitor will supply the load with the stored charge, as shown in Fig. 16.8. The duration between points A and B represents the charging period. After point B, the instantaneous value of the a.c. voltage drops below the capacitor's voltage, thereby reverse biasing the diodes. This allows the capacitor to supply the load during an interval represented by points B and C.

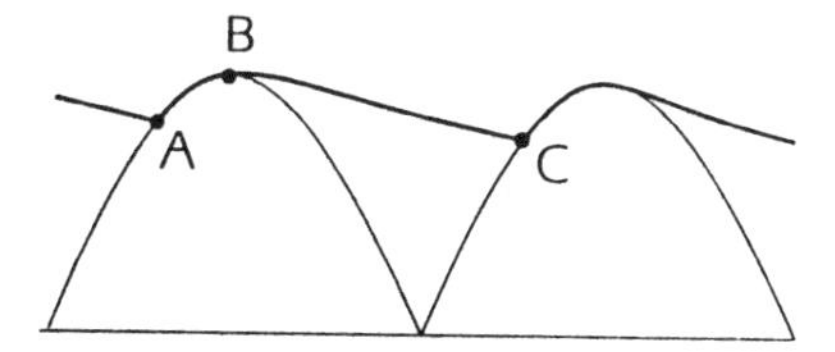

Fig. 16.8 Capacitor charge and discharge in a rectifier circuit.

Example 16.4: **In order to design a +5-V linear power supply, calculate the value of the capacitance C for a load current of 1 A and a frequency of the mains supply of 50 Hz. What is the voltage needed at the secondary of the transformer?**

In order to calculate the capacitance, let us allow for a ripple of 2.5 V. Rearranging Equation 16.3:

$$C = 1/(2 \times 50 \times 2.5) = 4000\,\mu F$$

One can then choose a 4700-μF capacitor, which allows for fluctuations in the mains voltage.

The 2.5 V allowed for the ripple needs to be added to a voltage drop across the regulator of 2.5 V. This implies that the d.c. voltage before the regulator should be:

$$V_d = 5 + 2.5 + 2.5 = 10\,V$$

The corresponding a.c. voltage can be found by dividing this value by the square root of 2. The result is 7.07 V. After allowing for the voltage drop across two diodes, a transformer of a 9-V secondary voltage can be chosen.

16.6.5 Battery backup

The operation of certain microcomputer-based applications should not be disturbed if the mains power disappears (this condition is known as 'blackout'), or if there is a temporary dip in its voltage (this condition is known as 'brownout'). Operation can continue from a back-up power supply. Its type and power capability depends on the application. Without a back-up, the smoothing capacitors of the power supply include stored energy which will supply power for a very short interval. If a shut-down is not acceptable, then a battery can be installed to provide the required power. An example of its connection to the power line is given in Fig. 16.9, where diodes D1 and D2 are employed to ensure the correct conduction of power. When the input voltage is high, D1 conducts, causing D2 to be reverse biased. No power is therefore delivered from the battery. If the mains voltage drops,

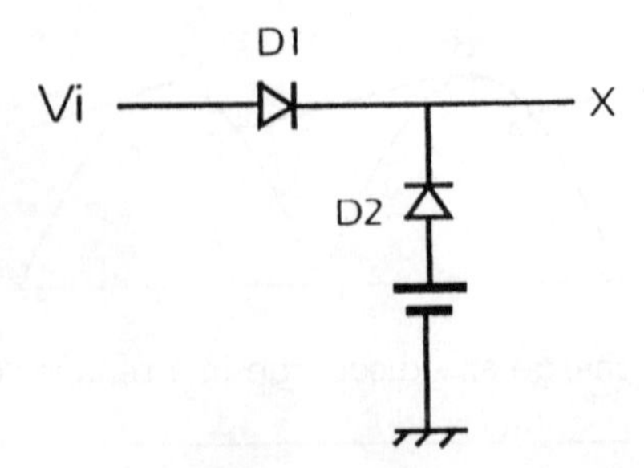

Fig. 16.9 Using a battery back-up.

then diode D2 starts to conduct, thereby reverse biasing D1. The use of D1 will ensure that power will only be supplied from the battery to the circuit and not to the mains. In some battery-powered applications, diode D2 is replaced with a thyristor to give the microcomputer the ability to control operation from the battery. This is shown in Example 11.5.

16.7 Design considerations for noisy environments

Electronic equipment incorporating microcomputers is used in a wide range of applications and in various environments. This section considers causes and effects of noise glitches on the operation of microcomputer-controlled equipment. The existence of such glitches and their effect depend on the environment in which the equipment is used. A designer should therefore take into consideration the type and magnitude of noise glitches which may occur, and their effects on the system under consideration. The designer can then decide on the extent of the protection needed for safe operation. Noise effects cannot be fully eliminated, but the possibility of their existence and the severity of their effects can be considerably reduced.

This section shows that a designer can introduce techniques to minimize the effect of the glitches and to provide the system with the ability to quickly detect incorrect operation and take a fast counteraction. Not all the presented techniques need to be included in every microcomputer-controlled system, but their utilization depends on the type of application and the degree of protection needed. The adoption of such techniques may lead to an increase in cost and/or a sacrifice in memory space and processing speed. The use of such techniques is justified when operating in a noisy environment.

It is important to take into consideration the effects of noise during the design stage. This is because when microcomputer-based industrial equipment fails on the field, the rectification of the fault may prove to be very costly for both the user and the manufacturer of such equipment. It may lead to users losing faith in the manufacturer of such equipment. The detection of the type of noise which has caused the failure is not likely to be a straightforward task. One of the main obstacles against fast detection is the randomness of the noise. It may be due to one of many causes, such as switching nearby heavy-current equipment. These and other factors need to be taken into account during the design of the system.

16.7.1 Effect of noise on microcomputer operation

Noise glitches may affect ICs within the system. This results in supplying the microcomputer with the wrong signals or affecting its internal operation. It leads to one or more of the following:

- Incorrect execution of instructions
- Attempting to fetch data from non-existing memory devices
- Resetting the system
- Altering the state of interrupt lines
- Executing an endless loop
- Altering the stack pointer. This leads to corruption of memory or incorrect operation
- Supplying wrong signals to external equipment controlled by the microcomputer. This may lead to permanent damage of the equipment
- The contents of the program counter may change, leading to a jump to a random location during the execution of the program. This leads to unpredictable results
- The state of memory elements may change. This results in corrupting data and/or program code

16.7.2 Causes of noise and methods of minimizing their effects

Switching heavy-current equipment off is one source of noise transients. Large transients are generated in the power lines as such equipment tries to maintain the current flow. Furthermore, loading or unloading of such equipment is usually associated with a large change in the magnitude of current, leading to the formation of similar transients.

A different type of noise results from the effect of electric field coupling. It can be minimized by taking two actions. The first is to place an earthed metal barrier between the source of noise and the circuit. The second action is to introduce a ground plane within a PCB. Such actions also help in minimizing the effects of magnetic and electromagnetic fields. Multilayer PCBs are expensive, but the increase in cost can be considered as an investment when operating in a noisy environment.

The designer should pay attention to the layout of the system. It is, for instance, good practice to keep circuits that generate unwanted noise signals compact and away from sensitive circuits. Glitches generated by power-switching circuits depend on the rate of change of the current. Consequently, the use of flywheeling diodes across inductive loads reduces the rate of change of the current and helps in reducing the magnitude of related glitches. Protection against noise can be improved by employing isolation transformers.

It does not need a heavy current to generate noise glitches, but when digital circuits operate they generate short current pulses (glitches) at each change of logic state. This can cause changes in the supply voltage which will therefore be delivered to each circuit supplied by the power supply. It may alter the operation of their circuits. Consequently, decoupling capacitors can be employed to supply the required current spikes. The capacitor will quickly charge to the supply voltage and will be ready to deal with future glitches.

Two types of decoupling capacitors can be utilized, IC-decoupling and board-decoupling capacitors. The first type provides the spike current to an IC, and is therefore placed close to the IC. Its magnitude should be selected so that it will be charged by the power supply at a rate of dI/dt less than that needed for the spike, thus preventing the power supply voltage from suffering from the spikes (0.1-μF capacitors are widely used as IC-decoupling capacitors). The second type of decoupling capacitor is the board-decoupling type. It supplies the current needed to charge the IC-decoupling capacitors as well as reducing any noise received from the power supply. It is desirable to reduce the path area between the IC-decoupling capacitors and the board-decoupling capacitors. This can be achieved by supplying power to the board through an additional layer in the PCB. Employing such a layer introduces a further increase in the cost of the board.

Let us now consider another type of noise, called ground noise. There are two types of ground: earth ground and the signal ground. There is a voltage difference between earth grounds at different locations. Consequently, if more than one ground point is used, then current from one circuit flowing through the ground wire generates a voltage difference due to the impedance of the wire. This can and should be avoided by selecting a single point as a common ground point to which all points are connected (in a star connection). When a microcomputer is interfaced to analogue circuits, digital and analogue grounds should be kept separate and jointed together at a single point.

In addition to the protection given above, a glitch (which could possess a high voltage peak) may reach the circuit components. The hardware should therefore be protected against over-voltage. This can be achieved by identifying those inputs which are likely to be the source of such glitches and protect them by placing a suitable protection device, e.g. zener diode, between each input and ground. Another zener diode can be placed across the supply rail.

In some systems, the information contained in the RAM is so valuable that its corruption may lead to disastrous effects. Thus, during the operation of the system, the valuable information is duplicated or triplicated to form two or three independent storage areas. The information of interest is compared before being used. If an error is detected in the case of a duplicate system, then a flag is set and an alarm may need to be generated. In the triplicate system, however, an area can be considered faulty when it differs from the other two. The information in that area can therefore be ignored and a flag can be generated indicating the type of fault. The system can continue to operate by using information from the other two areas.

Clearly, the difference between information obtained from two identical inputs does not always correspond to a fault. For instance, in devices such as A/D converters (Chapter 7) the results fluctuate and the comparison needs to include a hysteresis region. This, however, increases the processing time needed to perform the comparison.

16.7.3 Fault detection and recovery

The previous section presents methods of reducing the probability of occurrence of faults due to noise glitches. Let us now consider methods of protecting the system against the occurrence of faults. During the interval between the occurrence of a fault and taking an action, unacceptable signals may be generated from the output lines of the microcomputer. Consequently, a detection circuit can be introduced to recognize unacceptable combinations of the output signals. The circuit may consist of few logic gates. A decision on whether the use of such a circuit is justified depends on the magnitude of the expected noise glitches and on the effect of the faulty combination on the system. Clearly, economical reasons limit the extent of protection included within such a circuit. It therefore does not provide a complete protection against every possible combination.

In some systems, an external source is utilized to generate a periodic interrupt. The service routine of this interrupt ensures that a group of predefined tasks have been executed since the last time this interrupt occurred. If an error is indicated, then the interrupt-service routine will act to recover from the fault, reset the system, or shut the system in a controlled manner. The type of fault can be written in a reserved area of memory for later inspection. This area will not be overwritten by other parts of the program, nor cleared by a reset action.

Even when introducing the techniques described above, software execution may still be affected as a result of operating in a noisy environment. Detection at an earlier stage is important as it permits immediate action to be taken to prevent the consequences. This can be achieved by introducing software techniques such as:

1 Fill unused program memory with NOP instructions. They can help the program counter to point to a valid instruction after a faulty jump to an operand.
2 Checksums can be obtained for sections of the program and/or the data. Calculating checksums, however, is a time-consuming task and should be used only when necessary.
3 When receiving a value from an input, or supplying it to an output, a test can take place to determine if the value is within acceptable limits.
4 A location can be used to contain the number of bytes stored in the stack. Whenever a push to, or a pop from, the stack takes place the contents of that byte is adjusted. Ensuring correct operation is based on a test which takes place before each push or pop instruction. The test compares the contents of that byte with the contents of the stack pointer. If they do not agree, then a flag can be set to indicate the fault and the system can be brought to a halt. Clearly, performing such a test increases the execution time.

Chapter 17

Limitations, upgrading and development tools

17.1 Introduction

The continuous development in microcomputer technology has led to continuous changes and improvements. This makes microcomputers more attractive to use, but at the same time makes it more difficult for many users to keep up with the latest developments. The technological changes will continue to deliver more powerful microcomputers at affordable prices capable of running a wide range of very exciting new applications. We have already seen in the last few years a large increase in the number of vendors, device architectures and peripherals. To many inexperienced individuals, the name of the main processor in a microcomputer defines the performance. This is not always true, since it is possible to have a processor with fast peripherals outperforming a faster processor with slower peripherals. When choosing a microcomputer, one has to choose between functionality, performance, quality, price and after-sales support.

The provision of additional features, the need for faster sampling, the use of higher resolution calculations and representations and the increase in the use of graphics representation by modern programs create the need for higher processing power and storage capabilities. For instance, the use of higher screen resolution implies more information to be transferred from the processor to the screen. A faster rate of screen update means more demand for higher processing power. The increase in the use of multitask real-time systems requires the ability to respond to various commands and transfer quickly between tasks.

Since the requirements of various applications are not the same, and since the demand to keep up with new technology is ever increasing, one should have upgradability in mind when purchasing a microcomputer. Obviously, not everybody needs to do this. Many users are happy with a simple text editor and a modest storage capacity with a simple display. Nothing is wrong with that, as long as the microcomputer is employed for low-demand applications, e.g. as a text editor, or as a simple controller. This is different from the application of a designer who has to use the latest software packages, or is looking for the best performance for an interface circuit. The use of a slower microcomputer during the development time of a project may mean that one has to wait for the microcomputer to calculate. This is wasted time, which implies lower productivity and a loss of income. The use of a slower microcomputer for on-line monitoring or control means a slower sampling rate, limited control algorithms and probably the need for additional hardware and intelligence in an interface circuit.

The previous chapters have included a large number of examples where microcomputers are incorporated as controllers in open-loop and closed-loop applications. When a microcomputer is used as a controller, it normally needs to respond to two types of command, the first from an operator and the second from the system it controls. The interface to the operator is often accomplished through a display, keyboard, potentiometers and/or switches. The display may be a group of LEDs, numerical digits, alphanumeric display, graphic display or monitor. Numerous microcomputer-controlled systems employ menu-driven software as a means of communicating with the user. Others offer a graphical user interface which displays the elements of the system, their interconnections, and an up to-date display of the information transferred between them. These features increase the user-friendliness of the system. They not only allow an operator to observe values, but they can also be employed to permit the operator to choose the mode of operation, to configure the system and to alter its parameters. In many systems, however, changes of configuration and the values of parameters are not allowed without providing a password. In this way, only authorized personnel can introduce changes. Although a graphical or a menu-based user interface eases the usability of the system and allows operators to introduce changes without having to acquire knowledge of programming, it can be costly in terms of

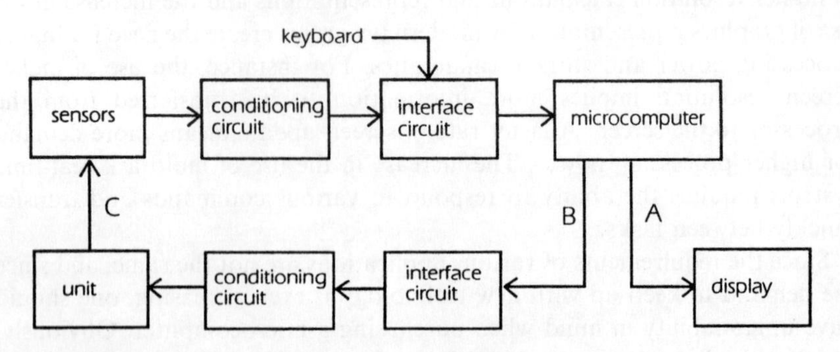

Fig. 17.1 A microcomputer-based controller.

processing time and associated hardware (graphical display, additional memory, etc).

Figure 17.1 is a block diagram representing a microcomputer-based controller. Three paths are shown; their use determines the type of controller and its associated interface circuit. An open-loop controller can be based on the use of path B without the need for paths A and C. In such a controller, the microcomputer sends signals to an interface circuit to be conditioned (amplified, isolated and/or filtered) before being applied to control the required hardware. A closed-loop controller, on the other hand, uses paths B and C, where the latter employs sensors to detect the instantaneous magnitude of signals and/or the state of the equipment. Signals from the sensors are conditioned before being interfaced to the microcomputer through a suitable interface circuit. Finally, path A is employed when monitoring is required. The use of the microcomputer as a data-collecting device employs paths C and A of the block diagram. This is explained in the following section.

17.2 Data-monitoring applications

Chapter 7 considers applications where analogue signals are supplied to microcomputers through single-channel and multichannel A/D converters. Examples of monitoring digital signals are given in Chapter 4, where signals are supplied to a microcomputer through its input ports. Once received by the microcomputer, the data can be stored, displayed, printed, transferred to other devices, or processed by the microcomputer. The microcomputer-controlled data-collection process can be arranged to reduce the involvement of the operator. This provides the operator with more time for examining and analysing the results. The work of the operator can be further simplified by employing a suitable software package to categorise the results and present them in a form suitable for test and analysis. One should, however, pay attention to the storage capability of the microcomputer, particularly in applications requiring the collection and storage of large blocks of data.

Example 17.1: **Consider an application where an analogue signal is to be monitored by a microcomputer employing a 12-bit A/D converter. The microcomputer stores a series of conversion results in a specified memory area. The storing commences following the detection of a logic-high signal supplied by an external device. The data-collection process continues until the specified memory area is full. How can such a task be implemented by the microcomputer of Fig. 6.1(b)?**

The result produced by the A/D converter can be supplied to the microcomputer through input port 2 (see Fig. 6.1(b)). The 12 digital-output lines of the converter can be connected to lines 0 to 11 of port 2. The EOC line of the converter and the line carrying the event indication can be connected to lines 12 and 13, respectively, of port 2. With the program-controlled input method, the microcomputer begins monitoring line 13 repeatedly, seeking a change of state from low to high. Once the event occurs, the microcomputer terminates the waiting mode and enters the data-collection mode, which is in the form of a software loop. The execution of this loop is repeated until the specified memory area is full.

The solution can be represented by the following steps, where 'p' is an address pointer, 'y' is a 16-bit word, and the '&' character represents a software AND:

```
1  LET p = TOP OF STORING AREA
2  READ PORT 2 AND STORE READING IN y
3  IF (y & 0010 0000 0000 0000) = 0
       THEN GO TO STEP 2
4  START CONVERSION
5  READ PORT 2 AND STORE READING IN y
6  IF (y & 0001 0000 0000 0000) = 0
       THEN GO TO STEP 5
7  STORE (y & 0000 1111 1111 1111) AT p
8  IF p < END OF SPECIFIED AREA
       THEN INCREMENT p AND GO TO STEP 4
9  INDICATE THE AVAILABILITY OF A NEW SET OF RESULTS
```

In this procedure, the microcomputer monitors the event-indicating line in steps 2 and 3 until the event is detected. Following this, the microcomputer starts the A/D conversion by issuing the SC signal (the generation of such a signal is the same as steps 1 to 4 of the procedure given in Example 7.1). By repeatedly executing steps 5 and 6 of the above procedure, the microcomputer detects the end of conversion as soon as it is reached (it assumes that a logic-high state represents the end of conversion). Step 7 extracts the 12-bit digital result and stores it at the location defined by the pointer 'p'. The pointer is then incremented and the loop repeated only if the memory area is not full. Note that, in step 8, the pointer is incremented by 2 bytes, since each 12-bit result is stored in two memory bytes.

Instead of waiting for the A/D converter to digitize the signal, the procedure can be modified to increase the rate of reading the signal. This can be achieved by starting the conversion after step 6 (as well as in step 4). The result is:

```
 7  START CONVERSION
 8  STORE (y & 0000 1111 1111 1111) AT p
 9  IF p < END OF SPECIFIED AREA
        THEN INCREMENT p AND GO TO STEP 5
10  INDICATE THE AVAILABILITY OF A NEW SET OF RESULTS
```

In this way, the digitization process is carried out while the microcomputer is storing the result and testing for the end of the available storage area. If the time taken by steps 8 and 9 is longer than a single conversion period, then there is no need to test for EOC indication.

No matter what method is adopted, the new set of results contains digital samples stored in sequential RAM locations in the order in which they were acquired. They can, for example, be displayed or printed for examination by the operator. The time spent on collecting all the results depends on the length of the specified area and on the conversion time of the A/D converter.

Instead of terminating the data-collection process when the memory area is full, some applications require continuous monitoring of data. For instance, when a microcomputer is utilized as an oscilloscope, it collects samples, stores them in a specified RAM area sequentially, and supplies them to the monitor to be displayed. At any instant of time, only the contents of the specified memory area are required, say 1000 values. Consequently, when that area is full, data storage continues by over-writing the oldest data. In this way, the available information consists of only the newest 1000 readings. The software pointer ('p' in the above procedure) identifies the location of the newest reading. The next location (in the case of a continuous data-collection process) contains the oldest available value. Consequently, the process of displaying the data starts by setting a second pointer 'q' to point to location 'p + 1' (where both 'p' and 'q' are double-byte numbers). By using pointer 'q', data is read and sent to be displayed sequentially. Let us start by assuming that the data-collection process will stop during the display of a single trace. This implies that the display process will start from the point of interest and continue until reaching the location pointed at by the software pointer 'p' (the process is likely to include reaching the end of the table and adjusting 'q' to point to the start of the table as it continues the display process). Since the display of a single trace has a limited use, then the procedure can be altered to allow a continuous update of the display. This imposes the requirement for the microcomputer to continue with the data-collection process and the storage of the new data to replace those being displayed. One should keep in mind that in order to keep the process continuous, the rates of collecting and displaying data should be the same.

Such a simple oscilloscope can easily be provided with a storage feature. One way of achieving this is to allow the microcomputer to start the data-collection process when a certain event occurs and continue until all the locations within the reserved memory area are updated. Alternatively, the occurrence of the event can be used to stop the data-collection process. The latter approach allows an operator to observe what happened prior to the occurrence of the event, whereas the former approach shows what happens after the event has occurred. The event can, for example, be a rapid transient in the sampled signal. This creates the need for the microcomputer to examine the collected data and search for a large change in value. In this case, the displayed data correspond to the last 1000 values received before the occurrence of the event. This can help an operator to examine the behaviour of the signal and, possibly, use the information to improve the performance of the system, or detect faults. If the value of the input signal is outside the range of the A/D converter, then resistors can be connected to form a voltage divider in order to bring the amplitude down to within the allowable range. Zener diodes can be employed to protect the converter from values exceeding its range. If the input signal was of a very high value, or is derived from a noisy environment, then isolation devices such as transformers or optocouplers are strongly recommended to provide electrical isolation.

The capabilities of such an oscilloscope can be extended by digitizing multiple analogue signals (Section 7.7). This effectively allows the use of the microcomputer as a multichannel oscilloscope. For instance, it gives the operator the ability to store and display several waveforms simultaneously. The processing power of the microcomputer can be employed to carry out mathematical operations on the received values and display the result. Clearly, the more operations carried out by software, the slower is the on-line update of the displayed information. In some fast data-gathering applications, the display of the data can be arranged to take place when the collection process is complete and the collected data are processed. An operator can enter a command to stop the process and to display the most up-to-date results. Alternatively, the operator can command the microcomputer to store the results on a disk. Clearly, the display or storage of the results does not have to commence from the oldest available reading, since an operator can command the microcomputer to begin from any point within the collected data. Although this approach is attractive in principle, its practical implementation is limited by the processing power and the storage capability of the microcomputer.

Example 17.2: **A test engineer needs to examine the behaviour of a system at different voltage levels (below 300 V). How can a microcomputer be employed to allow the engineer to set these levels, alter them by software, and examine the behaviour?**

A microcomputer can be employed to read the analogue signal through an A/D converter. Let us use a 10-bit unipolar A/D converter whose full scale corresponds to +5 V. The analogue-input signal will need to be scaled down to fit within the allowable range. In other words, 300 V of the input signal should be represented by 5 V. Since a 10-bit A/D converter is employed, this level will give a digital number of 1023 when digitized by the A/D converter. If, for instance, the operator wants to monitor what happens when the voltage level exceeds 250 V, then the data-collection process starts when the instantaneous value of the analogue signal exceeds an equivalent digital number of 853. Let us consider two methods of implementation.

Approach 1

This is based on the use of hardware to assist in fulfilling the requirement. It uses a D/A converter whose output is connected to an analogue comparator. The microcomputer delivers a digital value corresponding to the required level. It will be converted to an analogue value and compared to the input signal by the analogue comparator. A signal at a logic level will be delivered to interrupt the microcomputer as soon as the specified level is violated. The microcomputer only samples and stores the analogue-input signal when interrupted by the comparator.

Approach 2

This is based on fulfilling the requirement by the use of software without the need for additional hardware components. The microcomputer monitors the data continuously and compares each digitized value with a digital number equivalent to the level of interest. Once this is exceeded, the microcomputer starts the data-storage process.

Comparison

Each of the two approaches has its own advantages and drawbacks. Approach 1 uses additional hardware which may need to be put together especially for the test. The required circuit is, however, simple and of a low cost. On the other hand, approach 2 does not impose the additional hardware requirement, but demands continuous monitoring from the microcomputer, which is acceptable if the microcomputer is dedicated to the test.

17.3 Limitations of microcomputer storage capability

In many applications, microcomputers are incorporated as collection tools for a large amount of data. After collection, such data are normally stored in a specified portion of the RAM area of the microcomputer. Clearly, if the reserved area is large enough for collecting the data, then the storage will be a straightforward task, as in Example 17.1. Unfortunately, some applications impose the requirement of storing data of a greater length than the available RAM.

Many microcomputers allow the expansion of memory to meet the requirements of various applications (Section 17.5.1). Let us assume that a test needs to be carried out using an existing microcomputer without expanding its memory. The test involves collecting samples whose number exceeds the memory area that can be allocated for the storage of the results. In such an application, one needs to decide on what to do when the RAM area is full. The decision depends on the type of application and on the object of the data-collection task.

Many microcomputer-controlled systems (such as the one presented in the previous section) are operated in a data-collection mode, where the interest is in collecting the highest possible amount of data per unit time. Such systems require a fast microcomputer interfaced to one or more fast A/D converters. In most such systems, the microcomputer is permitted to over-write the oldest values when the memory is full, thereby losing the oldest information. In some systems, however, over-writing data is not allowed, as all the collected data need to be stored for subsequent categorization and examination. They can then be presented to the outside world, processed, or imported to other files for computation, analysis or inclusion in documents. This may mean arranging the software to command the termination of the data-collection process when the specified storage area is full. The program can then store the collected data on a disk before restarting the data-collection process. With this approach, a large amount of data can be collected and saved in different files on the disk. Each file corresponds to a certain operating condition, or is a result of using a different version of software. One can therefore say that the process of collecting data in such systems consists of two steps: the first is the data collection and storage in RAM, and the second is the transfer of those data to the disk. The approach can be different in applications where the storage time of the disk is lower than the rate of reading the data, e.g. a slow change in temperature. One is provided with the option of performing a continuous data-collection and data-storage process. It has the advantage of achieving the collection of a large amount of data with minimum intervention from the operator. Since the storage time of the disk is in milliseconds, the microcomputer may have sufficient processing power to execute other tasks besides handling the slow data-collection process.

17.4 Operating systems

The examples provided in the previous chapters have illustrated ways of executing tasks in a microcomputer. It is explained that tasks can be placed within a continuous loop and executed sequentially, or interrupts can be used to request the execution of tasks at certain instants of time. Dealing with interrupts and other processor-specific tasks requires gathering various machine-related information. The result is a system which uses the resources of the microcomputer to fulfil the requirements of the application efficiently. Software may be tailored to a specific application, or it may be provided with flexibility, allowing its use in a variety of applications.

An operating system is a program which handles various machine-related tasks and manages the allocation of the system's resources. A microcomputer with a good and flexible operating system removes from the user the need to know the detailed operation of interrupts, timers, initialization procedure, etc. The user can therefore employ the resources to implement tasks with minimal knowledge of the hardware. In other words, an operating system provides user interface, device interface and applications interface.

The continuous introduction of processors with higher speed of operation, better computation power and higher memory-addressing capabilities has increased their use in multitasking applications. In such applications, it is important to divide the processor's time between the tasks in the best possible way. The examples presented in the previous chapters emphasize the importance of achieving the best performance without overloading the processor or introducing the need for additional hardware. A compromise is often reached between solving the problem with software or hardware means. An operating system makes the best use of resources to execute tasks and transfer information between them efficiently. Certain resources may need to be shared between tasks. This creates the need for the operating system to keep a record indicating the availability of various resources at any instant of time. If, for example, a task requires the use of a resource which is already in use, then the execution of that task may need to be delayed until that resource is available.

Many operating systems are machine-specific, written in assembly language which is specific for a particular processor. Other operating systems are written in a high-level language, making them less dependent on the processor. The UNIX® operating system, for instance, is written in the C programming language, making it easier to use in another machine operating from a different type of processor.

One of the most widely used operating systems is MS-DOS® (Microsoft® Disk Operating System), which was included in the IBM PC. Newer versions have been developed to make the best use of technological advances in microcomputers. Another very popular operating system is OS/2®, which is used in the IBM PS/2® machines and is attractive to use in multitasking systems. UNIX is another very popular operating system capable of running multi-user programs.

Graphical user interface (GUI) has developed to become one of the most popular working environments. Microsoft Windows™ is a graphical environment which allows operations within various applications to be displayed graphically and can be run when selected with a mouse (or with the press of keys). It is a multitask environment which has a built-in graphics interface and allows the exchange of data and control between tasks.

Such features increase the demand for higher performance, disk storage capability and memory (RAM) capacity. The use of Microsoft Windows simplifies the use of computers and has opened the door to a vast number of new applications. Consequently, a very large number of application programs have been developed, making this type of environment even more attractive and easier to use.

17.5 Capabilities and upgradability

Many users are satisfied with a low-performance, few-year-old microcomputer for an on-line measurement and control application. Others accept waiting for a few seconds, or even many minutes, for an off-line microcomputer to calculate. Then, one day, they observe the performance of a faster microcomputer, or purchase a new software package, and they realize the limitations. They start to complain about how slow their microcomputer is, how difficult it is to execute certain tasks, or how introducing newer technology can help to improve productivity and performance. For example, many users of PCs have installed Microsoft Windows and other popular GUIs on their PCs. GUIs simplify the use of computers, but demand high processing power as well as a large storage area. More and more GUI products are being introduced for various applications, including monitoring, interfacing and real-time control. A user can, for instance, be provided with the ability to configure a microcomputer-controlled interface circuit, change the path of flow of data, alter the setting of temperature, define the movement of a robot arm, or select signals for monitoring. All these can be done with on-screen windows or pull-down menus, without the need for programming skills or additional user-interface hardware components, e.g. switches or potentiometers. One should take into consideration the required processing power and storage capacity which may dictate the utilization of a dedicated microcomputer for the application. An existing microcomputer may not be capable of meeting such requirements. The need for a higher processing power and storage capacity may be met by the purchase of a suitable microcomputer, or the upgrade of an existing one. The latter needs careful consideration in order to determine the optimum upgrade path. One needs to understand the exact capabilities of the existing microcomputer and the additional requirements of the new application. Also, allowance should be made for future needs.

Let us consider what happens when one selects a pull-down menu or opens a window. The menu or the window covers an area of the screen. The information present in this area needs to be stored in a temporary memory area and retrieved when the menu or the window is closed. The speed at which the information is stored and retrieved has to be high, to minimize the operator's waiting time. Dragging a window on the screen follows a similar approach. The need to open several windows simultaneously increases the stored information and creates the need for a bigger storage area. This can cause limitations in use and can create the need for upgrading the microcomputer in order to meet the requirements efficiently.

Many microcomputers are designed for a specific application. In such microcomputers, the type and speed of the processor, the capacity of memory, the resolution of counters and timers and the number of input and output lines are optimized for that specific application. They normally offer very little in terms of upgradability. This may take the form of adding or replacing memory devices or modules. Some microcomputers include empty sockets for the inclusion of upgrade ICs such as a math-coprocessor, or additional input and output devices. This is not the case for all microcomputers, as some (such as PCs) offer various upgrading paths. They give the user the ability to upgrade those parts that cause a bottleneck in the application. This is a very attractive feature, since one's needs change with time, and the computer industry changes fast and continues to offer more. It is therefore very attractive for one to have the freedom to update one's microcomputer and gain the best performance without having to replace the whole microcomputer. Not all PCs, however, offer the same upgrading paths. Some offer a more flexible path than others.

Some important reminders are as follows:

1 Certain parts of microcomputers have high voltage and this can result in an electric shock if repair or upgrading is attempted by inexperienced users.
2 Alteration or modification of a microcomputer can invalidate the warranty given by manufacturers.
3 It is always advisable to keep back-ups of software, especially when attempting to upgrade a microcomputer.

Let us take the IBM-compatible PCs as examples of upgradable microcomputers. Since such microcomputers have formed a standard, and since newly developed processors for such machines are backward compatible, such computers are no longer disposable machines. When such a computer gets older or its processing power can no longer satisfy the user's requirements, the user has the option of replacing one or more of its parts rather than the whole machine. The feasibility of this upgrade depends on what is forming the bottleneck and is greatly dependent on the type of PC used. Some lower-cost PCs combine many of the computer parts on the same board, making the machine good value for money in the first place, but could make some upgrades more expensive, if not impossible.

The availability of hundreds of different types of PCs provides a wide choice, but can make it very difficult to select the optimum machine, especially when there is a high rate of introducing new features and add-ons. There is a large difference in price which is dependent on many factors, such as built quality, manufacturer's name, reliability, upgradability, features, bundled software, type of warranty and after-sales support.

The reason for upgrading or changing a PC is not always one's inability to use it. It is often that one needs more speed, a better user interface, or the ability to make better use of newly developed software applications; in many cases the upgrade may be only for cosmetic reasons. Before spending the money, one should understand what one is getting and whether there is a better or more economical way to arrive at the same or a better result. For instance, upgrading the processor or changing the motherboard provides a higher performance which can greatly enhance the operation, but may not be the best value-for-money solution. This is because the user's requirements may be better served by the inclusion of additional memory, a hard-disk cache card or a video-accelerator card. Furthermore, one cannot decide on performance by comparing the speed of processors of different families. For example, a 33-MHz Intel386 DX processor is slower than a 25-MHz Intel486 SX processor. It is therefore extremely important to understand what each part does, and what upgrade paths are available. This may save one from buying the wrong part, or unnecessarily replacing the whole PC.

Without defining the requirements and without having a clear idea of what each upgrade path provides, one could buy something one does not really need. For example, it is explained in Section 1.2.2 that the inclusion of a math-coprocessor can speed some applications but not others. Understanding the facts and defining the requirements are therefore the most important steps in achieving the optimum upgrade, which may be in some cases the purchase of another PC. This is because upgrading an older machine may mean changing several parts which becomes more expensive than purchasing a new machine. Suppliers may be able to help one define the right upgrade path for one's particular machine. It is, however, extremely useful to provide the supplier with the full details of the machine. This does not mean only the type and speed of processor, but also the type of bus and what additional parts are included, e.g. a hard-disk cache or an interface hardware. This can be very useful in defining the exact upgrade without having to replace unnecessary parts.

Normally, the cost of an upgrade is more than if the required parts were purchased with the computer in the first place. If one, for instance, needs a certain capacity of a hard disk at the time of purchasing a PC, then one should seriously consider if there will be a need for a higher storage capacity in future. It is a difficult decision, especially when faster, cheaper and smaller parts for PCs are continuously being introduced. One should keep in mind, however, that it was not long ago when users were satisfied with home microcomputers possessing 1 kilobyte of memory. Nowadays, a single computer game reserves well over 1 megabyte.

A typical PC consists of a motherboard, one or two floppy disk drives, hard disk, power supply, monitor, keyboard, case, speaker, video controller, and a single board containing hard and floppy disk controller, serial communications ports and a printer port. A typical motherboard includes a processor, BIOS ROM, interface peripheral ICs, static RAM (cache memory), expansion slots, SIMM sockets (some or all of which are occupied), and connectors for keyboard, speaker and power supply. Let us consider a few upgrade paths.

17.5.1 Upgrading memory

In a large number of applications, the performance of a microcomputer can be considerably enhanced by increasing the memory capacity. A higher memory capacity can, for instance, remove the need for using the hard disk for temporary storage, thereby speeding the storage and retrieval of information. The increase in the use of RAM has led to a reduction in its cost, making it a very attractive upgrade. But before attempting to upgrade, one needs to know the type of memory required (e.g. whether ICs, SIMMs or an add-on board). Some microcomputers offer a choice. For example, an IBM-compatible PC with an ISA bus can have a memory upgrade in the form of SIMMs on the motherboard, but one can also add a memory board through an ISA expansion connector. It may be possible to purchase the board unpopulated or partly populated in the first instant and add memory to it if and when the need arises. Since the SIMMs are located on the motherboard, a fast processor can communicate with them at a much higher rate than is possible through the ISA bus. Communication with the add-on memory board, however, is much faster than writing to a hard disk and can be attractive in many cases; for example, it can be used as a RAM disk (Section 1.3.2). One should keep in mind, however, that information in RAM will be lost if the power to the computer is switched off, or if one is forced to activate a reset after the machine locks up. It is therefore advisable to save to the hard disk frequently, e.g. every few minutes. Some PCs offer a third choice in the form of dedicated slots (connectors), allowing a compatible add-on board to communicate with the processor at the highest possible rate.

When upgrading memory, it is important to get devices with the correct type and access time. One should either inspect the devices available within the microcomputer, or check with the supplier before attempting to purchase the devices. Some types of microcomputers, or add-on boards, include jumpers or DIL switches which need to be set in order to use the upgrade. Other microcomputers test the memory devices to determine the available capacity, while others require an update of the software. The number of additional devices needed for the upgrade depend on the type of computer. For instance, it is explained in Section 1.3.1 that upgrading the memory of a 32-bit IBM-compatible PC means adding or replacing SIMMs in groups of four.

Many modern software packages require a large area of memory (RAM) for their operation. There is a minimum limit of memory for each software package. This is stated in the manual associated with the package. Some state that if the available memory is not large enough, then one can install a simplified version of the package. Furthermore, if, during operation, the available memory is not sufficient to hold all the information, then some packages (such as Microsoft Windows) use a part of the hard disk for temporary storage. This will slow the operation, because the access time of a hard disk is much higher than that of RAM. It is therefore preferred to include as much RAM as possible if serious work is to be done with such software packages.

Let us assume that one decides to improve the performance of a PC by increasing the capacity of the memory. After deciding on the optimum capacity, one can check the number of slots available on the motherboard for the inclusion of additional SIMMs. One obviously wants to insert the additional RAM without having to throw away the existing SIMMs. Many PCs are sold with additional empty SIMM sockets. If these sockets are already in use, then one may be forced to replace the existing SIMMs with others of higher capacity. Some suppliers can offer good deals, as they will part exchange SIMMs.

17.5.2 Upgrading the processor or the motherboard

Older PCs are not designed to accept a direct replacement of new processors, but provide an empty socket for a math-coprocessor. The more recent PCs are designed with processor upgradability in mind. They offer such options as the inclusion of a specific higher-performance processor. They also include local-bus connectors to permit upgrading through the inclusion of fast interface boards to achieve communications with peripherals at the highest possible speed. Upgrading an older type of PC can therefore be achieved by changing the motherboard to give a significant improvement in performance. When changing the processor or the motherboard of a microcomputer, the realized percentage improvement depends on the type of processor to be replaced and what it will be replaced with. In the case of a new motherboard, further improvements can be realized depending on the difference in the clock frequency and on additional features, e.g. cache memory and the number of empty slots for future upgrades.

The cost of replacing the processor board may not be low, but it normally offers a cost-effective solution and can permit future upgrades (e.g. through an empty socket for a second processor or through local bus connectors). Before making the decision to upgrade, one should carefully consider when one will need the next upgrade and whether it is more economical to buy that upgrade now rather having to upgrade again soon. This decision is becoming more difficult, because the cost of processors and motherboards is continuously reducing, especially those of the highest performance (mainly due to the introduction of even faster and more powerful processors).

Buying a motherboard with a 'downwardly compatible' processor means that older programs will run on the new processor. The execution speed of programs will increase, which will reduce the operator's waiting time while the processor executes the instructions. Increasing the execution speed of some programs is not always favourable. For instance, delays introduced as software loops will become shorter, and some programs will run at a much faster rate, making them difficult if not impossible to use. Some computers, however, permit changing the clock speed of the processor from the keyboard.

Before attempting to replace a motherboard, one should consider the following:

1 What parts are included in the motherboard? Some motherboards contain more things than others. In most PCs, serial and printer ports, disk controllers and video controllers are on transferrable expansion boards. In other PCs, some or all of these are placed within the motherboard and cannot be transferred.
2 What can be salvaged from the board (e.g. memory devices)?
3 Will the supplier accept the old board in part exchange?

It is important to keep upgradability in mind when purchasing a computer. One can enquire on how easy it is to upgrade the processor board in future and clarify if the motherboard is available from more than one manufacturer (not all motherboards are of the same size, nor do they have the holes in the same positions). If, at the time of the upgrade, the manufacturer is no longer in business and no other motherboard will fit, then one may be able to purchase a new standard motherboard and a standard case and transfer the rest of the parts to the new case. Transferring everything else to the new case is not the easiest job for the inexperienced. Do not be alarmed, however, since many of the shops selling the parts will fit them for a small charge. This can save time, and enables one to see the computer functioning correctly before paying for the upgrade.

17.5.3 Upgrading disk drives

This section covers both floppy-disk drives and hard-disk drives. They are normally fitted in bays within the PC case. The number of bays depends on the type of case (e.g. desktop, mini tower, full tower). If all the bays are occupied and an additional drive is needed, then there is a choice between upgrading to a bigger case, or purchasing an external drive. In the case of a hard-disk drive, one has the option of replacing the drive with another of a higher capacity, rather than including a second drive.

The function of a disk controller is to control the transfer of data between the processor and the disk drives. A fast controller can therefore improve the performance of the PC. When one is looking inside the PC, the hard-disk controller board can be easily identified by the two ribbon cables connected to the hard-disk and floppy-disk drives. There are a few types of disk controller for PCs. In some PCs, the controller is designed as a part of the motherboard. Some of these PCs provide a means of disabling such a controller. This allows one to replace the controller in case it develops a fault at a later date, or to introduce a faster operation by adding a controller with a disk cache. The majority of PCs, however, include a disk controller in the form of a board occupying one of the expansion slots in the PC. The technological advances and the high volume of production made it possible to include into this board the controller for hard disks and floppy disks plus two serial ports and a printer port. This single board is sold at an amazingly low price. The inclusion of all these devices on a single board offers the additional advantage of leaving more expansion slots available for add-ons.

Writing information to and retrieving it from a hard disk depends on the type of disk, the type of controller, and whether a disk cache is used or not. Before defining the path for upgrading a hard-disk drive, one needs to define whether the requirement is for a higher storage capacity or speed. The requirement of a higher storage capacity may be satisfied economically by the use of 'disk-compression software'. This alters the way information is stored, thereby reserving a much lower area on the disk. Alternatively, an additional or a replacement drive is required. If this is the case, then one needs to understand the type and capability of the disk-drive controller, e.g. whether it possesses the ability to operate another drive, the type of drive it accepts, and the maximum drive capacity (keep in mind that newer hard disks are likely to be faster than older ones and may require a different type of controller). If, on the other hand, a higher speed is required, then one has the option of replacing the controller with another containing a hard-disk cache. This can introduce a considerable improvement in performance, particularly in applications involving frequent storage and retrieval of information. Even with this solution, one still needs to know the type of hard disk used in the machine to ensure that it is compatible with the new controller.

If one decides that the optimum solution is to use a second-hand hard disk (whether purchased alone or as a part of a second-hand PC), one should be careful of computer viruses, which are widespread. Buying from a reputable place with assurances may considerably reduce the risk. There are many programs which will test for viruses to ensure that there will be none of the viruses they check for.

17.5.4 Upgrading the monitor and video controller

The increase in the use of graphics-based applications has created the need for higher-performance processors, higher storage capacities and higher-resolution monitors capable of delivering high-quality pictures with a large number of colours. The requirements have been satisfied by the introduction of standards to make monitors compatible with PCs. This eases the task of specifying and changing monitors. However, changing from one type of monitor to another is likely to involve the replacement of the video controller, which is normally a board inserted in one of the expansion slots. This is not always the case, since in some PCs the controller is included within the motherboard. Some of the more recent PCs are designed for faster operation. They include, or allow the inclusion of, a compatible controller card in a local-bus slot. This permits the transfer of information at the highest possible rate, avoiding the bottleneck formed by limited transfer rate of the expansion bus.

Depending on the type of application, the performance of the whole PC can be affected by the type of video controller. During its operation, the processor executes programs and generates data at a much faster rate than the monitor can handle. Thus, video RAM is provided to act as a buffer. This approach enables the processor to dump information into that buffer and carry on with other activities while the information is being transferred from the buffer to the display. A low-performance controller includes a low-capacity buffer which (depending on the application) may mean more frequent writes from the processor, which may need to wait until information is transferred to the monitor. The higher the resolution of the screen and the higher the number of colours, the greater is the requirement for a high-capacity buffer. Depending on the type of computer and video-interface board, increasing the capacity of the video RAM may be performed by adding memory devices to the video-interface board. Performance can be improved if some of the work is done by the video card itself. This can be achieved by the use of an intelligent card, i.e. one which includes a special processor dedicated to the video operation.

The above shows that, in order to speed the operation of graphics-intensive applications, one has several options. One option is to use a high-capacity buffer-type video controller. This can be in the form of an add-on card placed in an expansion slot within a PC. The performance is limited by the capacity of the buffer and by the rate of transfer of information (because the rate of exchanging information through the expansion bus may not be high enough). Better performance can be realized if the video controller uses the local bus. This will increase the speed of transfer of information, leading to a much better performance. Improvement in performance and in the number of colours can be realized by the use of an intelligent graphics accelerator card. Thanks to the increase in demand, such cards are now available from many sources, offering different features and capabilities at a high performance-to-cost ratio. Before upgrading to such a card, one has to be sure of the type and features of the monitor to ensure compatibility.

17.6 Repair and warranty

The PC market has recently seen a dramatic price reduction as well as the introduction of better and faster PCs and add-ons. Suppliers are offering very attractive deals, e.g. free software, reduced price, better support, or a trial period and money back if not satisfied.

PCs are normally sold with a 12-month warranty for parts and labour. Some come with extended cover, or with an option to extend the cover at a specified charge. Some users prefer to have peace of mind by getting a yearly renewable maintenance contract. No matter what type of cover is offered, it is in one's own interests to check the terms of the cover. An important factor is where the repair will take place. Many people are not bothered with this at the time of purchase, because they assume that a new computer will not go wrong. But this is not always true. Things can go wrong and it is good to be prepared.

Many suppliers offer a return-to-base warranty with the option of an on-site warranty for an additional charge. Other suppliers offer the on-site warranty at no extra charge. Before accepting a return-to-base warranty, it is important to check where will the PC be taken for repair. Is it to the dealer, or to the manufacturer? One needs to consider the cost of sending it through a third party, since it is best sent insured. It is worth checking who will pay for the shipment and whether it will be sent back insured. If not, then is one prepared to offer to pay for insurance? One may decide to avoid these details and take it oneself. This can be easy if the supplier is local. But, in many cases, the best deal may be obtained from a supplier located 200 miles away, and it may not be possible to have the computer fixed on the same day.

The easiest approach will be to take out an on-site warranty, especially if the supplier is a long way away. This may be offered free or at a relatively small charge. Keep in mind that it costs the supplier to send a repair person. It is therefore to everyone's benefit to identify the problem, or obtain as much information as possible, before contacting the supplier. This will help them to get the right part and avoid an additional journey. Clearly, not all faults can be fixed on site; there is the possibility that the PC will have to go to the supplier for repair.

The reduction in the price of new PCs has led to the reduction in the price of second-hand PCs. But this is not always the case, especially with privately advertised PCs. They may have been purchased a few years ago, and the owner may not be aware of the recent price reductions and the introduction of faster and better machines at relatively low prices. It is not unusual to find in the local paper a private advertisement asking more for an older PC than the price of a brand-new identical machine advertised in one of the computer magazines.

17.7 Development tools

The continuous increase in the processing power offered by microcomputers has led to the development of more sophisticated programs. A large number of programs dealing with interfacing circuits are multitasking, where the processor's power is shared between different tasks. In a real-time control system the interference between tasks does not happen at exact intervals, but depends on external events, some of which may take place simultaneously. This should be taken into account when selecting the tools used in developing the product. Choosing the right tool can shorten design cycles considerably. This will not only reduce development cost and improve productivity, but, at a time when manufacturers are competing to introduce new products before their competitors, shortening the design cycle can have the advantage of getting a larger market share and maintaining customers' confidence.

Without thorough testing of programs, development timescales can increase and/or software of poor quality can be generated. The effect of the latter may be noticed at a later date when certain events coincide, for example, when interfacing the product to certain external devices, or when a later version of the software is produced.

17.7.1 Programming languages

One of the major decisions in the design of a microcomputer-based system is whether to use a low-level or a high-level language. The former involves working close to the machine and understanding the effect of every instruction on the operation of the system. This helps to achieve a very efficient code and can lead to high performance while making the best use of the machine resources. On the other hand, a single instruction of a high-level language is the equivalent of a number of low-level instructions. This makes programming and debugging much easier, which helps to reduce the development time. It also simplifies the task of understanding software written by other programmers. This is particularly useful to companies where one programmer modifies or continues the work of another.

Low-level language

Assembly code is a low-level language where the instruction set is specific to the processor. Once written, it can be translated into a 'relocatable code' by the assembler software. Software is normally divided into modules of manageable size. Each module is assembled separately. After generation of all the relocatable modules, a 'linker' program is run to join the modules and produce a code in a form suitable for loading into a memory device. Depending on the development stage and method, the code can be downloaded to a programming device, e.g. an EPROM programmer, or to a development tool such as an in-circuit emulator (Section 17.7.3).

High-level language

One of the most widely used high-level programming languages is C. It is efficient to use and produces a portable code. This makes the software less dependent on the processor, particularly if processor-related activities are combined in a separate software module. The need to run the software on a different processor may arise for many reasons, such as the availability of better processors, or a change in the market requirements. A high-level language allows programmers to use 'libraries', which are a group of functions or subroutines provided to simplify some of the programming tasks. For instance, a programmer may write an instruction such as y = sin(x); by the use of a maths library, it will be replaced by a call to a function containing the code needed to calculate the sine of variable 'x'. Clearly, this is much simpler than using assembly language, where the programmer will have to write a code to perform the mathematical equivalent of the sine function. A program written in a high-level language is normally divided into modules, which are normally stored in different files. A 'compiler' is employed to convert each file of source code into a relocatable module. The modules are then linked and downloaded to a programmer or to a development tool.

Efficiency

The percentage difference in efficiency between using a low-level and a high-level language depends on the type of assembler and compiler used as well as the programming approach and the type of processor (some processors are more optimized for high-level programming than others). There are other factors that can affect the efficiency of the system. For example, the use of a high-level language makes the programmer less involved with some details of the processor. This simplifies the programming task, but at the same time may divert the attention of some programmers away from the capability of the processor and the system resources. This can lead to producing an inefficient or a more complex system. For instance, carrying out mathematical operations while writing in assembly language makes the programmer aware of the amount of processing involved in its implementation. Writing in a high-level language, a programmer may find it attractive to use long integers (a 4-byte representation of variables), or floating point calculations, which will use a large percentage of the processing power. It is therefore good practice (particularly for inexperienced programmers of high-level languages) to examine the equivalent assembly code in order to realize how the software is translated. (Compilers produce 'list files' containing high-level instructions. Each instruction is followed by lines of its equivalent assembly code.) This approach is also useful in detecting compiler faults.

When designing the hardware, it is not an easy task to estimate the amount of code in the final product. If the linked code is larger than the available memory area, then it may be possible to take certain actions to reduce the size of the code. The following are two examples:

- Sections of the code which are repeated in two or more parts of the program can be written as subroutines, or separate functions. They can then be called by various parts of the program, thereby saving code space. This, however, imposes an additional time penalty, since their use involves the execution of additional instructions (a jump, a return and saving and retrieving the contents of some registers). Frequent usage will therefore increase the execution time and can introduce a limitation in some applications, especially if implemented within frequently executed parts of the program.
- Certain parts of the program can be rewritten to reserve less code space. This is more likely to introduce a better saving in applications using a high-level language, since the programmer may not be aware of how the program is translated until memory is exhausted.

17.7.2 Off-line testing

Software tools are produced to help in the detection and correction of software faults. They simulate the operation in order to permit testing parts of the software in the host computer before being transferred to the target system. Individual tasks can be examined and single stepped in isolation and before being run at full speed. One should keep in mind that operation can only be correctly simulated when the effect of other tasks is taken into consideration. Nowadays, simulators offer a user-friendly interface where configuring, monitoring and operation can be easily carried out through a keyboard or a mouse.

A useful means of debugging is 'stepping' through a program by executing one instruction at a time. Debugging an assembler program means stepping through the low-level instructions and observing the effect by examining the contents of registers and selected memory locations. A vast amount of information may be displayed simultaneously. Depending on the type of development tool, debugging a high-level program is achieved by stepping through high-level instructions (the source code) or through an equivalent assembler. Obviously, stepping through the source code is easier and clearer.

Stepping through an equivalent assembly code, on the other hand, makes the programmer more aware of what each instruction involves. The task is simplified by making compilers produce list files providing both the source code and the equivalent assembler.

17.7.3 On-line testing

Although off-line testing helps in the detection of many errors, a large number of faults will only appear during real-time testing. For instance, the simultaneous occurrence of a few interrupts may cause the stack to overflow and overwrite RAM locations. The larger the number of tasks running simultaneously, the more difficult it is to isolate the problems. In addition to software means, hardware tools have been developed to provide debugging functionality in order to assist in detecting such faults. The task will obviously be made easier by writing well-structured, well-documented and easy-to-follow software.

Debugging without special tools

Without an emulation tool, one has to follow a tedious and a difficult procedure. After writing the code, one has to program an EPROM, insert it in the circuit and test. If the operation is not as expected, then, after some investigation, the EPROM needs to be removed from its socket and erased by exposure to ultraviolet light in a suitable eraser. Depending on the type of fault and the complexity of the program, the search for the fault can be an extremely difficult task. It can take many days and, for an inexperienced programmer, it can be like searching for a white rabbit in a snowstorm!

With practice, one starts to develop techniques to speed up the fault-detection process. Some of these techniques are as follows:

- Inserting a code to assist in diagnostics and testing. For instance, the state of a digital-output line can be set high at the start of an interrupt and reset to low before returning to the main program. This can help to show the time taken by each interrupt and the effect of one interrupt on another.
- Interfacing a D/A converter to an output port and supplying it with the value of one variable at a time. The output of the converter can be examined by an oscilloscope in order to understand how some of the variables are changing in real time. This can be a very useful tool in the detection of time-related faults. Once a faulty behaviour is observed, one can use the other channel of the oscilloscope to examine other signals and identify the source of the fault, e.g. noise due to closing a relay.
- A display is an extremely useful tool, especially if it is capable of displaying more than one variable. It can be utilized to display the value of variables,

the state of flags, or the contents of registers at selected instants, e.g. when the state of an input line changes, or at the end of a calculation cycle.

- Another useful tool is the serial port of a microcomputer. It can be employed to transmit the value of one or more variables to a PC for later inspection or (in the case of a slowly changing variable) to a monitor to be examined by the operator in real time.
- Part of the memory of the on-line microcomputer can be utilized as a temporary storage area. The value of one or more variables can be stored in real-time in the form of a table which can be examined after the operation is halted (by the use of a numeric display or a group of LEDs).

These and many other techniques can be introduced to detect faults. Their introduction, however, consumes development time and they can only provide a limited fault-detection capability. A general-purpose tester can be designed and used in different projects. It can include a number of switches, a D/A converter, a multidigit display, a serial interface, a group of LEDs and other useful devices. All the parts can be connected to digital ports for interfacing to an on-line microcomputer through one or more of its digital ports.

Using emulators

An emulator is an extremely useful development tool. It is employed for the verification of software in order to help resolve difficult debugging problems, particularly in real-time multitasking applications. A low-cost type is an EPROM emulator. It provides a header for insertion in a socket instead of an EPROM IC. A more useful and a more costly tool is an in-circuit emulator. It provides a header for the insertion in the processor's socket, e.g. to replace a microcontroller. In either type, the header is connected to the emulator unit by a multiway cable, e.g. a ribbon cable.

An emulator is either a stand-alone unit or a board which can be plugged into a computer. In general, the latter type is simpler, offers less functionality and has a lower cost. The use of emulators simplifies the development process considerably. The target software can be downloaded and run unchanged, eliminating the need to program and reprogram devices. The program will simply be stored in RAM and run by the emulator to test the on-line behaviour of the system. The operation of a real-time in-circuit emulator should not alter the timing of the system. The execution of the program is carried out as if the actual processor is in operation. All the monitoring and storage of data is carried out in parallel with the operation of the processor without introducing an additional overhead.

When using an emulator, debugging can start even before the availability of hardware, especially when microcontrollers are used. All a microcontroller needs for its operation is a power supply, a reset signal and a crystal (to permit the operation of the on-chip oscillator). Emulators can be set to provide such signals (if additional circuitry is interfaced to the microcontroller, then an additional power supply will be required).

What an emulator can do depends on its type, make, age and price. Many attractive features are available in some emulators but not in others. One such feature is the ability to watch how the contents of certain RAM locations are changing as the program is running. Another feature is allowing the user to alter the contents of RAM locations as the program is running. The ability to set breakpoints is another very useful feature provided by emulators. It permits the operator to stop the program at the part he or she wants to debug. The operator can then observe the contents of registers and/or step through the instructions and observe effects that follow. Halting the execution of the program by the use of breakpoints may not be allowed in some real-time applications, since many systems require following a suitable shutdown procedure. A sudden halt of the program could lead to disastrous effects. Furthermore, using breakpoints or stepping through instructions may not show the actual effect of real-time behaviour. Instead of using breakpoints, one can utilize the trace facility where the emulator will incorporate a buffer to store information following the occurrence of an event. The length of the buffer is limited and can be specified by the operator. Tracing can help one to understand the actual behaviour of the system during real-time operation and explain the cause of faults.

During a debugging process, a typical emulator displays a part of the code as well as the contents of registers while providing an on-screen working area for the operator to command the operation. One can, for example, observe code or contents of registers, single step through instructions, and start or stop the program in real time. The program may, for instance, be commanded to stop at a specific location by the use of breakpoints. One can then single step through the code (either in high-level or low-level language, depending on the emulator and the programming language). The system's behaviour can be understood by observing how registers and certain memory locations change with time, or as a result of altering conditions, e.g. load current, input voltage or frequency. An emulator makes altering parts of the program a simpler task, since all the operator needs to do is alter the contents of RAM locations where the program is stored. If the aim is to introduce minor alterations to the program, then it may be possible to perform such alterations directly from the emulator. If, on the other hand, additional code is to be inserted, or sections of the program are to be replaced by others, then one needs to end the emulation session, alter the code, and then compile or assemble the code before linking and downloading to the emulator to repeat the test.

In general, emulators are not cheap, but their use increases the error-detection capability and reduces the development time considerably, especially in real-time control applications, such as robotics and motor drives. Technological advances will help to increase their user-friendliness and cost-effectiveness.

They will, however, continue to be regarded as a high-cost investment which can be justified by the saving in development time and the ability to use it in more than one project. A large number of emulators, however, are designed for a specific processor. The requirements of future projects and/or the continuous technological development may create the need to use a different processor. It is therefore attractive to invest in an upgradable unit. In general, lower-cost emulators are designed for a specific processor. Higher-cost emulators, on the other hand, may consist of a few boards and permit the replacement of one board in order to support another processor. The choice depends on the requirement. For example, a consultancy firm designing control systems according to the requirements of customers is likely to be interested in a flexible emulator which can be configured to any of several processors.

Since the price of emulators is high, it is desirable to have a good understanding of the operation and functionality of the emulator prior to purchase. Many software developers prefer to rent an emulator only to solve difficult problems. The difficulty with very short-term renting is the time needed to learn how to use the equipment. The increase in the user-friendliness of emulators, however, reduces the learning time considerably, especially for people who have used another emulator previously. Some emulator-producing companies simplify the task of purchasing an emulator by permitting the rental of their emulators and offer a discount if the same emulator is purchased within a specified time interval. In this way, one can familiarize oneself with the emulator before being committed to its purchase.

Appendix

Logic-gate ICs:

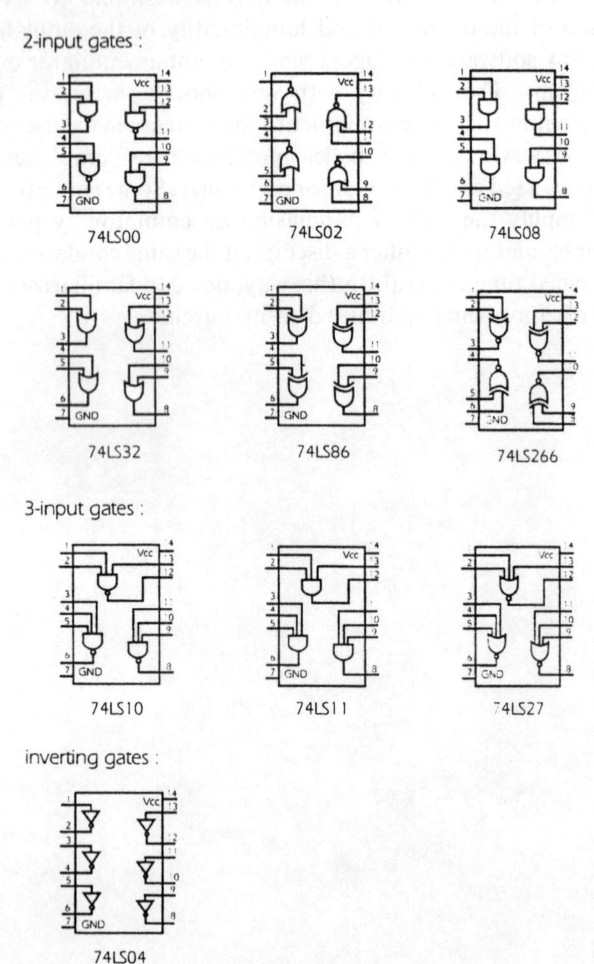

Index

Index

Index

Index

Index